P9-ARM-260

CARTOGRAPHIC DESIGN:
Theoretical and Practical Perspectives

International Western Geographical Series

editorial address

Harold D. Foster, Ph.D.
Department of Geography
University of Victoria
Victoria, British Columbia
Canada

Since publication began in 1970 the Western Geographical Series (now the Canadian and the International Western Geographical Series) has been generously supported by the Leon and Thea Koerner Foundation, the Social Science Federation of Canada, the National Centre for Atmospheric Research, the International Geographical Union Congress, the University of Victoria, the Natural Sciences Engineering Research Council of Canada, the Institute of the North American West, the University of Regina, the Potash and Phosphate Institute of Canada, the Saskatchewan Agriculture and Food Department, and the B.C. Ministry of Health and Ministry Responsible for Seniors.

CARTOGRAPHIC DESIGN: Theoretical and Practical Perspectives

edited by
CLIFFORD H. WOOD and C. PETER KELLER

JOHN WILEY & SONS
CHICHESTER • NEW YORK • BRISBANE • TORONTO • SINGAPORE

Baffins Lane, Chichester,
West Sussex PO19 1UD, England

National 01243 779777
International (+44) 1243 779777

Other Wiley Editorial Offices

John Wiley & Sons, Inc., 605 Third Avenue, New York, NY 10158-0012, USA

Jacaranda Wiley Ltd, 33 Park Road, Milton, Queensland 4064, Australia

John Wiley & Sons (Canada) Ltd, 22 Worcester Road, Rexdale, Ontario M9W 1L1, Canada

John Wiley & Sons (Asia) Pte Ltd, 2 Clementi Loop #02-01, Jin Xing Distripark, Singapore 0512

Library of Congress Cataloging-in-Publication Data

A catalogue record for this book is available from the Library of Congress

British Library Cataloguing in Publication Data

A catalogue record for this book is available from the British Library

ISBN 0 471 96587 1

Produced from camera-ready-copy supplied by the editors
Printed and bound in Great Britain by Bookcraft (Bath) Ltd
This book is printed on acid-free paper responsibly manufactured from sustainable forestation, for which at least two trees are planted for each one used for paper production.

Contents

The Contributors

Jacqueline M. Anderson is an associate professor at Concordia University, in Montréal, Canada. She received her doctorate from the University of Wisconsin (Madison) after completing graduate work at the Universities of Alberta and Glasgow and a year as an assistant map research officer with the Ministry of Defence (U.K.). The focus of her research is graphic literacy, her present studies investigating the mapping abilities and design needs of young school children. Currently, with Dr. Regina Vasconcellos (Brazil), she is initiating the formation of a formal working group of the International Cartographic Association on "Cartography and Children."

John A. Belbin has been head instructor of the Mapping Department at the College of Geographic Sciences in Nova Scotia for the past 15 years. He has been an instructor in cartography at the technician and technologist levels in Nova Scotia and Ontario for 20 years.

Barbara P. Buttenfield received a B.A. from Clark University, an M.A. in geography from the University of Kansas, and a Ph.D. in geography from the University of Washington. She is presently an associate professor of geography at the University of Colorado at Boulder. Her research interests are in analytical cartography, map generalization, visualization, and expert systems in cartography.

Rex G. Cammack is assistant professor in the Department of Political Science and Geography at OldDominion University. He received his Bachelor's degree in geography from the University of Nebraska at Omaha in 1989 and his Master's and Ph.D. degrees in geography from the University of South Carolina in 1991 and 1995, respectively. His research interests include spatial cognition, map communication, cartographic animation, and connectionist theory. Rex is a member of AAG, ACSM, and NACIS.

Kenneth J. Gilhooly is a senior lecturer in psychology at Aberdeen University. He is a graduate of Edinburgh University (1967) and obtained his doctorate at Stirling University in 1972. His main research interests are in the cognitive psychology of thinking and expertise. He has been treasurer and chair of the British Psychological Society's Cognitive Psychology Section.

Patricia P. Gilmartin is a professor of geography at the University of South Carolina. Prior to joining the faculty at South Carolina, she taught cartography for four years at the University of Victoria in British Columbia. Her research interests include cognitive aspects of map reading, map design, and navigation, geographic education, and women explorers. She is past president of the Cartography Specialty Group of the AAG, former associate editor for Cartography and GIS and the Annals of the AAG and currently serves as National Councillor of the AAG and chair of the AAG Publications Committee.

Nikolas H. Huffman studyied cartography, social theory, and visual language as a master's student in geography at Pennsylvania State University. He has an undergraduate degree in geography and philosophy from the University of Kansas, with concentrations in cartographic design and the philosophy of language. He has been working as a freelance cartographic designer for four years, including two and a half years with the National Geographic Society, working on projects for the 27th International Geographical Congress and the Society's scholarly publication *Research & Exploration*.

C. Peter Keller is an associate professor at the University of Victoria, British Columbia, Canada. He received his undergraduate education from Dublin University, Trinity College in Ireland and his graduate degrees from the University of Western Ontario, London, Canada. He has been teaching cartography, GIS and spatial analysis since 1984. His research interests are in cartographic design focusing on map user needs, the advancement of GIS for better analysis and decision making, and teaching innovations. In the past he has been chair of the Pacific Institute of Cartographers Society and the Canadian Cartographic Association. He currently is chair of the Canadian National Committee for Cartography.

John B. Krygier taught cultural geography at Penn State University during the 1994/95 academic year. He has a M.S. from the Department of Geography at the University of Wisconsin-Madison and a Ph.D. in geography from Penn State University. He is working on an Atlas of the Molly Maguires, hypermedia teaching tools for introductory Earth Science, the geography of garbage in Pennsylvania, and Rails to Trails projects in Pennsylvania.

Mark Kumler received a B.A. from Dartmouth College, an M.A. in geography from Michigan State University, and a Ph.D. in geography from the University of California at Santa Barbara. He is presently an assistant professor of geography at the University of Colorado at Boulder. His research interests are in map design, map projections, and digital terrain models. This particular work was inspired by a statement he made in his first year in the classroom: "studies have shown that women don't read maps as well as men, but we're not sure why." He regrets making the statement.

Robert Lloyd is a professor in the Department of Geography at the University of South Carolina. He received his Ph.D. in geography from Pennsylvania State University.

Matthew McGranaghan is an associate professor of geography at the University of Hawaii-Manoa. He holds undergraduate and Masters degrees from SUNY-Albany and a Ph.D. from SUNY-Buffalo. He teaches cartography, remote sensing, and geographic information systems.

William Mackaness has a geography degree and a Ph.D. in knowledge-based techniques applied to automated map design. In various guises, he has been a lecturer in Information Systems at Otago University, NZ, a post doctoral researcher for the NCGIA at State University New York-Buffalo, a consultant for an architectural practice in London, a post doctoral and assistant research professor in surveying engineering (NCGIA) at University of Maine. More recently, he has returned to the UK from Australia where he was a researcher at Murdoch University, researching the role of GIS in Native Title Land Claims. Currently he is a lecturer in geography at the University of Edinburgh, and is course director of the MSc in GIS.

Janet E. Mersey joined the Department of Geography at the University of Guelph in 1985, after completing her Ph.D. degree from the University of Wisconsin-Madison. She teaches courses in cartography, GIS, remote sensing, and aerial photography, and manages the digital mapping facilities. Her research interests are in the areas of thematic map design, symbolization techniques, and data visualization with GIS. Dr. Mersey is active in the Canadian Cartographic Association and is book review editor for the journal *Cartographica.*

Mark Monmonier is a professor of geography in the Maxwell School of Citizenship and Public Affairs at Syracuse University. He received a B.A. in liberal arts and mathematics from Johns Hopkins in 1964 and a Ph.D. in geography from Penn State in 1969. His current research interests are dynamic cartography, hazard-zone mapping, and the ethical use of geographic information. Monmonier is the author of *Maps with the News* (1989), *How to Lie with Maps* (1991), *Mapping It Out* (1993), and *Drawing the Line: Tales of Maps and Cartocontroversy* (1994).

Phillip C. Muehrcke is a professor of geography at the University of Wisconsin-Madison where he has taught since 1973. He earned his Ph.D. from the University of Michigan in 1969. Prior to coming to Wisconsin, he was in the geography faculty at the University of Washington. He is the author of numerous scholarly articles published in a wide variety of geographic journals. He is also the co-author of the well-known text, *Map Use: Reading, Analysis, Interpretation*, now in its third edition.

Elisabeth S. Nelson is a lecturer in the Department of Geography at San Diego State University. She received her Bachelor's degree in geography from the University of North Carolina at Charlotte in 1987 and her Master's and Ph.D. degrees in geography from the University of South Carolina in 1989 and 1995, respectively. Her research interests include map learning and encoding strategies, cognitive map structures, and map reading processes. Elisabeth is a member of AAG, ACSM, and NACIS.

David K. Patton is assistant professor in the Department of Geography and Environmental Science at Slippery Rock University in Pennsylvania. He received his Bachelor's degree in English from Northern Michigan University in 1988 and his Master's and Ph.D. degrees in geography from the University of South Carolina in 1993 and 1995, respectively. His research interests include spatial cognition, category theory, and map communication. David is a member of AAG, SEDAAG, and NACIS.

Elzbieta Rostkowska-Covington is cartographer in the School of Public Health at the University of South Carolina. She received her M.S. degree in geography from the University of South Carolina.

Theodore Steinke is a professor in the Department of Geography at the University of South Carolina. He received his Ph.D. in geography from the University of Kansas.

D.R. Fraser Taylor earned his Ph.D. in geography from the University of Edinburgh. Currently he is a professor of Geography and International Affairs at Carleton University in Ottawa, and also holds the administrative posts of vice-president international and director of Carleton International in the Faculty of Graduate Studies and Research. Research interests include the application of computer-assisted cartography to the understanding of socio-economic issues, the theory of cartography, and cartographic education. His interests in cartography and international development issues are often inter-related. He has published widely in both the international development and cartographic fields.

Regina Vasconcellos, past president of the ICA, Ph.D. in geography, is an assistant professor in the Department of Geography, University of Sao Paulo, Brazil. She is coordinator of the geography Education Laboratory at the University of Sao Paulo. She is vice-president of the ICA and a full member of the ICA Commission of Tactile Cartography and Low Vision Maps. Dr. Vasconcellos organized the IV International Symposium on Maps and Graphics for the Visually Impaired, on behalf of the International Cartographic Association, held in Sao Paulo, Brazil from February 20-26, 1994. Her areas of research are thematic cartography, tactile mapping and geographic and cartographic education.

Irina Vasiliev teaches cartography and environmental geography at the State University of New York at Geneseo.

Roger Wheate has been a technician and lecturer in cartography, GIS and remote sensing at the University of Calgary since 1981. In August 1994, he took up the position of GIS Lab coordinator/lecturer at the new University of Northern British Columbia at Prince George. He has been a cartophiliac almost from birth and has a Master's in cartography from Queen's (Kingston) and an undergraduate degree from St. Andrews (Scotland) where he is also enrolled in a part-time Ph.D. program. He was secretary and then manager of the Canadian Cartographic Association from 1986-1993.

Clifford H. Wood is a professor of geography and director of the Memorial University of Newfoundland Cartographic Laboratory where he has taught since 1977. Prior to joining the faculty at Memorial University, he taught at the University of Idaho and at the University of Maryland-College Park. He earned the B.S. and M.S. degrees at the University of Idaho and the Ph.D. at the University of Wisconsin-Madison. His research interests revolve around cartographic design issues and involve the perceptual/cognitive aspects of map analysis using eye movement techniques as the investigative tool. He has served as president of the Canadian Cartographic Association, president of the Canadian Institute of Geomatics, and chair of the Canadian National Committee for ICA.

Michael Wood is a senior lecturer at Aberdeen University. He completed his graduate studies in 1964 in the University of Glasgow under the direction of J.S. Keates and since that time his main research interests have centred on the design, creation and reading of maps and cartographic panoramas. He has been president of the British Cartographic Society and is currently president of the British Society of Cartographers. He is also a member of the UK Committee for cartography and president of the International Cartographic Association.

Series Editor's Acknowledgements

This volume represents a major step forward for the Department of Geography, University of Victoria. In the early summer of 1995, the University established the Western Geographical Press. Negotiations were conducted with both John Wiley and Sons, Ltd. and UBC Press. As a result, the Western Geographical Series was split into two components. The International Western Geographical Series is edited under the auspices of Western Geographical Press, and published and distributed by John Wiley and Sons, Ltd. The Canadian Western Geographical Series is published by the Western Geographical Press but distributed by UBC press. This is the first volume of the International Western Geographical Series.

Several members of the Department of Geography, University of Victoria co-operated to ensure the successful publication of this volume. Special thanks are due to members of the technical services division. Diane Macdonald undertook the very demanding task of typesetting. Cartographic work was performed by Ken Josephson, who also designed the cover. Their dedication and hard work is greatly appreciated.

University of Victoria
Victoria, B.C., Canada
10th February 1996

Harold D. Foster
Series Editor
Western Geographical Press

Preface

A great deal has changed since the Symposium on the Influence of the Map User on Map Design was held at Queen's University, Kingston, Ontario, in September of 1970. Organized by Henry Castner and Gerald McGrath, the first Canadian symposium dedicated to map design explored two aspects. First was the role of the variables the cartographer must cope with in designing a map that facilitates the exchange of graphic information. The second element, an issue at the time which had received little attention, was the influence the map reader exerts over the communication process. While both of these issues have been the focus of much research since 1970, questions continue to remain unanswered. Most cartographers today, however, do recognize the importance of the map reading audience as a design factor that must be considered. Contemporary cartographers are aware of the need for precision and clarity of thought as they set about the task of designing a map. This requirement exists whether the map is intended to communicate information, or its primary purpose is data exploration, as in scientific visualization.

One of the most significant changes that has occurred since the first symposium was held has been the widespread acceptance and reliance on digital methods in cartography. With the exception perhaps of printing, no other technological change has had more of a profound influence on cartography than that of the computer. A great deal of time and effort has been placed on realizing the potential and the methods of digital automation of the practice of cartography, and few would dispute that the digital revolution has brought great benefits, However, several cartographers, including the symposium organizers, share a concern that one element of cartography has been overlooked, or at least somewhat neglected in the move to digital automation, namely "map design". It was this apparent lack of attention to design that spawned the idea to organize another symposium dedicated solely to that one issue.

The result was the Symposium on Cartographic Design and Research held on the campus of the University of Ottawa, Ottawa, Ontario on 7-8 August 1994. Sponsored by the Canadian Institute of Geomatics (CIG), the Symposium attracted a total of 84 registrants from Austria, Brazil, Canada, Finland, the United Kingdom, and the United States. Although the program consisted of several topics, the principal organizational structure revolved around two design perspectives - theoretical and practical. A total of 16 papers on these two themes were presented and were anchored at both beginning and end by an opening keynote address by D. R. Fraser Taylor and a closing keynote by Phillip C. Muehrcke.

The organizers wish to acknowledge with gratitude the support of the sponsor and the invaluable assistance of Ms. Susan Pugh of the CIG head office in Ottawa. Thanks are also due David Douglas of the University of Ottawa who oversaw arrangements on the University of Ottawa campus. The success of the Symposium and of this volume, in large part, is directly attributable to a number of people who have helped us in several capacities, including session moderators, session discussants, or reviewers of manuscripts. They include, in alphabetical order: Alberta Auringer Wood, Christopher Board, Henry Castner, Michael Coulson, Borden Dent, David DiBiase, Matthew Edney, Grant Head, Diana Hocking, Jon Kimerling, Brian Klinkenberg, Gail Langran Kucera, Robert McMaster, Judy Olson, Tom Poiker, Michel Rheault, and Carolyn Weiss. Special thanks also go to our editor in Victoria, Dr. Harold Foster, and his production team consisting of Diana Macdonald, who exhibited patience and poise in deciphering manuscripts in many different formats, and Ken Josephson, without whose understanding of the graphic requirements and a calm approach in solving all of them, the volume would not be what it is.

Clifford H. Wood
C. Peter Keller
Symposium Co-organizers

1

Design: Its Place in Cartography

Clifford H. Wood

Department of Geography, Memorial University of Newfoundland

C. Peter Keller

Department of Geography, University of Victoria

Cartographers frequently use the term "design" when referring to one of the processes involved in preparing and producing a map, whether in an automated environment or in the traditional sense. While most would admit to having some notion about the meaning of the term, many likely would have some difficulty in defining precisely what design is and what the design processes are. Most would agree, however, that design is truly important in cartography. DeLucia (1974) called design "the most fundamental, challenging, and creative aspect of the cartographic process".

If one accepts the premise that design has this pre-eminent role in cartography, how is it that cartographic design appears to have been taken for granted, or treated rather superficially; why has design been neglected, poorly appreciated, or even ignored by many cartographers? In order to answer these questions, we must first understand what design is—what we mean when we say that we are going to "design" something. Secondly, we must agree on the purpose of graphic design. Third, we must understand the processes of graphic design. Finally, we must understand and learn to work within limitations imposed on design by technology, the media, and operational logistics. Achieving some understanding of these elements may assist in perceiving not only where design fits into the discipline of cartography, but also where emphases currently are being placed in addressing unresolved design issues of today's cartography. The Symposium on Cartographic Design and Research was organized and held in August of 1994 to articulate many of these design issues.

What is meant when we say that we are going to design something? From many definitions of the verb "to design" one that appears rather comprehensive is that of Professor L.B. Archer of the Department of Design Research at the Royal College of Art in London. He defines design as:

> *. . . to conceive the idea for and prepare a description of a proposed system, artifact, or aggregation of artifacts. No distinction is drawn between architectural design, engineering design, graphic design, and industrial design. . . . the design act is logically identical in all these*

fields. Professional designers understand that [the] essential element in the definition of the verb design is the notion of conceiving in the mind a plan or scheme of something to be done (Archer, 1969, 1.6, 1.4).

The one element that seems to be missing in Archer's definition relates to designing something for a *purpose.* One could argue that knowing what the purpose will be for what is being designed will exert a tremendous influence over the design process. Blumrich (1970) approached defining the term as a noun and concluded that design is a solution to a problem not before solved, or a new solution to a problem previously solved a different way. These definitions are rather broadly based. One definition more directly related to cartography is that from Richard Taylor's book, *A Basic Course in Graphic Design,* in which he states: "Graphic design, usually thought of as a derivative of painting, has emerged as a problem-solving design activity. The graphic designer, like any other designer, is a specialist who is concerned with solving other people's visual communication problems" (Taylor, 1971). If one narrows the scope to cartography, perhaps design can be defined as the planning of a solution to a problem in graphic communication—the solution itself.

In reviewing these and other definitions, a number of common characteristics emerge. It is clear that mental activity, i.e. thinking, is a fundamental element of design. Other elements, according to DeLucia (1974) that taken together form the essence of design include: problem-solving, systematic process, solutions, and product or discipline independence. DeLucia integrated these characteristics into the following definition of design: "Design is a systematic thought process which yields solutions to a wide range of human problems. The process is logically independent of the specific nature of the systems or products required to solve those problems" (DeLucia, 1974).

Although there may be disagreements over the precise definition of design in a cartographic context, there should be little doubt regarding its purpose. "The purpose of graphic design is to facilitate human thought and communication. Success in graphic design is achieved when diverse design principles are manipulated and adapted to produce an image with a high degree of readability" (Wood, 1992). If one accepts the notion that cartography is functional, then the design of a map must fit some previously selected concepts or ideas around which design decisions are made. These decisions must be made by a designer mindful of the defined purpose for which the map is intended. As MacEachren (1994) states: ". . . the map is primarily a presentation device. It presents an abstract view of some portion of the world with an emphasis on selected features . . . . When most map users, even trained cartographers, approach the task of map symbolization and design, they typically assume a presentation goal for which a single map must be selected."

In order to arrive at a functional design that fits the intended purpose of the graphic, a logical progression of steps must be followed. The methodology of the design process consists of stages, as discussed by both Dent (1993:243) and Tyner (1992:48-51). A simple model of the cartographic design process would include an initial stage at which ideas and concepts are generated. This initial stage is the personal, imaginative and creative one in which, among other things, the purpose of the graphic is articulated. A second stage would involve the development of the graphic structure in which the geographic and thematic information is assigned to various visual levels. And the third stage would include the preparation of the worksheet and the specifications for the cartographic marks that would convey the intended cartographic message. It is at this stage that the technical competence of the cartographer assumes a greater importance. A more elaborate structure for the design process was promoted by DeLucia whose model included five stages as follows: 1) problem identification and definition, 2) search for solutions, 3) evaluation of alternative solutions, 4) selection of the optimum solution, and 5) production (DeLucia, 1974). Regardless of the complexity or the number of stages in the design process, the end result should be the same—a graphic that fits the defined purpose. Few empirical studies are available that articulate the stages in the design process, but one of the more entertaining is that of Ommer and Wood (1985) who divulge the author-cartographer interaction in a map design exercise. The study amply demonstrates the reliance on a set of steps that lead to a final conclusion in cartographic design—a functional map.

Design has always had a place in cartography, but the relative importance attributed to it has changed through time; it has ebbed and flowed with a constantly changing and evolving discipline. During the late 1960s and early 1970s, for example, research cartographers (Kolacny, 1969; Koeman, 1971; Ratajski, 1973; Morrison, 1974, 1976) produced theories of cartographic communication. Their efforts were largely an attempt to formalize the cartographic process, but also to further the acceptance of cartography as a scientific discipline. Considerable doubts were raised, however, about the validity of the underlying assumptions of cartography modelled as a communications system (Petchenik, 1975; Guelke, 1976; Salichtchev, 1978). Of particular concern in the communication models was the apparent reduction of cartographic tasks to merely minimizing the loss of cartographic information while ignoring the acquisition of knowledge during the map analysis operation (Keates, 1982). Of no less concern was the failure in the communication models to account for the extraction of cartographic information (map reading) and the cognitive processes involved in interpreting the map image. As Morrison (1976) suggested, attempts at producing a simplistic model of cartographic communication based on system input equalling system output fail to address the question of interpretation.

Cartographic communication, as a model, has largely faded from the literature. But, one aspect that gained considerable attention when the notion of cartographic communication was at its zenith, has not. That aspect, **cartographic design** and the need to address many of the questions that would result in a successful transfer of map information to the map analyst, continues to figure very prominently today.

Some may argue, perhaps, that cartographic design and communication are synonymous, or at least so inextricably linked that differentiation is difficult. Regardless, cartographic design has continued to occupy a research niche. A number of research avenues have been pursued, including individual subjective judgment (Robinson, 1980), verbal protocols and user-reactions (McGrath, 1971; Bartz,

1969; Greenberg, 1971; Stanley, 1975), and psychophysical methods (Flannery, 1956; Wright, 1967; Castner and Robinson, 1969; Kimerling, 1975; Cox, 1976). Petchenik (1975, 1983), Morrison (1976), and Muehrcke (1982) lament the fact, however, that much of the design research has not produced the desired results. Petchenik (1975:183) stated: ". . . no coherent and comprehensive whole theories [for map design] emerged." Morrison (1976) attributed this failure to the lack of research structure. Muehrcke (1982) placed the blame on map design problems that were too narrowly defined. He stated: "Analytical test procedures simply do not capture the essence of the map design process and, therefore, are of limited use" (Muehrcke, 1982:115). The lack of significant advances may be responsible for the general malaise that seems to have affected research efforts in map design. Recognition of the role that cognitive processing of cartographic information (Petchenik, 1977; Thorndyke, 1981; Steinke and Lloyd, 1981; Castner and Eastman, 1984, 1985; Peterson, 1985; Blades and Spencer, 1986; Lloyd, 1989; MacEachren, 1991) plays in cartographic design and communication may also be a contributing factor, and may be leading map design research in slightly different directions, including learning strategies (MacEachren, 1991), visualization (MacEachren and Ganter, 1990), and the pursuit of a rule-based expert system for map design.

Cartographic design has, to be sure, undergone periods of benign neglect and redefinition. During the time that cartography was undergoing reformulation as a truly scientific discipline, automation was also sweeping cartography as the newest, evolutionary and revolutionary technique that offered many advantages over the more traditional approaches to map preparation and production. The advantages that automation offered cartography were soon consumed by the need to prescribe the various operations in a rigorous, systematic manner. Addressing these needs relegated cartographic design to the back seat while the emphasis was focused on understanding how to make the technology perform the necessary cartographic operations and writing the algorithms to accomplish these ends. During the last two decades or so, automation, or more accurately our fascination with technology, seems to have absorbed our interests along with the need to satisfy our thirst for technological advances in the discipline.

How has this fascination with automation affected design in cartography? One cannot deny that time, money, and manual labour can be saved in cartographic production by relying on automated data manipulation. The strengths of the computer are numerous. First, graphics software now duplicates and supports all the capabilities of traditional design. Second, the digital environment allows for timely and cost effective modifications to a design, something that was very time consuming and expensive using the old media. Third, the computer supports independent digital storage of images, designs and text that can be accessed and combined to form composite images at a whim. These flexibilities allow for exploration of significantly greater design alternatives without adding to the overall cost of time or preparation.

But what is the role of the designer in this age of automation? The computer suffers by comparison with the human mind in two critical areas of the design process. First, there is the stage of problem definition and structuring. In order to define or structure a problem, a computer algorithm must enable the selection and interpretation of data in a problem-oriented context. The second area in the design process in which the human mind is superior to that of the computer is in the evaluation and decision-making stage. A digital computer must receive its instructions and data in a discrete form, bit-by-bit. Programming instructions that would allow an automated evaluation of a graphic image seem difficult due to the plethora of variables that must be considered. The assessment of aesthetic qualities of a particular design, for example, would create a special problem in attempting to substitute computers for tasks better performed by the human designer. The methods by which this information may be imparted to computers fall under the aspirations of researchers pursuing the elusive "expert system", or the "rule- or knowledge-based system" for map design. Although strides have been made in this area, much work remains, and it is unlikely that a computer will ever replace the creativity and imagination that a skilled cartographer brings to map design.

One of the results of automated or digital cartography has been the emergence of analytical cartography. Automation permits the manipulation of large geographic data sets from which the cartographer can extract any number of different parameters, separate or in combination with other factors. Automation has also provided the analytical tool that has fostered the derivation of geographic information systems (GIS). Although analytical cartography and GIS rely on the same technology, GIS goes well beyond automated mapping. Using GIS technology, it is possible to perform a number of different, complex operations which include: encoding of geographic (geocoding) features into the geographically-referenced data base, the freedom to manipulate and integrate data, the ability to spatially cross-reference data files within the data set or with other related sets of data, to name a few. Clearly, the strength of a GIS is in its analytical power.

Interest in GIS has grown exponentially. This growth has not come about unexpectedly. On the contrary, it is a natural outgrowth of 1) the focus on dealing with the environment in an integrated, holistic manner (Muehrcke, 1990), 2) our ability to collect vast amounts of geographic information using many advanced techniques on a variety of platforms, 3) the need to grasp an understanding and make sense of this wealth of information, 4) the opportunities that automation provided in increasing the potential for managing the data, and 5) the flexibilities of the geographic information systems in displaying the many combinations of characteristics inherent in the collected data for a particular region. The impact of GIS on cartographic design is summarized by Muehrcke:

> *. . . the ability to generate graphic iterations easily encourages the cartographer to explore alternative design concepts before making final design decisions. But the role of GIS tools in the conceptual process is clearly limited. They may have wonderful, almost magical powers to offer, but in the foreseeable future at least they will not replace the human element. Ideas come from people* (Muehrcke, 1990).

In addition to GIS, interest in visualization, more specifically, scientific visualization, has gained momentum recently, underscored by the publication of two new texts (MacEachren and Taylor, 1994; Hearnshaw and Unwin, 1994). Visualization is a computing method in which computer graphics and image processing technology are used in data-intensive scientific applications to transform the symbolic into the geometric, thereby enabling the researcher to observe their simulations and computations (McCormick et al., 1987). Visualization permits the researcher to produce quickly a number of images with various combinations of variables from a number of data sets. Visualizing how the data appear in different interactions can increase the researcher's understanding. As Fisher et al. (1993) point out, the strength of scientific visualization lies in the development of ideas, not the presentation of them in a traditional cartographic sense. Although design may be of some importance in visualization, that is, colours must be sufficiently different to facilitate discrimination, it is not the primary emphasis.

The need to discover new methods of data management and to produce the technological means to carry them out has had, what we consider, a deleterious effect on cartographic design. The result appears to be a lack of concern for the impacts that good design has on the readability of graphics, and maps in particular, a point made by van Elzakker (1991) at the Bournemouth ICA meeting. Are we in cartography so caught up with technology that we are overlooking design?

It seems only now that map design is beginning to emerge from a rather prolonged neglectful period. Cartographers are beginning to realize the impact that automation is having on map design. It is commonplace that people untrained in cartography and unfamiliar with principles of graphic design, for example, now have ready access to a variety of mapping programs. The results of this access are maps that fail in one way or another to communicate cartographic information in the most effective and efficient manner. Until these programs contain a built-in design sense, or users learn the theory, concepts and rules behind the software, even greater incidence of graphic design failure can be anticipated. Perhaps the increasing reliance on graphics in GIS and in scientific visualization has sharpened our awareness of how important "graphic" is in graphic design.

Whatever the reason, many cartographers now recognize the niche that design occupies in cartography. It is a niche that is in danger of being lost. This recognition prompted a re-examination of design in a number of areas at the Symposium on Cartographic Design and Research, including the place of design in geography and cartography, as well as the relevancy of cartographic design. Of concern to every cartographic designer is the intended audience for whom the map is targeted. Basic questions are investigated in three areas, the special mapping considerations of young elementary school children, what we know about gender differences in map reading abilities, and the design of maps for the visually impaired. The increasing complexity of data often raises special problems for cartographic representation. Two such topics, the cartographic treatment of time and the third dimension are viewed from the designer's standpoint. In our attempt to formulate a rule-based expert design system, not only must we understand the linkages between automation and the human paradigm, but we must also determine what we currently know about cartographic design. Both approaches are documented in the following chapters. Using the knowledge of linkages between computers and humans, we are then in a better position to investigate some areas of applying design in an automated environment that present interesting challenges. Two such areas are: 1) the requirements for symbolizing maps in a microcomputer-based GIS and 2) how statistical maps can be designed for more effective displays on monochrome monitors. The concluding chapters address a number of wide-ranging experimental design issues. Without research and experimentation, cartographic design would progress little beyond the intuitive stage. Such topics as quantitative, multivariate point symbols, feature matching, the effects of task type and map complexity on statistical maps, and the application of Gestalt theory to type placement are examined.

A greater body of knowledge gained from research and experimentation will contribute toward understanding the complexities of cartographic design. But this goal will not be realized unless cartographic design remains high on the research agenda. We know that today many maps fall below their thought-provoking potential because of design failure. It is clear that cartographers need to become proactive about design issues if we are to reduce the incidence of poorly designed maps.

REFERENCES

Archer, L.B. (1969). *The structure of design processes.* London: Department of Design Research, Royal College of Art.

Bartz, B.S. (1969). Search: An approach to cartographic type legibility measurement. *Journal of Typographical Research,* 3(4), 387-398.

Blades, M., and Spencer, C. (1986). The implications of psychological theory and methodology for cognitive cartography. *Cartographica,* 23(4), 1-13.

Blumrich, J.F. (1970). Design. *Science,* 168, 1551-1554.

Castner, H.W., and Eastman, J.R. (1984). Eye-movement parameters and perceived map complexity - I. *The American Cartographer,* 11(2), 107-117.

Castner, H.W., and Eastman, J.R. (1985). Eye-movement parameters and perceived map complexity - II. *The American Cartographer,* 12(1), 20-40.

Castner, H.W., and Robinson, A.H. (1969). Dot area symbols in cartography: The influence of pattern on their perception. *Technical Monograph No. CA-4.* Washington, D.C.: Cartography Division, American Congress on Surveying and Mapping.

Cox, C.W. (1976). Anchor effects and the estimation of graduated circles and squares. *The American Cartographer,* 3(1), 65-74.

DeLucia, A.A. (1974). Design: The fundamental cartographic process. *Proceedings of the Association of American Geographers,* 6, 83-86.

Dent, B.D. (1993). *Cartography: Thematic map design.* Third edition. Dubuque, Iowa: Wm. C. Brown Publishers.

Fisher, P., Dykes, J., and Wood, J. (1993). Map design and visualization. *The Cartographic Journal,* 30(2), 136-142.

Flannery, J.J. (1956). The graduated circle: A description, analysis and evaluation of a quantitative map symbol. [Unpublished Ph.D. dissertation]. Madison, Wisconsin: University of Wisconsin-Madison.

Greenberg, G.L. (1971). Irradiation effect on the perception of map symbology. In E. Arnberger and F. Aurada (Eds.), *International yearbook of cartography* (pp. 120-126). Zurich: Orell Fussli Verlag.

Guelke, L. (1976). Cartographic communication and geographic understanding. *The Canadian Cartographer*, 13(2), 107-122.

Hearnshaw, H.M., and Unwin, D.J. (1994). *Visualization in Geographical Information Systems*. Chichester: John Wiley.

Keates, J.S. (1982). *Understanding maps*. London: Longman.

Kimerling, A.J. (1975). A cartographic study of equal value gray scales for use with screened gray areas. *The American Cartographer*, 2(2), 119-127.

Koeman, C. (1971). The principle of communication in cartography. In E. Arnberger and F. Aurada (Eds.), *International yearbook of cartography* (pp. 169-175). Zurich: Orell Fussli Verlag.

Kolacny, A. (1969). Cartographic information—A fundamental concept and term in modern cartography. *The Cartographic Journal*, 6(1), 47-49.

Lloyd, R. (1989). The estimation of distance and direction from cognitive maps. *Cartography and Geographic Information Systems*, 16(2), 109-122.

MacEachren, A.M. (1991). The role of maps in spatial knowledge acquisition. *The Cartographic Journal*, 28(2), 152-162.

MacEachren, A.M. (1994). *Some truth with maps: A primer on symbolization and design*. Washington, D.C.: Association of American Geographers.

MacEachren, A.M., and Ganter, J.H. (1990). A pattern identification approach to cartographic visualization. *Cartographica*, 27(2), 64-81.

MacEachren, A.M., and Taylor, D.R.F. (Eds.) (1994). *Visualization in modern cartography*. Oxford: Pergamon.

McCormick, B.H., DeFanti, T.A., and Brown, M.D. (1987). Visualization in scientific computing. *ACM SIGGRAPH Computer Graphics*, 21(6).

McGrath, G. (1971). The mapping of national parks: A methodological approach. In H.W. Castner and G. McGrath (Eds.), Monograph No. 2. *Cartographica* (pp. 71-76).

Morrison, J.L. (1974). A theoretical framework for cartographic generalization with emphasis on the process of symbolization. In G.M. Kirschbaum and K-H. Meine (Eds.), *International yearbook of cartography* (pp. 115-127). Chicago: Rand McNally and Company.

Morrison, J.L. (1976). The science of cartography and its essential processes. In G.M. Kirschbaum and K-H. Meine (Eds.), *International yearbook of cartography* (pp. 84-97). Chicago: Rand McNally and Company.

Muehrcke, P.C. (1982). An integrated approach to map design and production. *The American Cartographer*, 9(2), 109-122.

Muehrcke, P.C. (1990). Cartography and geographic information systems. *Cartography and Geographic Information Systems*, 17(1), 7-15.

Ommer, R., and Wood, C.H. (1985). Data, concept and the translation to graphics. *Cartographica*, 22(2), 44-62.

Peterson, M.P. (1985). Evaluating a map's image. *The American Cartographer*, 12(1), 41-55.

Petchenik, B.B. (1975). Cognition in cartography. In J. Kavaliunas (Ed.), *Proceedings* of the International Symposium on Computer-Assisted Cartography (Auto-Carto II); 21-25 September 1975; Reston, Virginia. Washington, D.C.: U.S. Department of Commerce - Bureau of the Census and American Congress on Surveying and Mapping - Cartography Division (pp. 183-193).

Petchenik, B.B. (1977). Cognition in cartography. In L. Guelke (Ed.), Monograph No. 19. *Cartographica* (pp. 117-128).

Petchenik, B.B. (1983). A map maker's perspective on map design research 1950-1980. Chapter 3. In D.R.F. Taylor, *Graphic communication and design in contemporary cartography* (pp. 37-68). London: John Wiley and Sons.

Ratajski, L. (1973). The research structure of theoretical cartography. In L. Guelke (Ed.), Monograph No. 19. *Cartographica* (pp. 46-57).

Robinson, A.H. (1980). Design research: A reply. *The American Cartographer*, 7(1), 78.

Salichtchev, K.A. (1978). Cartographic communication: Its place in the theory of science. *The Canadian Cartographer*, 15(2), 93-99.

Stanley, G.W. (1975). An investigation into certain aspects of spot depth numeral design on nauticalcharts. [Unpublished M.S. thesis]. Madison, Wisconsin: University of Wisconsin-Madison.

Steinke, T.R., and Lloyd, R.E. (1981). Cognitive integration of objective choropleth map attribute information. *Cartographica*, 18(1), 13-23.

Taylor, R. (1971). *A basic course in graphic design*. London: Studio Vista.

Thorndyke, P. (1981). Distance estimations from cognitive maps. *Cognitive Psychology*, 13, 526-550.

Tyner, J. (1992). *Introduction to thematic cartography*. Englewood Cliffs, N.J.: Prentice-Hall.

van Elzakker, C. (1991). Map use research and computer-assisted statistical cartography. In K. Rybaczuk and M. Blakemore (Eds.), *Proceedings* of the International Cartographic Association 15th Conference - Mapping the Nations; 23 September - 1 October, 1991; Bournemouth, England. London: ICA 1991 Ltd., 2, 575-584.

Wood, C.H. (1992). The influence of figure and ground on visual scanning behavior in cartographic context. [Unpublished Ph.D. dissertation]. Madison, Wisconsin: University of Wisconsin-Madison.

Wright, R.D. (1967). The selection of line weights for solid, qualitative line symbols in series on maps. [Unpublished Ph.D. dissertation]. Lawrence, Kansas: University of Kansas.

Challenge and Response in Cartographic Design

2

D.R. Fraser Taylor
Department of Geography, Carleton University

OPENING KEYNOTE ADDRESS TO THE SYMPOSIUM ON CARTOGRAPHIC DESIGN AND RESEARCH

Cartography, as with many disciplines in the information era, is undergoing dramatic and substantial change driven primarily by technological developments in both the computer and telecommunications fields. These changes, although driven by technology, affect the discipline in fundamental ways and involve changes in the conceptualization of the discipline itself and all cartographic processes including design. The response to the changes has been varied. Some have suggested that cartography might be dying (Loy, 1993), others predict that the need for cartographers will diminish and what few cartographers remain will function simply as design consultants (Rhind, 1993), whereas others see cartography as having a revitalized and promising future (MacEachren, 1994b). The answer to where cartography is going, in my view, depends on cartographers themselves. We are not passive victims of the factors affecting cartography—we are active participants and have a definite influence on our own fate. Arnold Toynbee, the English historian, has analysed history in terms of 'challenge and response' seeing great civilizations and empires emerging and declining over time largely as a result of their ability or failure to deal with challenges. Emergence and growth comes from a successful response to challenge; decline and, in some cases disappearance, result when societies fail to deal with new challenges in an adequate manner. Although Toynbee's analysis of history can be disputed by alternate paradigms and epistemological approaches, it has some merit and I have, therefore, chosen as the title of this chapter, "Challenge and Response in Cartographic Design". The future of cartography and of cartographers depends upon the nature of our response. A reactive and defensive response will lead to decline; a pro-active and offensive response will lead to renewed vigour for the discipline and the profession.

What then is the nature of these challenges? I should like to use as a graphic framework for my analysis the diagram in Figure 2.1, which is a revised version of one that I introduced some time ago. The three arms of the triangle represent three major aspects of cartography. The base of the triangle, which I have called 'Formalization', represents the cartographic production aspect where the emphasis is on production process and production techniques.

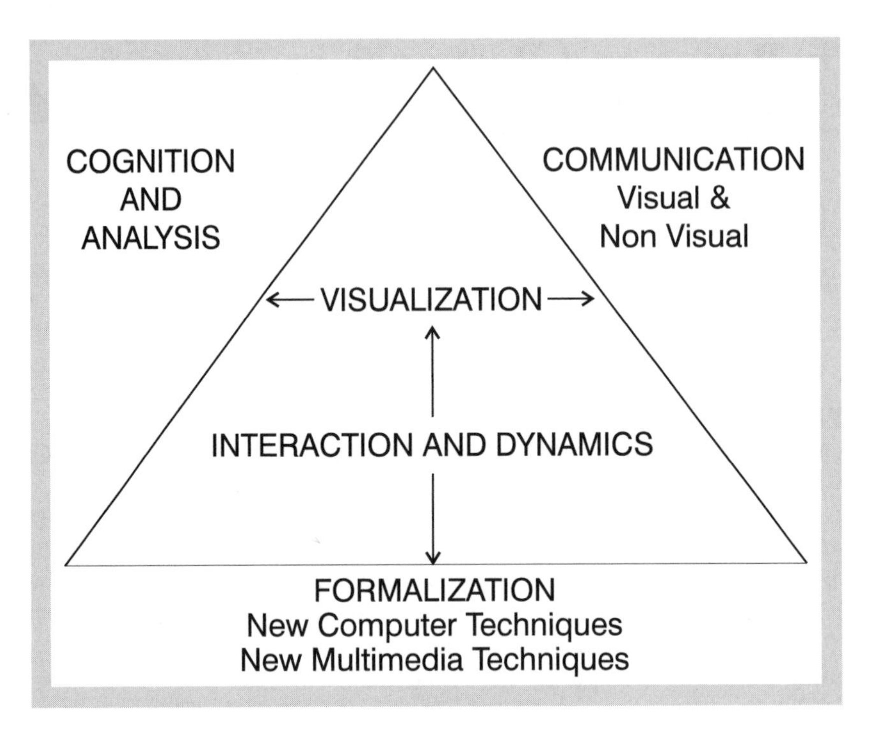

Figure 2.1 A conceptual basis for cartography.

Currently the cartographic design challenges are to come to terms with the emerging computer, multimedia and telecommunications developments which are taking place at an ever-increasing speed. Multimedia, for example, is exploding as a production tool, but like so many other technological developments, the needs of cartography are not the main driving force, which is the emerging market for entertainment, training, education and marketing. Cartographic designers have a great opportunity to utilize multimedia tools in new and exciting ways. Although there is evidence of this, much more needs to be done. It is not enough to decry the cartographic incompetence of some of the designers of the new multimedia products as we continue to do with some GIS programs. The criticism may be justified, but must be complemented by action and the introduction of new or improved cartographic products as an integral part of the emerging market.

In December, 1993, the American Dialectic Society chose as its new expression for the year the term "information superhighway". The technological convergence of computer, broadcast and telecommunications technologies and media will realize a long standing promise of the information revolution which is the interactive delivery of information directly into the home. I have noticed recently in both the Toronto *Globe and Mail* and the *Ottawa Citizen* newspapers that several of the syndicated cartoonists have picked up on the move toward the information superhighway.

One of the cartoon strips depicts the results when an individual resists joining in on the move to the electronic medium and explicitly remarks that some reluctant users will be run over in the rush to it, thereby becoming "roadkill". The message the cartoon conveys is especially relevant. Cartography and cartographers must become relevant players in the emerging information superhighways, or we may indeed become roadkill! Cartographic information is already being transmitted over the new data highways, but at present, this is only a tiny fraction of the information flow. The highways can also be used for cartographic and geographic research, and Michael Goodchild has drawn attention to the innovative use of INTERNET for this purpose (Goodchild, 1993).

What has been called EDUTAINMENT is a growing trend, and there are already geographic and cartographic examples such as the computer game Magellan, which is an interactive touchtalk globe that can be used either as a computer game, or as an educational tool. In this case, virtual keyboards respond to touch and give taped audio information (which can be updated by new tapes) on each country. The production of cartographic products on CD-ROM is increasing, and second generation electronic atlases in this form are beginning to appear. The National Geographic Society is already marketing CD-ROMs, and in Alberta, the Jean-Talon project is creating modular computer operated, multimedia resources on Canada's geography and history (CMMM, 1992).

There are many other technological developments that will affect cartographic design which time does not allow me to consider. It is clear, however, that the new technologies, the information superhighways, interactive video and TV, video on demand, and virtual reality all pose challenges to cartographic design to which adequate responses must be found.

Before turning to the other two arms of the triangle shown in Figure 2.1, I would like to emphasize a crucial point. The nature of recent technologies developments allows for a dynamic interaction between the user and the product, which has never been possible before , and to deal effectively with interaction is perhaps the greatest single challenge to cartographic design. This is much more than a technological challenge as it affects the balance of power and blurs the distinction between creators of cartographic products and users of cartographic products. The single map or single product solution is no longer viable, and process, including the epistemological bases and the nature of the questions and topics being addressed, is likely to be much more important than the cartographic products themselves. This is a fundamental and revolutionary change made possible by technological developments and which is much more significant than the technologies themselves.

A second arm of the triangle in Figure 2.1 is "Communication", which includes both visual and non-visual elements, such as sound and touch (Krygier, 1994; Vasconcellos, 1993). Communication has always been important to cartography; and in the 1970s, the communication model of cartography was dominant, although interest in it has declined. However, as some of my earlier diagrams indicated, it is very easy to get lost on the information superhighways, and navigation of them is far from easy. In addition, we are being inundated with increasing volumes of data, and the need to convert these into meaningful and useful information has never been greater. Many of the new interactive systems and databases offer too much, rather than too little, choice which can lead to confusion and serious under-utilization of the extensive information in the data bases. Cartographic products such as the map are ideal media for effective organization, presentation and communication of information on a wide variety of subject areas, and the importance of a better understanding of the cartographic communication process in the merging dynamic and interactive situation created by the new technologies may revitalize cartographic communication and design research.

There will, however, need to be a searching re-examination of some previous research approaches to cartographic communication which had serious limitations. There is the ongoing issue of ensuring that research results are fed back into production processes as well as the argument that research on individual map elements has serious methodological drawbacks as it ignores the holistic nature of images. There are additional problem areas. The new electronic products are different from the paper map, and the human brain's perception of electronic images is also different. Research in neuromechanics has shown that different parts of the brain respond to different kinds of visual input, and that variables such as shape, motion, size, and colour are dealt with in different parts of the brain with much processing taking place outside the primary visual cortex which was the focus of much earlier research (Sejnowski and Churchland, 1989).

Cartographic research on user reaction to the new products is still quite modest, and there are many unanswered questions (McGuiness, 1994). The products themselves are multidimensional, including variables such as sound about which cartographers know relatively little. As McGuiness points out, emphasis will also have to shift from stimulus to process and concept. Cartographers' interest in their own products often leads to an emphasis on stimulus to the neglect of other user and process variables. Recent physiological research (Booth, 1989) indicates that effective interaction between user and the product is key to learning and communication. The degree of interactivity possible between the user and the emerging range of new cartographic products, therefore, becomes of central importance to cartographic design as was argued earlier in this chapter.

The third arm of the triangle is cognition and analysis. Cartographic cognition is a unique process as it involves the use of the brain in recognizing patterns and relationships in their spatial context. This analytical function, as I have argued elsewhere (Taylor, 1991), cannot be easily replicated by GIS software which process information in an essentially linear fashion. It is in the cognitive and analytical area where scientific cartographic or geographic visualization has a central role to play. Alan MacEachren has defined visualization as "a human ability to develop mental images (often of relationships that have no visual form) together with the use of tools that can facilitate and augment this ability" (MacEachren, 1991: 12).

MacEachren et al. (1992) following DiBiase (1990) present four stages of visualization in the research sequence as shown in Figure 2.2. These stages are exploration, confirmation, synthesis, and presentation. These are conceptualized as a

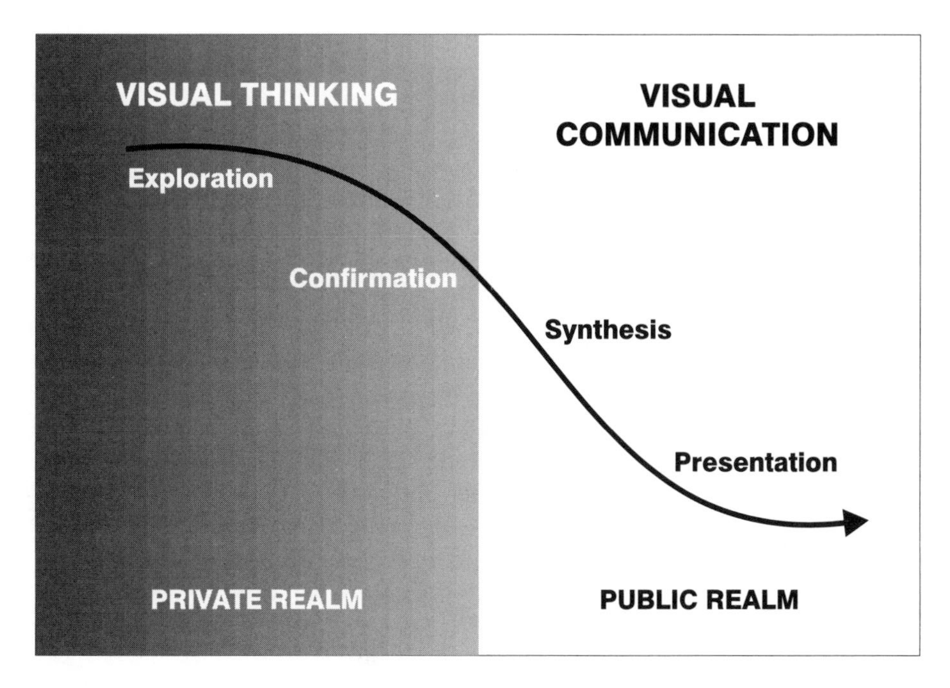

Figure 2.2 Four stages of visualization.

continuum with the first two being in the largely private realm of visual thinking and the latter two in the more public realm of visual communication. MacEachren's more recent work (MacEachren, 1994b) is again emphasizing the analytical and cognitive aspects of visualization, which he sees in terms of map use distinguished by the combination of variables along three continua, as shown in Figure 2.3: private to public, unknown to known, and high interaction to low interaction.

MacEachren draws a distinction between the use of visualization for analysis and its use for communication, although he sees this as a continuum. My own view is that the distinction may not be a clear cut one, and that the key is interaction. A highly interactive system, especially one which allows the user access on an individual terminal, even at the 'public' and 'known' end of the other two continua in 'cartography cubed', may result in the user analysing and interpreting in ways quite different from the perspectives of those who created the cartographic product being used.

Cartographic visualization will have importance both to the analytical and cognitive function of cartography, as well as to the communication function and should stimulate the whole field of cartographic design as MacEachren has indicated in his recent book on design, *Some Truths With Maps: A Primer on Symbolization and Design* (MacEachren, 1994a), which I recommend to you.

Modern cartographic visualization is an extension of spatial analysis and imaginative and creative data presentation which have been present in cartography

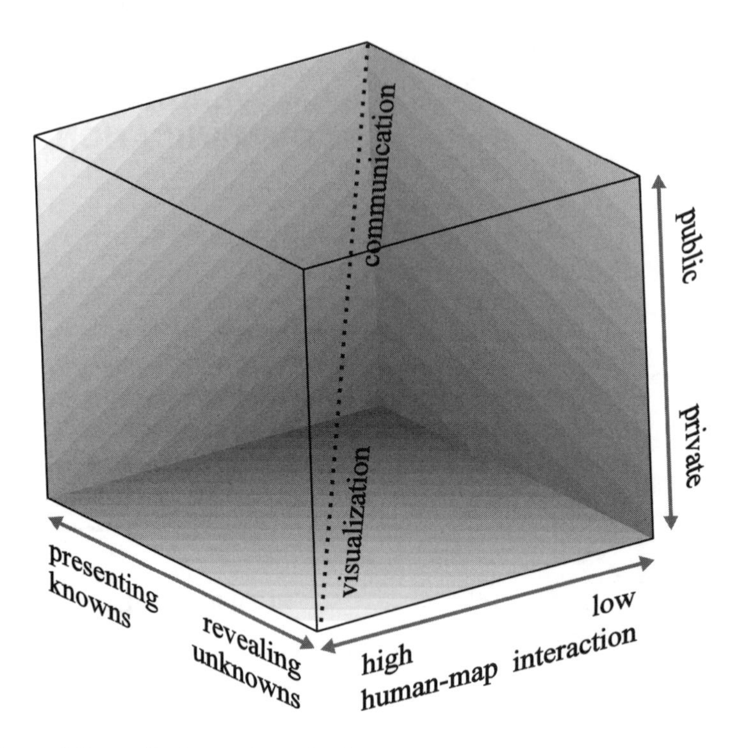

Figure 3.2 Cartography cubed.

for centuries. As computer-based techniques are used, this can be seen as a scientific methodology, but at the same time it involves the use of imagination, intuition and artistry, especially in the development of new multimedia products and in the exploding use of visual analysis in a whole range of new topics with which cartography has not been traditionally involved. The challenges and opportunities for cartographic design in this area are growing exponentially.

What is also growing is the recognition that cartography and cartographic design must broaden their horizons and re-examine their basic concepts. The normative, positivistic approach of many cartographers must be supplemented by an understanding of different epistemological approaches such as those of the humanists and the structural/realists. This will affect not only what we choose to represent, but how we choose to represent it. Here the new technologies that facilitate the introduction and integration of different kinds of cartographic products with pictures, sound, motion, movement, time, and abstract concepts are an advantage. We are also reminded by the work of Harley (1990), Wood (1993) and others of the importance of both social context and cultural variety. Cartographic designers are not value-free individuals utilizing neutral technologies. Both the social and cultural contexts of cartographic design deserve consideration.

To this I would add human emotion and values. I was struck by a comment that appeared on August 1 , 1994, in the Toronto *Globe and Mail* (A17) which repeats an editorial from the *Economist*: People are not thinking machines, (they absorb at least as much information from sight, smell and emotion as they do from abstract symbols), and the world is not immaterial. 'Virtual' reality is no reality at all. The weight on mankind of time and space, of physical surroundings and history—in short, of geography—is bigger than any earthbound technology it is ever likely to lift" (*Globe and Mail*, 1994, A17).

At the Cologne meetings of the ICA, there was a huge exhibit of the most modern computer-based technology, GEOTECHNICA, which attracted the attention of thousands of people together with impressive national and international map exhibits, but the display that attracted most attention was that of the children's maps entered in the Barbara Petchenik Map Competition. Twenty-seven member nations sent in several hundred entries. Of the 10 winning entries, two dealt with environmental issues: 'Greenland' by 13 year-old Yue Seiko of Japan, and 'No Smoking' by Agnes Horvath of Hungary, age 15. Six dealt with peace and harmony in the world: 'Peace in the World', Alfonsas Lekavicius, 11, of Lithuania; 'Children around the World', Jana Durankova, 13, Slovakia; 'Pomul Victu', Bratu Poud, 15, Romania; 'Damai Di Dunia' (Peace in the World), Trihedi Okatianko, 12, Indonesia; 'Stitching up the World', Andre Nguyen Machiaverni, 11, Brazil; and 'Greetings from the World', Henry Huang, 14, United Kingdom. The final two winners were: 'Head Ship and Tree' by Siskara Samanmali Ratnayake, 15, Sri Lanka, and 'The Horse Population of the World', Taiga Marthens, 11, United States. I had the privilege of seeing the several hundred entries that did not reach the final display and can tell you that there were some remarkably creative cartographic designs.

I started this chapter by posing a 'challenge and response' theory for cartographic design. My own view is a highly optimistic one. Cartography and cartographic design have a bright future. Computer technologies are the direct source of many of the new challenges, but the core of the response is the creativity, curiosity, imagination and enthusiasm of cartographers themselves, and I look forward to your contributions to the responses, to the challenges facing us.

REFERENCES

Booth, P.A. (1989). *Introduction to human computer interaction.* Hillsdale, N.J.: Lawrence Erlbaum Associates.

CMMM. (1992). The Jean-Talon Project. *The Canadian Multi-Media Magazine*, 1(5), 6-8.

DiBiase, D.W. (1990). Scientific visualization in the earth sciences. *Earth and Mineral Sciences*, 59(2), 13-18.

Globe and Mail (1994). August 1, 1994, p. A17.

Goodchild, M.F. (1993). Ten years ahead: Dobson's automated geography in 1993. *The Professional Geographer*, 45(4), 444-46.

Harley, J.B. (1990). Cartography, ethics and social theory. *Cartographica*, 27(2), 1-23.

Krygier, J.B. (1994). Sound and geographic visualization. In A.M. MacEachren and D.R.F. Taylor (Eds.), *Visualization in modern cartography* (pp. 149-166). Oxford: Elsevier.

Loy, W. (1993). Is cartography dead? Association of American Geographers, Cartography Specialty Group *Newsletter*, 14.

MacEachren, A.M. (1991). Visualizing uncertain information. *Cartographic Perspectives*, 13, 1-19.

MacEachren, A.M. (1994a). *Some truths with maps: A primer on symbolization and design*. Washington, D.C.: Association of American Geographers.

MacEachren, A.M. (1994b). Visualization in modern cartography: Setting the agenda. In A.M. MacEachren and D.R.F. Taylor (Eds.), *Visualization in modern cartography* (pp. 1-12). Oxford: Elsevier.

MacEachren, A.M., Buttenfield, B.P., Campbell, J.B., DiBiase, D.W., and Monmonier, M. (1992). Visualization. In R.F. Abler, M.G. Marcus, and J.M. Olson (Eds.), *Geography's inner worlds: Pervasive themes in contemporary American geography* (pp. 99-137). New Brunswick, N.J.: Rutgers University Press.

McGuiness, C. (1994). Expert/novice use of visualization tools. In A.M. MacEachren and D.R.F. Taylor (Eds.), *Visualization in modern cartography* (pp. 185-199). Oxford: Elsevier.

Rhind, D. (1993). Mapping for the new millennium. *Proceedings* of the 16th International Cartographic Conference, Vol. 1. Bielefeld, Germany: German Society of Cartography, pp. 3-14.

Sejnowski, T.J., and Churchland, P.S. (1989). Brain and cognition. In M.I. Posner (Ed.), *Foundations of cognitive science* (pp. 301-356). Cambridge, Massachusetts: MIT Press.

Taylor, D.R.F. (1991). Geographic information systems: The microcomputer and modern cartography. In D.R.F. Taylor (Ed.), *Geographic Information Systems: The microcomputer and modern cartography* (pp. 1-20). Oxford: Elsevier.

Vasconcellos, R. (1993). Representing the geographic space for visually handicapped students: A case study on map use. *Proceedings* of the 16th International Cartographic Conference, Vol. 1. Bielefeld, Germany: German Society of Cartography, pp. 993-1004.

Wood, D. (1993). Maps and mapmaking. *Cartographica*, 30(1), 1-9.

Geography and Cartographic Design

John B. Krygier

Department of Geography, Pennsylvania State University

INTRODUCTION

A peculiar mixture of malaise and exuberance characterizes attitudes about the status of cartography and cartographic design research. Bill Loy, chair of the American Association of Geographers Cartography Specialty Group asks "Is cartography dead?" as he relates a seeming shift of interest away from cartography (Loy, 1993). Patricia Gilmartin notes a precipitous decline in communication oriented map design research (Gilmartin, 1991). At the same time, excitement surrounds the possibilities of animation, sound, video, hypermedia, visualization, and geographical information systems (MacEachren and Taylor, 1994). The future of cartographic design research is, in part, dependent upon how cartographers choose to approach such new technologies and methods being used for visualization and analysis in geography. This chapter describes an approach to cartographic design driven by its relations to geography and the practice of geographic inquiry.

The first section of this chapter suggests that geographers conceptualize maps and mapping technologies as geographic methods and are interested in how maps and mapping technologies relate to and serve as means in the process of geographic research. Geographers have questioned the assumption that writing, numbers, and visual representations are specific, unproblematic, objective procedures which can be considered independent from the complexities of geography (Sayer, 1992). In general these critics argue that geographical methods are fundamental to what geographers do—how we construct geographic knowledge—and are embedded within and shaped by conceptual, theoretical, and philosophical assumptions. Methods are active, conceptually, theoretically, and philosophically determined and determining means by which we conceive of, think about, and present our understanding of the world and its inhabitants. As evidence for this, I suggest that debates in geography over theory and the nature of spatial structure produce critical commentary on maps and visual methods. Maps are often criticized in paradigm shifts in geography, or during times of haughty debate over the conceptualization of geographic entities. Because conceptual, theoretical, and philosophical assumptions vary immensely within geography, ignoring this diversity by treating maps and mapping technologies as objectified, generic techniques independent from geography is problematic. This conceptualization of maps and mapping technologies as geographic methods, I argue, suggests the possibilities of an explicitly geographic approach to cartographic design.

In the second section of this chapter I discuss three general cartographic design issues which arose in the context of a geographic case study undertaken to investigate the relations between geography and cartographic methods. My interest is in exploring how geographic understanding can guide and shape research on cartographic design and methods while drawing cartography closer to geography.

GEOGRAPHY AND CARTOGRAPHY

The intimate relation between maps and geographic conceptualization and theory is revealed in the relatively uniform view of maps in academic geography into the 1950s. We have, for example, Hartshorne's famous quote from *The Nature of Geography*:

> *So important, indeed, is the use of maps in geographic work, that, . . . if (the) problem cannot be studied fundamentally by maps—usually by a comparison of several maps—then it is questionable whether or not it is within the field of geography* (Hartshorne, 1939: 249).

. . . and that of Mill:

> *In geography we may take it as an axiom that what cannot be mapped cannot be described* (Woolridge and East, 1951: 64).

. . . and that of Ullman:

> *I recall an economist once telling me that the map was a theory which geographers had accepted* (Ullman, 1953: 57).

Hartshorne and Mill extol the map as a geographic method which *defines* the nature of geography. Ullman's economist friend, given the advantage of an outside view, notes the substantive relationship between maps and geographical theory. All three comments conceive of the map as defined by and defining of geographic thinking and conceptualization.

While the intimacy between maps and geography in all of the above cases is assumed, this relationship was seldom explored in detail. J.K. Wright was an exception, and his seminal paper "Mapmakers are Human" conceptualized maps and cartographic design as geographic methods and described their relationship to scientific integrity, judgement, and aesthetics (Wright, 1942). The lack of explicit rigor in map making noted by Wright gained attention during World War II and would serve as an impetus for the academic study of cartography and map design in the 1950s.

1950s

Robinson, in his 1952 book *The Look of Maps*, saw the discipline of cartography based upon a very rigorous and focused research agenda: "The development of design principles based on objective visual tests, experience, and logic; the pursuit of research in the physiological and psychological effects of color; and investigations in perceptibility and readability in typography are being carried on in other fields . . . such a movement in cartography cannot fail to materialize" (Robinson, 1952: 13-14). *The Look of Maps* was influenced by Wright's ideas in "Mapmakers are Human" and Robinson's experience during World War II, where he observed the lack of map design and analysis skills, the lack of guidelines with which to construct optimal maps for military needs, and gravely noted the threat of propaganda maps. Robinson's elemental and "scientific" approach to cartography inspired Jenks to speculate that "the scope of . . . cartography . . . is broad enough to justify the organization of independent departments" of cartography (Jenks, 1954: 321). Geography, thus, need not necessarily be part of cartographic research.

The adoption of a psychological foundation for cartographic research and the possibility of an independent academic cartography inspired criticism from geographers. It was not entirely clear what all the excitement about "visual tests" on the "physiological and psychological effects of color" had to do with geography, geographic conceptualization, and geographic thinking. Further, the call for "independent departments of cartography" set some geographers on edge. Mackay called for the preservation of cartography as a geographic method and asserted that "cartography by itself is sterile" (Mackay, 1954: 13). Beishlag forecast in 1951 that

> *. . . many of the new enrollees in cartography classes will not want to learn to be cartographers but to be better geographers. What happens to them in cartography classes may determine the future of relations of cartography and geography. If cartography teachers put these new students to learning hand-lettering or to constructing a series of different map grids from mathematical calculations, then good relations between cartography and geography may be jeopardized. Such training is neither interesting nor very useful to most geographers* (Beishlag, 1951: 6).

Thus the discipline of cartography began and developed into a somewhat uncomfortable relationship to geography (Muehrcke, 1981). Certain cartographers were convinced by their war experience, military and government support, and the availability of psychometric research methods that more attention should be focused on the map in and of itself as a communication device. That this research could improve the map as a geographic method was undoubtedly in the minds of these cartographers, but the *focus* of cartographic research was simply *not* on maps as geographic methods and means of thinking. Certain geographers, on the other hand, seemed reluctant to accept cartographic research that seemed so alienated from geography. They undoubtedly saw maps in the same light as Hartshorne and Mill—as inseparable from geography. What was of interest to these geographers was how maps related to geographic thinking and conceptualization, questions most academic cartographers and geographers chose not to explicitly address. Ullman's intriguing anecdote about the map as "a theory which geographers had accepted" remained (and remains) unexamined (Ullman, 1953).

1960s and 1970s

The cartographic communication model served as the structuring model and "theory" of cartography from its formal introduction in the mid-1960s (Keates, 1964; Kolacny, 1969). The rapid adoption of the communication model was due to two factors. First, the communication model was a formalized statement of what had already been the agenda of most cartographic research in the late 1950s and the 1960s. The second and more general reason for rapid adaptation of the communication model is that it reinforced and capped off a desire for cartography to be considered an objective science based on an explicit cartographic theory.

The apex and culmination of this science of cartographic design is best illustrated in a series of articles by Joel Morrison (Morrison, 1974a; Morrison, 1974b; Morrison, 1976; Morrison, 1978). Drawing from strict psychophysical theory, he deemed it the task of the science of cartographic design to devise a system of map symbols matched with the optimal "physiological and psychological responses of the map user" (Keates, 1982: 93). Morrison's behaviouralism assumed a mechanistic commonality among map user's perception, response, and actions when faced with a map, and the task of the science of cartography was to delimit this perceptual commonality. The end product of this work would be, in essence, a grammar of mapping which could be applied to all mapping situations regardless of context. Morrison predicted that the science of cartography would result in "the freedom to map abstractly and to develop methodology free of specific real world distributions." Morrison asserted that "cartography can operate like English and mathematics in relation to the social sciences" and that cartography "might easily become independent of geography, history, or any other discipline concerned with the mapping of 'real' distributions." "Cartography," Morrison proclaimed, "is indeed a science in its own right" (Morrison, 1974b: 9-10, 12).

While Morrison and other cartographic researchers were pursuing the psychometrically optimal map, the issue of the map as a geographical method and a means of geographical thinking had surfaced once again among geographers. Purveyors of three very different conceptual and theoretical perspectives in geography—humanism, Marxism, and positivism—all called attention to the map as a geographic method and the apparent lack of interest in this perspective among academic cartographers. Wood, for example, argued for a humanist-inspired "cartography of reality":

> *Unlike contemporary academic cartography, a cartography of reality must be humane, humanist, phenomonological, and phenomenalist.... It must reject as inhumanly narrow both the data base and subject matter of contemporary academic cartography ....* (Wood, 1978: 207).

From a very different theoretical and conceptual foundation, Lacoste, a Marxist, asserted that

> *It is important that we gain (or regain) an awareness of the fact that the map, perhaps the central referent of geography, is, and has been, fundamentally an instrument of power. A map is an abstraction from concrete reality which was designed and motivated by practical (political and military) concerns; it is a way of representing space which facilitates its domination and control. To map, then, means to formally define a space along the lines set within a particular epistemological experience; it . . . serves the practical interests of the State machine; it is a tedious and costly operation done for, and by, the State* (Lacoste, 1973: 1)

Bunge, an early purveyor of quantitative methods in geography, argued in the 1962 edition of his *Theoretical Geography* that

> *The reason that geography has always paid such respect to maps is that, besides their vital role as a device for storing facts areally for uniform regional geography, they have been the logical framework upon which geographers have constructed geographic theory. They have been to geography what mathematics has been to some other disciplines* (Bunge, 1962: 33).

If visual methods—primarily in the form of the map—were as closely related to geographical theory and understanding as these geographers claimed, and if there was dissatisfaction with what geography was or had accomplished from any of these often incommensurate perspectives, the map may be part of the problem. In his influential methodological tome *Explanation in Geography*, Harvey asserted that

> *The map, it must be understood, is a model of spatial structure. Before we can accept that map as a theory about actual spatial structure (and therefore act upon the basis of that theory) we require to show that the model is empirically realistic with respect to the phenomena it is designed to represent. The use of the map, like the use of any kind of model, poses a number of problems concerning inference and control. It is time, therefore, that these methodological issues were explicitly and comprehensively discussed* (Harvey, 1969: 376).

Geographers such as Wood, Lacoste, Bunge, and Harvey were obviously attuned to the same insight recorded by Ullman in 1953—mainly that "the map was a theory which geographers had accepted" (Ullman, 1953). All of these geographers recognized the map as a visual method imbued with conceptual and theoretical assumptions. The map, as Harvey asserts, is "a theory about actual spatial structure" and the conceptualization of spatial structure was at the centre of raging debates among humanists, Marxists, and spatial scientists. How did these fundamentally different conceptualizations of space and phenomena and humans and their relations relate to maps and visual representation? Again, such issues received little attention from cartographers and geographers.

1980s and 1990s

Academic cartography has now, in large part, forsworn explicit allegiance to a strict, perceptually-oriented notion of the communication model. Cartographers have increasingly focused on issues of map cognition based in cognitive psychology (Olson, 1979; Blades and Spencer, 1986; MacEachren, 1995), or on issues of analytical cartography and geographic information systems based in quantitative geography (Tobler, 1979; Hearnshaw and Unwin, 1994). While these approaches to cartography were developing, critical human geographers and historians of cartography fired yet another salvo at maps and cartographers, akin to critiques made in the 1960s and 1970s and noted in the previous section (Cosgrove, 1984; Harley, 1989; Wood, 1992). These critics focus on what they see as the naive understanding and use of maps and GIS (Miller, 1992). Maps and GIS, the critics argue, are imbued with particular theoretical and conceptual assumptions—particularly those of spatial analysis—which make them *antithetical* to other kinds of geographical work. Once again, debates in geography over theory and the nature of spatial structure have produced critical commentary on maps and visual methods.

For example, Pudup reiterates Harvey's sentiments of 20 years earlier when she asserts the problematical nature of the map as a geographical method for the "new" regional geography: ". . . map's methodological contribution and, just as important, their limitations are not explored" (Pudup, 1988: 374). Pile and Rose argue that mappable space "limits the possibility of critique by refusing to acknowledge other kinds of space" (Pile and Rose, 1992: 131). Rose argues that seeing and vision are "masculinist," implying control and domination, and suggests that visual representations of landscapes are used to control space just as pornography is used to control women (Rose, 1993). Harley asserts that "Much of the power of the map, as a representation of social geography, is that it operates behind a mask of seemingly neutral science. . . . maps are at least as much an image of the social order as they are a measurement of the phenomenal world of objects" (Harley, 1989). Smith argues that the problem with GIS "lies in the outlandish disciplinary ambitions, the radical exclusion of other perspectives, and the dangerous and self-defeating renunciation of an intellectual (as opposed to technical) agenda that too often accompany the programmatic advocacy of GIS" (Smith, 1992: 258).

In all of these cases, mapping and GIS are seen as problematical and limiting methods because the manner in which they shape and are shaped by particular social, cultural, and geographic conceptualizations is not understood or explicitly discussed. The issue, then, remains nearly unchanged from that posed by Harvey in 1969: maps are "a theory about actual spatial structure" (Harvey, 1969). The conceptualization of spatial structure is at the centre of current debates between spatial scientists, Marxists, and humanists, to which can be added the perspective of feminists, post-structuralists, and critical social theorists. How do these fundamentally different conceptualizations of space and phenomena and humans and their relations relate to maps and visual representation? The problem lies in the unarticulated relationship between visual methods and the various (and often incommensurable) theoretical, conceptual, and methodological contexts of geography. This suggests that differences in geography may imply differences in cartographic design and geographic visualization, and that an understanding of geography, and especially its theoretical and conceptual differences, may be used to guide cartographic design. There are, in other words, limitations to an a-geographic or geographically naive cartography.

Geography is not homogeneous. At the most fundamental and philosophical level, geographers make very different assumptions about what the world is like. Such assumptions shape the manner in which the world is conceptualized and understood (Sayer, 1992). Geography has at least three basic philosophical traditions: classical empiricism/positivism, which underpins quantitative geography; idealism, which underpins humanistic geography; and realism, which underpins Marxist, structurationist, and social theory approaches to geography. These three fundamental approaches to geography all conceptualize and theorize spatial structure and geographic phenomena very differently. Further, issues of difference based on gender, race, class, and culture increasingly dominate the literature of geography (Young, 1990). Cartographers have had little to say about how such fundamental differences in geographic philosophy and theory relate to maps and cartographic design. Nor have cartographers addressed the issue of why certain approaches to geography tend to use maps and other visual representations more than others.

Given the above issues, how, in particular, is this geographically informed approach to cartographic design to proceed and what design questions and issues does it raise? The second section of this chapter briefly reviews a case study focused on addressing the geographic and methodological issues raised by a realist-oriented approach to landscape geography (Krygier, 1995a). I explore three general cartographic design issues generated by the case study and provide an example of how geographic and methodological issues shape cartographic design.

INCORPORATING GEOGRAPHY IN CARTOGRAPHIC DESIGN

The first section of this chapter argues that geographers have long been interested in the relations between maps and geography, and specifically how maps relate to conceptualization and theorization in geography. Critics of maps have claimed that maps represent particular conceptualizations of spatial structure and that the methodological function of maps is not sufficiently understood. Certain geographers use maps and other visual representations extensively, others use few or none. This, and the fact that maps are often criticized in paradigm shifts in geography or during times of haughty debate over the conceptualization of geographic entities, suggests that more attention needs to be paid to the complex relationship between a multifaceted geography and cartographic design.

In this section of the chapter I briefly review a geographic case study undertaken as a means of exploring the relations between geography and cartographic design and discuss three general design issues which arose in the process of the research. The case study and the issues it raises suggests, I believe, that a geographic approach can open up new perspectives on cartographic design.

The case study is situated in what has been called the "new" regional or "new" landscape geography (Cosgrove, 1984; Pred, 1984; Soja, 1986; Pudup, 1988; Daniels, 1989; Sayer, 1989; Thrift, 1990). Current debates in realist-oriented landscape and regional geography have led to a methodological and theoretical imperative to use intensive studies of particular landscapes, regions, and places as a means for understanding the complex relations between society, culture, and space (Sayer, 1992). My case study is an attempt to undertake such an intensive study.

My research focuses on a seemingly marginal and remote landscape in rural Pennsylvania—the Quehanna Wild Area. Initial impressions of Quehanna are chaotic and paradoxical. A 16-sided polygon encloses the 50,000 acre state Quehanna Wild Area and its quiet woods, nuclear reactor, meandering nature trails, derelict missile test pads, clear streams, toxic waste dumps, Civilian Conservation Corps camp ruins, boot camp convicts driving trucks, Rare Reptile Preserve, and caved-in industrial buildings. This odd landscape provides a lens through which to discuss and explain the geographies of a landscape in its historical, social, and cultural context. The case study focuses in particular on the issue of landscape dereliction. How did the landscape of Quehanna develop? How was the development of the place related to the dereliction of that place? Who was involved in the development and decline of the place? How did the scale of incorporation at state, national, and international scales vary over time? All of these questions raise a further methodological question: how does one realize such an intensive, landscape and place oriented project? How do you represent Quehanna (visually and otherwise) so as to explicate the origins, functions, and meaning of its landscape? The geographical literature contains discussions of the methodological problems and difficulties involved in realizing such intensive, place and landscape-oriented studies (Sayer, 1989; Soja, 1989). My case study is an attempt to address the geographical, methodological, and, in particular, representational issues raised by this geographic literature.

The case study was chosen, in part, because of personal interest, but also because it was very different from the more quantitative case studies which tend to dominate cartographic design and visualization research. My hope was that by undertaking this particular geographic case study I would gain some insight into how theoretical and conceptual issues and differences in geography relate to cartographic and other visual methods. At the same time I hope to be able to address the geographic criticisms of maps and other visual representations as geographic methods. The rest of this section consists of three general cartographic design issues raised by my case study. The particular design strategies created to address these issues are illustrated in a prototype hypermedia application (Krygier, 1995b).

Visual Functions and Cartographic Design

A fundamental assumption of a geographic and methodological approach to cartographic design is that there is a range of functions which maps and other visual representations serve within the geographic research process. DiBiase has speculated that these generalized functions include exploration, where questions and hypotheses are sought; confirmation, where answers and hypotheses are checked; synthesis, where ideas are brought together; and presentation, where a final, optimal visual representation is designed (DiBiase, 1990). Thus, maps and other visual representations are conceptualized as means rather than as only ends in the geographic research process. Different design strategies, then, are appropriate at different stages in the process. Cartographic research has discussed and demonstrated the differences between exploratory and presentation graphics: the former entails design focused on looking at data in many different ways; the latter entails design focused on presenting a single, optimal view (MacEachren et al., 1992; Monmonier and MacEachren, 1992).

Yet it is also important to consider how the research process differs among different approaches to geography. For example, the manner in which a quantitative geographer goes about research is different from the manner in which a humanistic geographer goes about research. My case study raised several issues about the generalized research process model in relation to differences in the research process within geography.

My case study confronted me with a large quantity of "messy" data collected in the process of researching Quehanna: historical photographs, maps, text, along with statistical data, interviews, my own photographs, videos, maps, graphs, and diagrams. All of this "data" needed to be organized, linked together, and understood in tandem with the theoretical ideas defined in the literature of the 'new' regional geography and the questions my study sought to address. The voluminous "messy" data I collected required a form of *synthesis* before I could really begin to "explore." As the research proceeded, the important role of synthesis became evident: I needed to design visual methods to assist me in exploring relations and conceptual ideas, synthesizing and resynthesizing the "messy" data to attempt to understand the complexity of the landscape, its history, and its current status. I found hypermedia, in its ability to bring together different media and create non-linear linkages between visual materials and text, to be ideal for this synthetic function.

My case study revealed the importance of synthesis, then, in the research process and particularly for the kind of geography in which I was engaged. In a more general sense, synthesis is seen by geographers as one of geography's primary academic strengths and contributions (Turner, 1989). Some geographic research, such as my case study, is fundamentally based on synthesis. Other research, such as certain quantitative studies, are initially less dependent upon synthesis, but may depend on it at a later time to recontextualize the findings. For example, current trends in global warming research in geography indicate an interest in synthesizing quantitative and modeling research results with cultural and social issues. In either case, how much do cartographers know about designing visual methods for synthesis? My case study reveals that hypermedia is particularly adept at serving as a basis for synthetic visualization methods. This suggests that there are substantial relations between the form of hypermedia and geographic synthesis, and that hypermedia can be conceptualized as a substantial geographic method rather than just a flashy new technology (Krygier, 1995a).

Geographic Conceptualization and Cartographic Design

A second general consequence of a geographic approach to cartographic design is that there are differences in the manner in which the same geographic entities are conceptualized by different theoretical approaches to geography. In my case study, it quickly became evident that there are different and contested conceptualizations of region and landscape, and that these geographic differences implied differences in design. Thus, cartographic design varies as conceptualizations of the same geographic entity vary. Geographers have commented on this relationship between conceptualization and maps. Natter and Jones (1989) argue that "by combining the 'reading' of place with its deconstructed representation in maps we may come nearer to closing the circle on understanding place" (1989: 114). Lewis argues that cartographic representation and design has "lagged behind social theory" and urges geographers to develop cartographic methods more suitable to the spatial conceptualizations of current regional geography (Lewis, 1991: 621).

My general concern in the case study was how to design visual means for a research project situated in the "new" regional/landscape geography. As previously noted, recent geographic literature has discussed the importance of situating particular regions and landscapes within their larger contexts and exploring how such regions are simultaneously effected by and effect such contexts. This can be contrasted with a more traditional conceptualization of the region as a spatial unit with internal consistency, focused inward rather than outward. The literature on the "new" regional and landscape geography posed a series of representational challenges: First, how to represent time and change while avoiding problems of serial causality, for example, assuming that later events are necessarily caused by earlier events. This entailed designing visual means for organizing, representing, and exploring the data and information about Quehanna in both a chronological and non-chronological manner. Second, how to represent the elastic boundaries and changing scale of incorporation of a region over time. This entailed designing means for visually representing the different spatial scales of Quehanna over time, their interrelations, and connections to particular events, people, and artifacts in the Quehanna landscape. Third, how to represent the experience of different groups (class, gender, race) and their relations to the Quehanna landscape and each other over time. Fourth, how to represent the relations between abstract theoretical structures and concrete events and artifacts in the Quehanna landscape. This entailed designing and linking abstract diagrams to particular images and maps and text relating to particular events in the Quehanna landscape. Together, these (and other) issues define the focus and goals of the 'new' regional geography and can be used to guide the design of visual methods which aid in the process of this particular approach to regional and landscape geography.

In sum, a second general consequence of a geographic and methodological approach to cartographic design research is to stress the importance of understanding differences in the manner in which geographic entities are conceptualized.

Different theoretical and philosophical approaches to geography assume different conceptualizations and suggest different design strategies. A cartographic "language" independent of geography may not be cognizant of such differences. Nor can it keep abreast of changes and developments in geography which lead to new design challenges.

Visual Forms and Cartographic Design

A third general consequence of a geographic approach to cartographic design is that a focus on maps alone is not sufficient. This is not to argue that the traditional focus on maps has been misguided, but that cartographic design has to consider the role of other visual forms—such as images and abstract graphics—in the process of geographic research. My case study raised two general and related issues regarding a range of visual forms (including photographs, images, maps, graphs, and relational diagrams) and cartographic design.

First, geographers actually use a broad range of visual forms in the practice of geographic research because maps alone cannot capture the breadth of geographic phenomena that geographers study. My case study required dealing with photographs, video, maps, graphs, and abstract diagrams, and the manner in which they relate to text and each other. Cartographers have had little to say about the ends of the visual forms continua including photographs and images and graphs and abstract diagrams. For example, abstract diagrams are common and often central in academic geography (and cartography—witness the communication model diagrams), yet there has been little or no research on the design of such "conceptual maps." Further, and more importantly, many design issues are raised by the interrelation between the different visual forms and text. This subject has been dealt with to some degree in literature on atlas design, but is becoming increasingly important (and complicated) in the context of hypermedia. My case study, for example, required linkages between images, maps, abstract diagrams, and text. No single visual form alone could capture the range of data and ideas needed to understand the Quehanna landscape and the geographic questions it raises. It is important to note that such linkages are not superfluous, but are necessary and fundamental to addressing the representational and methodological issues raised in the 'new' regional and landscape geography literature.

Second, different visual forms can be used to reflect different conceptualizations of spatial structure; e.g., the visual forms are related to differences in conceptualization in geography. For example, Davis has claimed that humanistic (idealist) geographers conceptualize space and place in a manner which may be more adequately represented by photographs and images than by other visual forms:

> *As humanistic geography and landscape studies begin to recognize the shortcomings of strictly empirical modes of analysis ... creative photography may come to play a more important role in scholarly investigations of the cultural landscape* (Davis, 1989: 1).

Realist geographers (Marxists, structurationists) tend to stress abstract relational conceptualizations of space in their work and thus have argued that the use of abstract relational diagrams (rather than images and maps) is fundamental to their work (Pred, 1984: 292). Such abstract ideas and their visual representation and links to other visual forms and text were fundamental to my case study. Theoretical relations were represented in diagrammetric form, then linked (via hypermedia) to images, maps, and text representing less abstracted elements of the Quehanna landscape. Do different approaches to geography actually have a tendency to use visual forms differently? Do some approaches to geography use more maps, others more photographs, others more abstract diagrams? In any case, my argument is that an approach to cartographic design focused only on maps will exclude visual forms which may be appropriate to the entire range of philosophic, theoretical, and conceptual approaches to geography and geographic phenomena.

In sum, a third general consequence of a geographic and methodological approach to cartographic design research is to stress the need to carefully consider the design implications of a range of visual forms including images, graphs, and abstract diagrams, and how all these visual forms relate to each other and to phenomena, issues, and ideas in geography. Again, my case study suggests that hypermedia serves as an ideal means of exploring such relations. Attention focused on a range of visual forms acknowledges that geographers do not restrict themselves to maps, that maps are substantially related to other visual forms and text, and may enhance the use of visual methods by a broader range of different approaches to geographic research.

CONCLUSIONS

The first section of this chapter consists of an historical overview of questions geographers have raised about the relations between maps and geography. I argue that paradigm shifts in geography often lead to geographic critiques of maps, implying that geographers understand and conceptualize maps and other visual representations as geographic methods. Questions regarding the nature of maps and other visual representations as geographic methods provide the impetus to expand the purview of cartography and cartographic design. I argue that understanding differences within geography can inform cartographic design while simultaneously answering the criticisms of maps and cartography by geographers. Cartography and cartographic design, then, are understood as intimately related to debates, disagreements, different conceptualizations, and different assumptions about the world and its phenomena rather than as an objective language independent of the diversity of geography.

Three general consequences of such a geographical approach to cartographic design are discussed in the second part of this chapter. Three general cartographic design issues are raised, including the role of synthesis as a visual function, the effect of differing geographic conceptualizations on cartographic design, and the consideration of a range of visual forms and their interrelations within the context of cartographic design. All three of these issues involve the use of hypermedia which, I suggest, can be conceptualized as substantially related to fundamental issues in geographic research and representation.

I hope to have shown a few general ways in which cartographic design can be guided and enhanced by geographic understanding. At the same time, I hope that geographic research will be enhanced by such an approach to cartographic design. I have attempted to create a broad conceptualization of cartographic design and visualization which can encompass the methodological needs of the range of geographers and geographic work. By approaching cartographic design in terms of the complexities, contradictions, and multiplicity of geography, cartographers have the option to engage in current debates over differences in knowledge and representation in terms of social, cultural, racial, gender, political, and academic issues. At the same time, such an approach allows cartographers to consider the nature and possibilities of cartographic design as a contextualized and active participant in such fundamentally important debates.

REFERENCES

Beishlag, G. (1951). Aims and limits in teaching cartography. *The Professional Geographer*, 3, 6-8.

Blades, M., and Spencer, C. (1986). The implications of psychological theory and methodology for cognitive cartography. *Cartographica*, 23, 1-13.

Bunge, W. (1962). Metacartography. In W. Bunge (Ed.), *Theoretical Geography*. Lund Studies in Geography, Ser. C. General and Mathematical Geography No. 1.

Cosgrove, D. (1984). *Social formation and symbolic landscape*. Totowa: Barnes and Noble.

Daniels, S. (1989). Marxism, culture, and the duplicity of landscape. In R. Peet and N. Thrift (Eds.), *New Models in Geography: The Political Economy Perspective* (pp. 196-220). London: Unwin Hyman.

Davis, T. (1989). Photography and landscape studies. *Landscape Journal*, 8(1), 1-12.

DiBiase, D. (1990). Visualization in the earth sciences. *Earth and Mineral Sciences*, 59, 1-18. (Edited and reprinted in *Geotimes*, July 1991, 13-15.)

Gilmartin, P. (1991). A content analysis and comparison of three cartographic journals: 1964-1989. *Cartographic Perspectives*, 11, 3-22.

Harley, J.B. (1989). Deconstructing the map. *Cartographica*, 26, 1-20.

Hartshorne, R. (1939). *The nature of geography: A critical survey of current thought in the light of the past*. Lancaster, PA: Association of American Geographers.

Harvey, D. (1969). *Explanation in geography*. New York: St. Martins Press.

Hearnshaw, H. and Unwin, D. (Eds.) (1994). *Visualization in geographical information systems*. New York: Wiley & Sons.

Jenks, G. (1953). An improved curriculum for cartographic training at the college and university level. *Annals of the Association of American Geographers*, 43, 317-331.

Keates, J. (1964). Cartographic communication. *Abstracts of Papers, the 20th International Geographical Congress, London.*

Keates, J. (1982). *Understanding maps*. New York: John Wiley & Sons.

Kolacny, A. (1969). Cartographic information—A fundamental term in modern cartography. *Cartographic Journal*, 6, 47-49.

Krygier, J. (1995a). Visualization, geography, and landscape. [Unpublished Ph.D. dissertation]. University Park, PA: Pennsylvania State University.

Krygier, J. (1995b). HyperLandscape: The visual geographies of Quehanna. [Unpublished Hypermedia Application]. University Park, PA: Pennsylvania State University.

Lacoste, Y. (1973). An illustration of geographical warfare. *Antipode*, 5, 1-13.

Lewis, M. (1991). Elusive societies: A regional-cartographical approach to the study of human relatedness. *Annals of the Association of American Geographers*, 81, 605-626.

Loy, W. (1993). Is cartography dead? *Cartography Specialty Group Newsletter*, 14, 1.

MacEachren, A. (1995). *How maps work*. New York: Guilford Press.

MacEachren, A., et al. (1992). Visualization. In R. Abler, M. Marcus, and J. Olson (Eds.), *Geography's inner worlds: Pervasive themes in contemporary American geography* (pp. 99-137). New Brunswick: Rutgers University Press.

MacEachren, A., and Taylor, D. (Eds.) (1994). *Visualization in modern cartography*. Oxford: Pergamon.

Mackay, J. (1954). Geographic cartography. *The Canadian Geographer*, 4, 1-14.

Miller, R. (1992). Beyond method, beyond ethics: Integrating social theory into GIS and GIS into social theory. *GIS/LIS Proceedings*, 2, 585-593.

Monmonier, M., and MacEachren, A. (Eds.) (1992). Special content: Geographic visualization. *Cartography and Geographical Information Systems*, 19(4), 197-272.

Morrison, J. (1974a). A theoretical framework for cartographic generalization with an emphasis on the process of symbolization. *International Yearbook of Cartography*, 14, 115-127.

Morrison, J. (1974b). Changing philosophical-technical aspects of thematic cartography. *The American Cartographer*, 1, 5-14.

Morrison, J. (1976). The science of cartography and its essential processes. *International Yearbook of Cartography*, 16, 84-97.

Morrison, J. (1978). Towards a functional definition of the science of cartography with emphasis on map reading. *The American Cartographer*, 5, 97-110.

Muehrcke, P. (1981). Whatever happened to geographic cartography? *The Professional Geographer*, 33, 397-405.

Natter, W., and Jones, J. (1989). Response to J.B. Harley's article deconstructing the map. *Cartographica*, 26, 113-114.

Olson, J. (1979). Cognitive cartographic experimentation. *The Canadian Cartographer*, 16, 34-44.

Pile, S. and Rose, G. (1992). All or nothing? Politics and critique in the modernism/postmodernism debate. *Environment and Planning D: Society and Space*, 10, 123-36.

Pred, A. (1984). Place as historically contingent process. *Annals of the Association of American Geographers*, 74, 279-297.

Pudup, M. (1988). Arguments within regional geography. *Progress in Human Geography*, 12, 369-390.

Robinson, A. (1952). *The look of maps*. Madison: University of Wisconsin Press.

Rose, G. (1993). *Feminism and geography*. Minneapolis: University of Minnesota Press.

Sayer, A. (1989). The new regional geography and problems of narrative. *Environment and Planning D: Society and Space*, 7, 253-276.

Sayer, A. (1992). *Method in social science: A realist approach*. London: Routledge.

Smith, N. (1992). History and philosophy of geography: Real wars, theory wars. *Progress in Human Geography*, 16, 257-271.

Soja, E. (1986). Taking Los Angeles apart: Some fragments of a critical human geography. *Environment and Planning D: Society and Space*, 4, 255-272.

Soja, E. (1989). *Postmodern geographies*. London: Verso.

Thrift, N. (1990). For a new regional geography I. *Progress in Human Geography*, 14, 272-280.

Tobler, W. (1979). A transformational view of cartography. *The American Cartographer*, 6, 101-106.

Turner, B. (1989). The specialist-synthesis approach to the revival of geography: The case of cultural ecology. *Annals of the Association of American Geographers*, 79, 88-100.

Ullman, E. (1953). Human geography and area research. *Annals of the Association of American Geographers*, 43, 54-66.

Wood, D. (1978). Introducing the cartography of reality. In D. Ley and M.S. Samuels (Eds.), *Humanistic geography: Prospects and problems* (pp. 207-219). Chicago: Maaroufa Press.

Wood, D. (1992). *The power of maps*. New York: Guilford.

Woolridge, S., and East, W. (1951). *The spirit and purpose of geography*. New York: Hutchinson's University Library.

Wright, J.K. (1942). Map makers are human. *Geographic Review*, 32, 527-544.

Young, I. (1990). *Justice and the politics of difference*. Princeton: Princeton University Press.

4

You Can't Get Here from There: Reconstructing the Relevancy of Design in Postmodernism

Nikolas H. Huffman
Department of Geography, Pennsylvania State University

YOU CAN'T GET HERE FROM THERE

This mixed-up cliché expresses the particular conundrum that I find myself in as a young cartographic designer confronted with the contemporary critical work in cartography. "Here" is the world of cartographic design, in a sense, the very setting of the design conference. Like the other designers I know, I am interested in good design, that is, planning and creating objects that I believe are well-suited to their purposes. Because I am not prepared to offer a full theory of design in this chapter, I hope it will suffice to say that "good design" might be characterized by maps that we can be proud of as professional designers and cartographers; maps that are attractive and useful to the people who want and need them; that live up to our own professional and aesthetic standards, as well as our academic, geographic and political standards as they have developed through our personal and professional experiences.

"There," on the other hand, is the contemporary critique of cartography, which I will place cavalierly under the rubric of "postmodernism," in spite of the fact that many of these critics forthrightly would deny the label. Although most frequently associated with contemporary French philosophy and social theory, "postmodernism" is a complex and convoluted label that means many different things to different people, and is a far cry from a unified or consistent philosophical position.

In recent years, the work of numerous "postmodernists" has been adapted to a critique of cartography, most notably by Denis Wood and Brian Harley, but also by Richard Helgerson, Barbara Belyea, and Robert Rundstrom, among others. As a whole, their work has raised compelling arguments challenging the traditional foundations of the Anglo-American cartographic discipline, and questioning perceived problems associated with its claim to an apolitical, scientific status for cartography. These critiques have disputed the way that language and the production of meaning have been theorized in cartographic research on maps and mapping, and introduced new ways of understanding how we interact and communicate with and through maps. They have raised questions about the politics of representation: What are maps and who profits from them? Who makes maps and who is being mapped? Who speaks for maps and to whom do maps speak? They have also pointed towards a broader sociology of mapping in which maps and mapping can be understood as artifacts within our social and material culture, and have demanded that greater attention be paid to issues of representation, politics, and social action.

While these critics are personally concerned with establishing a more equitable ground for the cartographic enterprise, what they have not addressed is how we, as cartographic designers (perhaps) interested in responding positively to these critiques, should be rethinking how maps are designed and created. Their critiques have focused primarily on the interpretation of maps as cultural artifacts, and have done precious little to engage cartographic design and production as an object of critique. In fact, Wood, for one, mockingly has argued for the irrelevance of cartographic design (Wood, 1985), dismissing cartographic design research as shallow and vapid, if not entirely useless because of its determined reliance on the "scientific" nature of maps and map communication (Wood and Fels, 1992). But if one accepts the concept of a cartographic process as proposed by Rundstrom (1991), then design needs to be considered as an integral part of maps and mapping in this critique of cartography. And, by implication, cartographic design theory, as it is taught and practised today, has been undermined by this critique for its persistent claims to be an objective science, its apolitical rhetoric, and its shallow and mechanistic treatment of human visual language.

But even as the cartographic enterprise is restructured in this digital, post-industrial world, our responsibilities will continue in our (currently) unavoidable role as designers and design theorists within the established social and political organization of mapping. Academic cartographers will continue to write specifically for other cartographers, and map designers will continue to work as map designer *per se*, even as we encourage more people to engage in the design, production, and use of their own maps. The task of responding to the "postmodern" critique, then, is our own, and, given the dishevelled state of cartography and cartographic design theory today (Petchenik, 1983), we absolutely need to embark on a vigorous dialogue to reconstruct cartographic design theory in response to the contemporary critique of cartography as a visual language, a sociopolitical structure, and an academic discipline. Admittedly, this chapter is little more than a call to arms in this project of critically re-evaluating design theory, boldly declaring that we need to earnestly engage the full range of issues raised by the "postmodern turn" in a serious effort to recreate a cartography that can flourish as an academic discipline and as a public and private enterprise.

In order to adequately address the "postmodern" critique, I will begin by outlining four different definitions of "postmodernism," and how they relate to the various cartographic critiques. The definitions of "postmodernism" are also relevant to contemporary issues in the nascent field of design studies. In closing, I will comment on some of the major issues that I feel are important in addressing the project of reconstructing cartographic design in "postmodernism," namely the nature of visual and cartographic language, the role of technology in the future of cartographic design, and the possibility of a cartographic ethics. Re-evaluating cartographic design theory in response to these postmodern critiques does not mean that we should all emerge as full-blown "postmodernists," but that the "postmodern" debate and challenge should be engaged with the full integrity and depth of consideration that it deserves. These ideas are challenging, but we stand to gain much as a discipline by rising to the challenge and staking out a new, more robust ground for the cartographic enterprise.

POSTMODERNISM, MAPS, AND MAPPING

In geography and cartography, as in academia in general, the term "postmodernism" is used so loosely and unproblematically that I think it is important to actually explain "postmodernism" by drawing a distinction between the different modes of thought that fall under the rubric of "postmodernism." The key feature of all "postmodernism" is, of course, a distinct opposition to, or break from, "modernism." But this is really an useless definition because neither term has been adequately explained, and both stand for a variety of things in different contexts. In defining the term "postmodernism," I have divided its many uses into four basic categories: the postmodern style in architecture, art and literature; postmodern social theory; the political economy of late capitalism; and poststructuralist philosophy. These four categories all have very different implications that can not be mutually compared without also clarifying their differences. However, theorists often contend that their particular position provides the account which explains all of the other categories, a conviction that will be more or less true depending on your own theoretical proclivities. More importantly, confusion over the term "postmodernism" is often strategically employed to dismiss the most devastating claims of "postmodernists" by conflating them with its more vapid aspects (Strohmayer and Hannah, 1992).

Mapping and the Postmodern Style

Although the term "postmodern" can be traced back to 1870 (Best and Kellner, 1991), it entered the academic discourse primarily through discussions of the postmodern architectural style that emerged in the late 1960s in response to the worst dehumanizing tendencies of the "modernist" International Style (Berman, 1988). By reviving the previously scorned ornamentalism in an odd-ball blend of historical styles, postmodern architects claim to extol emotion and living experience over the faceless rationality and pure functionalism of the International Style. However, this mixing of styles is also derided as a mere cosmetic pastiche of serious architectural style, a charge also levelled against the complex and self-reflexive parody of the postmodern style in the arts (Hutcheon, 1989). This style is hyper-aware of the perplexities of language and representation, and uses parody to simultaneously build up and undermine the artist's desire to work within the limitations of language, rather than trying to overcome, deny, or ignore them.

To the best of my knowledge, no cartographer has openly adopted a postmodern style of cartography, which is hardly surprising given the (alleged) scientific fetishism of the discipline. But the interesting question remains: What would a stylistically postmodern map look like? One possible example is the Australian (upside-down) map of the world; or any other map that self-consciously uses unusual map elements to openly confront the reader with ideas that question how we relate to maps, and to the world through maps. The ultimate goal is not to confuse or disorient the reader, but to encourage them to read deeper into the map and the mapping process, and to challenge the objective and scientific mystique of the map as mirror of the world (Wood and Fels, 1992).

Commenting on another type of style, Rundstrom is critical of the stylistic politics of inclusion he attributes to ideologically postmodern academics (Rundstrom, 1991). He rightly argues that in spite of their expressed desire to include the politically disenfranchised, they have a hypocritical and "ironic tendency toward exclusion," particularly with respect to adversarial "modernist" ideologies. However, Rundstrom's counterclaim that no representation can be totally inclusive conflates the exclusion or "generalization" necessary in all language with the ideological blindness of the master subject position criticized in feminist theory (Haraway, 1991). This blindness, a consistent feature of First World discourse that is neither confined to "modern" nor "postmodern" representations, excludes by its failure to even recognize the existence of the excluded Other (Said, 1978).

Cartography and Postmodern Social Theory

This postmodern concern for the politics of style is in many respects a purposeful response to postmodern social theories. One aspect that consistently runs through these theories is an overwhelming concern for the politics of representation. Lyotard, and Deleuze and Guattari glorify the contemporary cacophony of voices, advocating for difference and dissensus over the "modern" consensus politics that they claim silences voices with its rational discourse of universal rights (Best and Kellner, 1991). Lyotard also argues strongly in favour of partial, situated, and contingent social theories, over and against totalizing or universalizing discourses that claim to provide complete theoretical descriptions.

The postmodern critique of cartography similarly has concerned itself with people speaking out for themselves through maps. Wood has argued against the map as a fixed or comprehensive statement about the world, promoting an exploratory narrative style that fosters fluid mappings that are continuously open to interpretation over a totalizing cartographic narrative which claims only to communicate scientific knowledge (Wood, 1987). While Wood has been openly promoting this cartographic multiplicity as a broad cultural program (Wood, 1993a; Wood and Fels, 1992), Harley focused more generally on the social, political, and ethical aspects of government mapping programs (Harley, 1992), a program that

Rundstrom follows, although for reasons he places outside of the "postmodern" (Rundstrom, 1991).

These multiplicitous theories of postmodernism have lead contemporary modernists to argue that this critique of "modernity" is itself a totalizing discourse (Best and Kellner, 1991), and that the social realities are far more complex than the postmodernists have acknowledged (Berman, 1983). While conceding to some of the negative consequences of "modern" reason, the modernists are not willing to so easily abandon the Enlightenment project of social transformation through traditional rational discourse (Berman, 1988; Habermas, 1983). Social theorists in other fields, such as feminism, post-colonialism, and post-Marxism, have also taken up arguments against "postmodernism," but have situated themselves most commonly between the foundationalism of modern science and the multiplicity of postmodernism (Graham, 1992; Nicholson, 1990; Said, 1978).

Another concern of postmodern social theorists that has strongly influenced cartography is the means by which power works in and through society, particularly in the well-known social critique of Foucault (1969). His diverse historical investigations probed the development of the social and linguistic structures which manifest social relations of power and define us as individuals within that society. Foucault was explicitly anti-modern and critical of the Enlightenment's new forms of rational and scientific domination, but never openly identified himself as postmodern, and in later years, moved towards a position of reconciliation with modernist rationality (Best and Kellner, 1991).

This critique was adopted by Harley in the development of his self-proclaimed postmodern critique of cartography. Concepts like "power-knowledge," "episteme" and "discourse" were applied to the role of maps in creating and sustaining social and political power within a given society (Harley, 1988a; Harley, 1988b; Harley, 1989). In a recent commentary, Belyea takes Harley to task over his proposed "epistemological shift," arguing that he does not fully embrace the more extreme postmodern conclusions of Foucault's social theory. Her reworking of Foucault and cartography, reminiscent of Latour's recent work in the sociology of science (Latour, 1990), considers how concepts of reality are created through social relations which empower scientific discourse (Belyea, 1992).

This work is also reminiscent of Wood's claim that maps are actually "weapons of power" that create the territory desired by those empowered to make and enforce maps (Wood, 1992). This directly opposes the traditional (modern) cartographic claim to be reporting objective facts about the real world, and while I don't believe Wood would bother to call himself a postmodernist, many of his ideas have that same flavour. Wood's maps work much like simulacre in the social theory of Jean Baudrillard, which proposes that a new political economy of the sign is emerging under the influence of commodity marketing and the incorporation of the mass media and telecommunications into our daily lives (Best and Kellner, 1991). As a simulacrum, the sign of the map, in place of the land itself, stands for the desires of the cartographers and planners codified in the map sign which becomes that reality.

Late Capitalism and the Economy of Mapping

The contemporary political economy that gave rise to Baudrillard's simulacre is the focus of the third type of "postmodernism," which has been called alternately late capitalism by theorists such as Jameson and Harvey, and post-industrialism by Bell among others (Bell, 1973; Harvey, 1989; Jameson, 1984). Here, the concern is for analysing and making sense of the staggering structural changes in the global economic, social, and political fabric. Although they dispute "postmodernism" as a radical break from the present historical moment, the various Marxist critiques have advanced the most highly developed account of these structural changes, while most other theorists have simply pointed to these changes without analysing possible causes.

Mapping in the late capitalist economy poses many new and interesting problems for cartography. One serious political concern in late capitalist society has been the emergence of a "surveillance society" as legal and cultural definitions of privacy have not kept pace with the collection and manipulation of information about our personal and electronic lives by corporate information systems which track and record our daily interactions (Lyon, 1994; Review, 1972; Rule, 1974). Along more moderate lines, Harley has argued that the current drive towards all-digital public mapping programs is politically objectionable because it does away with the historical record of map artifacts that has been the main focus of his critical work (Harley, 1992).

While Bolton argues that the technological and economic imperatives of late capitalism have reduced architecture to mere advertising, it could well have the opposite effect on cartography, creating a new place for mapping in a culture increasingly responsive to and dependent upon the potential of maps to transform the world around us. Access to computers has allowed many people to become their own cartographers, and small-scale map production operations have flourished like never before. We all stand to benefit from increased access to high-end GIS and geodemographics systems to fulfil our own geographic and cartographic needs, whether that simply means reading the *USA Today* colour weather maps, or downloading maps on-demand off the Net at work or home.

Late capitalism is definitely changing the face of cartography, and microcomputers will be increasingly instrumental in bringing mapping tools to the general public. Professional cartographers presently have a golden opportunity to play a critical role in this so-called "democratization" of cartography. In the final pages of *The Power of Maps*, Wood openly invites his readers to become part of the vanguard of the personal cartographic revolution (Wood and Fels, 1992), and other cartographers have expressed a concern for more personal cartography as well (Monmonier, 1991).

Cartography and Poststructuralism

Perhaps the most profound "postmodern" critique of cartography has come from poststructuralist philosophy of language, commonly accepted as the source of the bewildering "crisis of representation." In a grave oversimplification, the poststructuralist position can be summed up in two major tenets. First, all human language and meaning is based on the interpretation of signs that are ultimately indeterminate and always open to further analysis. Second, and more importantly, all human knowledge, and I mean that quite literally, is mediated through language structures that shape our perspectives on the world around us, as well as constituting such personal concepts as self and subjectivity. In cartography, this work has focused on the language of maps as complex social and cultural objects, and how maps construct and are constructed by society and language.

As the premier poststructuralist philosopher, Derrida (1967) has conducted a critical analysis of philosophical works, showing how their subtle rhetorical and metaphorical assumptions favour an account of language as unmediated thought. Derrida, on the other hand, insisted on the necessity of *writing*, which in the strict Derridean sense, fixes thought into an artifact that can and must be subjected to the vagaries of free and open-ended interpretation (Norris, 1982). Following Derrida's philosophical critique, adherents have applied the deconstructionist analytical tool box at different levels of interpretation in their political and literary critical analyses. In cartography, Harley adopted this position in a political critique of cartographic "texts" that seeks to reveal the underlying political interests and prejudices embodied in maps (Harley, 1989), and Belyea pushes this analysis even deeper into a reading of the philosophical assumptions of the language of mapping (Belyea, 1992).

The work of Wood, on the other hand, follows from a deep engagement with the ideas of Roland Barthes, whose career is emblematic of the history of poststructuralism itself. Barthes began writing as a structuralist semiotician, but developed an increasingly poststructuralist position to tease out the deepest consequences of the original structuralist philosophy of language. Although neither Barthes nor Wood are specifically aligned with the deconstructionist movement, their work often has a similar feel to it, digging beneath the surface of signs to reveal hidden political and philosophical assumptions (Harley, 1989; Norris, 1982). Barthes' concept of "mythology," which describes how one sign-function is used to stand as the signifier for another, higher-order sign-function, has been used by Wood to expose the broader social significance of maps and mapping (Wood and Fels, 1992). Wood is particularly critical of the mythology of scientific mapping whose "rhetorically orchestrated denial of rhetoric" buttresses the claim to an objective view on the world and disguises the power of maps to define our world.

On the other hand, Wood's most recent work is a plausible sign that poststructuralism and structuralism, which is the study of the structures, or sets of relations, that determine the meaning of signs and the function of the social order within a particular cultural context, are not inherently antagonistic as the names seem to imply, but are in fact mutually inclusive (Caws, 1988). Wood proposes a structuralist social theory to elucidate a sociology of mapping that can account for all the different types of maps and mapping found throughout the world, from USGS topo sheets to the Inuit navigational charts scratched in the Arctic sand (Wood, 1993b). But, Wood persistently has also been critical of claims to a fully scientific status for cartography based on strict structuralist models, such as the communication and psychophysical models that theorize meaning as solely and unambiguously mediated by linguistic or perceptual structures.

If this discussion of poststructuralism has made anything clear, perhaps it is that the key to this impossibly complex paradox of language is that while language always structures our society and ideology, structure can never fully account for the interpreted meaning of the representations we rely on in our daily lives. Human language is an exceptionally powerful and complex force shaping our lives, but language and meaning can never be reduced to a simple projection of thought from one person to another, and is always open to further interpretation and analysis at any level.

What I do hope is abundantly clear by now, though, is that the confluence of "postmodernism" and cartography is by no means a clear cut one. A proper treatment on the complexities of "postmodernism" is crucial to a fully adequate response to the contemporary critiques and philosophical positions that have imperilled traditional cartographic design theory. This is not to imply a total capitulation to "postmodernism" by any stretch of the imagination, but our response to these critiques is critical to the reassertion of the relevance of our work in the present and future political economy of mapping. One excellent model for forging just such a diverse and responsive cartographic design theory is the current debate going on in the nascent field of design studies.

POSTMODERNISM AND CONTEMPORARY DESIGN THEORY

"Postmodernism" has also affected other areas of design theory besides cartography, similarly calling traditional assumptions into question and creating new opportunities for designers. Recognition of these fundamental problems for all designers has prompted efforts to overcome the isolation of designers in their normally separate fields, such as architecture, industrial design, and graphic design. The new field of Design Studies developed to integrate the various branches of design into a broader theoretical perspective, and new journals, such as *Information Design Journal, Design Issues, Visible Language,* and *Design Studies* have served as a forum for contemporary designers and design theorists from different fields to establish a common ground for a general theory of design which unites the common elements of these specialized design activities (Margolin, 1989). Cartographic designers, also feeling the pinch of the "postmodern" critique, would do well to engage this new field of study, and could have much to contribute to the contemporary problem of reconsidering traditional design theory, adapting to changes in the political economy, and developing new design theories.

Reconsidering Design Theory

Functionalism, as the guiding principle of "modern" design theory in the 20th Century, concentrated on maximizing the utility of design by creating objects that were industrially efficient and well suited to their intended purposes (Lemoine, 1988). While most theorists recognize the inherent teleology of design, that constant concern for "changing existing situations into preferred ones" (Simon, 1969), in moving beyond "modernism," contemporary design theorists have argued that utility is hardly the sole quality of a designed object. Design studies have further questioned what the purpose of design is, and what impact these designed objects, and thus our designs, have upon the society around us. For Margolin, an editor of *Design Issues*, the design studies project is bounded by design theory, criticism, and history, and considers the broader social theory of planning as part of the impact of design (Margolin, 1989).

While not explicitly "postmodern," Vitta's interest in a "culture of design" certainly moves beyond the limits of functionalism to consider the totality of questions that will help us to understand our relation to the built objects that define, and through which we define the social order around us (Vitta, 1989). Positing a preferred social condition and acting to achieve it both imply that one is theorizing about society. In this sense, no design theory can exist without an accompanying social theory, in effect a social theory of design that considers how designs alter society. At this level of analysis, design theory and social theory merge in that the designer can both read the social values in designed objects, as well as write those values into objects.

Such a social theory also implies a critique of the existing social order and affirms that design plays an integral role in the creation and maintenance of the social structures that surround us. The critical and historical analysis of the social values of design also contributes to this social theory of design by analysing values embodied in our material culture. Regardless of the critical perspective taken, whether postmodern, poststructuralist or feminist, a critical history of design reveals the values which are privileged or excluded by a particular design theory. For example, Buckley argues that the elevation of functionalism and technological development in modern industrial design played a significant role in the denigration of women's productive roles as artisans involved in traditional craft trades (Buckley, 1989). In looking at cartographic design, Wood has commented on the bias towards developers in the design of USGS topo sheets (Wood and Fels, 1992), and Harley pointed to the value systems inherent in the symbolization of the historical maps he studied (Harley, 1988b).

Post-Industrial Design

Design values, like social values, are integrally related to the political economy in which they occur because the means and modes of production structure the design and production possibilities available to designers. "Modern" design theory developed out of a concern for efficient industrial production, and in a more contemporary vein, Diani calls for an "integrative science of design," which goes beyond the design of physical objects to include the design of services and processes as part of the complex interplay of humans and machines (Diani, 1989).

However, advances in flexible manufacturing technology have made small-scale production increasingly efficient, and presaged unprecedented levels of personalized design to suit the needs and desires of individual consumers, prompting Lemoine to speculate: "What is to stop consumers themselves from designing what they consume?" (Lemoine, 1988). Building upon this post-industrial technology, many design theorists have promoted design concepts which extol personal expression and self-fulfilment over function (Branzi, 1989; Burkhardt, 1988; Moles, 1989). This trend toward personalized production is, of course, occurring in cartographic circles as well, where the rapid advances in print technology are forging the way for other products. The push for all-digital, national mapping programs is one example, but the much vaunted availability of mapping software will be most instrumental in promoting interest in and access to personal mapping systems.

This trend towards flexible production has also fostered a redefinition of "design as process," that is, a shift away from the "product-thinking" that only recognizes the user in relation to the product, to a concern for "process-thinking," considering the user's work processes or needs, rather than the product itself, as the object of design (Jones, 1988). Ironically, this postmodern design theory is quite similar to Rundstrom's pre-modern model of "process cartography," which situates the map product as an artifact in the map making process, and interprets the map making process in a context of cultural dialogue (Rundstrom, 1991).

Their concern for process in design is reiterated by Lemoine, who argues that designers can neither fix, nor fully anticipate all of an object's possible functions, so it is better to focus on the production of an adaptable object than on the optimization of a limited set of functions (Lemoine, 1988). This process design is also compatible with the definition of cartography as a visual method proposed by Bertin and celebrated by Wood for its focus on answering questions and solving problems with maps (Wood, 1985). This recent focus on doing things with maps, rather than on making general purpose maps is integral to the concerns of visualization in contemporary science (Lynch and Woolgar, 1990). Muller has reported on Bertin's general theory of graphics that could serve as a model for investigating the role of visual methods in scientific practice (Muller, 1981). And, in Latour's sociology of science, visual methods play a prominent role in the rhetoric of scientific discourse that creates and utilizes graphic arguments to establish and verify scientific results (Latour, 1987).

Design as Rhetoric

For Buchanan, design is a form of rhetoric in which the main function of a particular design is to persuade the user to accept some belief or statement associated with various features of the design (Buchanan, 1989). In his design theory, the objective expression of function is replaced by the persuasiveness of the design as conveyed in the three elements of the design argument: logos, ethos, and pathos. Logos is the technological reasoning of the design, or how the object is made technically functional and relevant to the users needs. Ethos is the character of an object that gives it credibility and allure for the user, and pathos is the emotional quality of a design that stirs the feelings of the user. Buchanan's design theory reduces functionalism to a rhetorical position which claims to be the pure expression of an object's use, masking the object's ethos and pathos under the guise of the logos.

Cartographic design theory has, and continues to focus primarily on the technological functionalism of the design, while the consideration of character and emotion in map design rarely have been considered explicitly as an aspect of design theory. The continued expediency of the logos is evident in the recent *Cartographic Journal* special issue on "Map Design and the Cartographer" (Keates, 1993). The majority of articles focus on technological and visual issues of design logos, but Dorling's work on British census maps and cartograms, for instance, works from the logos of the census cartogram to the pathos of the census maps to make the conclusions drawn from the maps more credible and accessible to the user or researcher (Dorling, 1993). Given the technological upheaval we face today, this technological fixation is not entirely bad, but we also need to begin to focus our attention on the other objectives behind design rhetoric. As in other forms of design, the production and use of maps and mapping technologies have a major influence on how people interact with the world around them, and we need to join with other designers in considering our role in shaping the world around us, and how we can involve others in the work of designing their world.

CARTOGRAPHIC DESIGN THEORY IN POSTMODERNISM

Of course, it is very easy for me to rattle off these critiques, especially since I have already admitted that I refuse to address the real issue, proposing new cartographic design theories. What I am willing to do, though, is discuss some of the issues that I believe are particularly relevant to the reconstruction of cartographic design. One central issue is the question of who this reconsidered design theory will address. Will it be solely for other academic and professional cartographers? For scientists and navigators? For political activists? Ultimately, the answer must be that these design theories should offer something helpful to everyone, because the technological impact of digital cartography could be that broad as well. To make maps as accessible and useful as possible for everyone, we need to consider the importance of visual literacy for using maps as tools, the politics of designing technology in society, and the ideological and ethical conflicts inherent in the cartographic enterprise.

Writing In Visual Language

Visual language already plays a vital, if perhaps under-theorized role in our society, and this is one area where cartographers should be particularly well qualified to engage important issues in such a social theory. Visual methods have an extensive history in "modern" science (Wood, 1994), and are becoming an increasingly common part of the average scientific researcher's tool box. Visual methods are also important in our daily lives, as Wood repeatedly demonstrates in the cartographic tales inspired by his family life. But the use of visual methods is analogous to learning and using any other language effectively, for as Wood points out, sign systems are not intuitive, but learned (Wood, 1994). We endure years of training and practice at home and school to become effective English speakers, but for the most part, this kind of graphic education is missing in our culture.

Writing, in the strict Derridean sense, is as necessary in visual language as in any other, and could play a useful role in offering an account of the reader's interaction with the author's *written* map. While this perspective inherently focuses on reading and interpretation, I, for one, am not ready to totally abandon the human factors approach to graphic design, either. I agree that design theory has focused too heavily on cognitive and psychophysical issues, but I still believe in the importance of human information processes in the interpretation of visual language. It is for good reason that Bertin's perceptually structuralist visual variables have played a central role in cartographic design theory since the early 1970s, and this in spite of the criticism that they are tantamount to "graphic fascism" according to one anonymous GIS specialist.

However, if this is true, and I am perfectly willing to accept it as such if for no other reason than to make my point, then it is also true that grammar in our written and spoken language is fascism as well. Grammar structures language according to a set of common rules, determining the conditions of interpretation so that we can understand what we are saying to each other. It is perfectly acceptable to violate those rules, but then it usually takes more interpretative effort on the part of the reader to still be understood. In visual language, there is a perceptual grammar that is a function of the human visual information processing system, and it is this grammar that is one of the key features of Bertin's graphic semiotics (Bertin, 1983).

Bertin's graphic information processing is a visual scientific method that relies on human visual perception to process data graphically by matching the graphic presentation of the data to the type of data being analysed. By following his perceptual rules for constructing data images, the user will be able to analyse the data by *seeing* it in Bertin's strict sense. For Bertin, to *see* the data means to analyse the data using the human perceptual system directly instead of through conscious thought, for instance when the eye *sees* the relative difference in symbol sizes before trying to *read* the data relationship between the symbols. *Seeing* is the backbone of his visual methodology, and is the principle that makes it such an expedient way of analysing complex data.

But this strict perceptual *seeing* is not the only factor that determines the meaning of the data, and in the broadest cultural sense, perception can never be more important than *reading* the data. In Bertin's sense of the term, *reading* involves both visually processing and mentally interpreting the data image as the reader meticulously pours over the image making constant mental notes and comparisons about the perceived relations in the image. This most certainly involves the normal interpretative activity of making sense of signs, but the *seeing* is vital to the visual method, as well. *Reading* and interpretation are utterly inescapable, but while it is always possible to "learn the code" (yet another anonymous quote) when one is *reading* an image, I don't believe that we can freely learn to *see* graphic code and to physically perceptually process information as we see fit.

The relationship between *seeing* and *reading* is a very complex question, but it is also a vital one if maps are to be used productively in the way that Bertin suggests. Wood accepts this perceptual structuralism without comment in his review of Bertin's manual on graphic information processing (Wood, 1985), but is also highly critical of cartography for its determined reliance on psychophysical research and Bertin's visual variables in map design theory. This kind of duplicity is not really a problem for me, but is instead symbolic of the complex nature of visual language and the fact that graphic images are, but are not totally arbitrary. Meanwhile, I remain confident that there can be a more robust cartographic design theory that can account for both the importance of interpretation and the mediation of meaning through perception. I am, however, not holding my breath in anticipation.

Mapping and the Politics of Technology

In the end, the real issue in the debate over visual language is making maps work for us, assuring that the visual methods actually work toward their purposes. And if our goal is to help people learn to use maps more extensively in their personal lives, then teaching visual grammar, visual language, and visual methods will play an important part in advancing the visual literacy that cartography depends on as a method. Technology plays a double role in this effort. It is because of the prevalence of new graphics technologies that more people need to be able to use these skills effectively, but the technology also offers new tools to teach graphic skills more efficiently and to help overcome the socially privileged status of the mechanical arts that restrains many people from using graphics for themselves.

Of course, cartographers will continue to make maps and write and teach about cartography in the traditional media, but these nascent technologies will greatly expand our resources for cartographic education and training. Hypertext help systems and mapping tutorials to accompany the cartographic software might be a very efficient and cost-effective way to help people get the most out of their new mapping tools. Cartographic design expert systems may have the potential to play an active role in helping people teach themselves about graphics by offering advice and pointing to relevant issues where appropriate, but the dubious social and technological limitations of machine knowledge systems calls their decision-making abilities into question.

It will become increasingly vital for cartographers to have a greater impact in the development of commercial cartographic systems and new mapping technologies. Our expertise could be useful in designing mapping packages that are effective without years of cartographic training, but also affect that training while using them. This is one of the main concerns of design in the "Human Age," where the hardware and software limitations no longer override the user's requirements (Mitchell, 1988). This can be affected by involving ourselves more directly in the software design process, and by interesting other designers in cartographic issues. This concern also applies to the development of adaptable visualization platforms, geographical information systems, and cartographic data networks, and the problem of making these tools available and functional to people in academia, government, and business, as well as the general public.

And that brings us back to the underlying theme of this chapter, the politics of mapping and technology. If the "postmodern" critique of cartography has demonstrated anything, it is that these issues surrounding our work and technology are deeply political. Maps bear political messages (Harley, 1988b; Pickles, 1992), maps are part of the political game (Harley and Zandvliet, 1992; Helgerson, 1986; Wood and Fels, 1992), and maps are weapons in the struggle to exert and enhance political power and social control (Said, 1994; Wood, 1992). An important part of the way that maps exert their power in contemporary society is by denying the very existence of their power, the "rhetorically orchestrated denial of rhetoric" exposed by Wood (Wood and Fels, 1992).

In any case, technology is not objective and neutral (Kinross, 1989), and although power is rarely open and obvious, the political power system builds tools for itself, or co-opts new technologies that reinforce and expand its power. However, the success and spread of technologies means that they become available to more people and can be used for many different purposes. These may be aligned with the interests of the empowered in a number of ways, or they may subversively challenge the interests of the social powers. Regardless of how the power of maps is employed, what is clear is that within this broad political spectrum there is considerable room to act out one's personal interests. And within this open, but not quite level playing field, there is enough freedom to act that we need to consider the question of a cartographic ethics. How can we do the right thing with maps?

On the Frustrations of a Cartographic Ethics

The question of a cartographic ethics is an important one, and I doubt that many cartographers would deny that ethical issues should be addressed. I agree wholeheartedly with Harley's declaration that "ethics cannot be divorced from questions of social justice" (Harley, 1991), particularly in response to the "postmodern" critique of how political power relations work in and through cartography. But to extend this critique of the power of maps to the concerns of map designers and a general social theory of design in cartography is another matter altogether. Ethically accounting for the power relations of a particular map would require some normative framework for evaluating that design's intended and actual results in much the same way that we evaluate other, less political design criteria. This recognition of the political component of design is a positive step forward because it undermines the rhetoric of neutrality and objectivity, and injects at least the illusion of social and political responsibility into the work of designers.

The problem with social justice and ethics is that issues of right and wrong are deeply personal, and commonly based on social, cultural, political, and religious ideologies that are themselves inherently subjective. Additionally, design activity, except in the most private situations, does not occur in a social vacuum but within a cultural network that strongly obliges us to attend to the politics of those around us, such as our bosses, patrons and publishers. Within this social framework, there are too many voices and too many interests for Harley's general ethics of cartography to be tenable (Harley, 1991). And besides, this call for a normative cartographic ethics is quite "modern" in that it relies on a rational and totalizing concept of social justice. One problem with such ethical systems is that they tend to be socially conservative and most unethical for the people excluded from deciding what justice means.

This is not to say that we should not base the evaluation of our work on ethical principles; or that we should not engage mapping from a political perspective; or that we should not engage in political cartographic work. Those things are exactly what we should be doing, but we should not be searching for some universal, guiding principle that can determine the rightness or wrongness of the maps we design and produce. Wood constantly beseeches people to involve maps in their politics, to use the cartographic "weapons of power" for themselves. He extends this overture to people of all political persuasions, but his progressive tendencies occasionally shine through his rhetoric, and when necessary and appropriate, he openly declares his own political interests in his work (Wood and Fellowes, 1994). Similarly, Rundstrom's political activism in Canadian and Nunavut mapping is an excellent example of the coincidence of politics and cartography (Rundstrom, 1991). Their approach is a much better model for doing ethical cartography, for designing and producing ethical maps, because it is neither totalizing nor dogmatic and leaves the power and discourse of ethics open to debate.

In proposing a general social theory of design, one would also need to consider the impact of personal theories of design ethics, and how they shape decisions about the political and social ramifications of a design. This is complicated by the fact that no matter how powerful an object's design rhetoric, the ultimate use of that object is beyond our control, and so even our best intentions are often thwarted. As designers, we participate in a constant critical dialogue of praise and reproach between the intentions and uses of the objects we design. Following this model for an ethics of design, there are at least four design responses that a cartographer could take in response to the political critique of power relations in cartography. However, this model definitely reflects my own political interests, so many of you may see things a little differently.

A *conservative* approach is the first response to the political critique of cartography. This approach implies making "corrections" to a map in direct response to a specific critical analysis, thus silencing the revealed rhetoric of the map and re-inscribing its mythology and power as a result of the critic's attempt to open or reveal that hidden power. According to Wood, this is exactly the response taken by the North Carolina Department of Transportation in response to his deep reading of the North Carolina highway map (Wood and Fels, 1986). The *reactionary* response is more insidious still, in that this approach co-opts the critical perspective for itself in the early stages of design in order to pre-empt future critical readings and more deeply entrench the mythology of the map.

A *liberal* response would acknowledge the criticisms and try to incorporate them into a new socially-responsive map, but only as a mask that lends the appearance of having resolved the conflict or problem that was criticized. However, beneath the surface of the map it is clear that this approach has simply removed the visible object of criticism without addressing the power inequalities underlying the social system.

The *progressive* approach would acknowledge the value of the criticism and incorporate it into a broader social effort to foster democratic participation and access to power. This could be done by creating maps that are themselves critical, and by encouraging people to think critically about maps and the worlds these maps construct and represent.

This model extends Buchanan's (1989) theory of rhetoric as design into the political spectrum, and provides a framework for discussing political intention and action in the context of cartographic design. The consideration of the politics of mapping is an important part of the effort to encourage people to really take up cartography to fulfil their own needs and political interests. This concern for politics, as well as the discussions of visual language and technology, should have a significant impact on cartographic design theory in responding to the "postmodern" critique and encouraging greater social interest in the power of maps.

WHERE ARE WE NOW?

Returning to the original cliché, it should be clear that the original concern for "good design" under consideration here is itself a personal political position that will vary among design frameworks. Seen from there, we might ask ourselves: Are we doing good design? The answer, I think, is no, and yes, but more importantly that it just doesn't matter that much. What does matter is that we could be doing better design and design theory.

The contemporary "postmodern" critique of cartography has questioned the most basic foundations of cartographic design theory, but it has also proposed many new theories that could be useful in reconsidering the fundamentals of our practice and discipline. Similar reconsiderations are currently under way in the new field of design studies, and as designers we could do well to participate in this new field of research. Perhaps the most profound result of all this is the recognition of the broader social situation of the cartographic enterprise.

The cartographic discipline would benefit significantly from greater participation in the social issues that affect our work, just as society would benefit from participation in mapping, whether on a personal or public level. Cartographic design theory will be reconstructed in some fashion, but whether it becomes a robust and widely accessible theory is dependent upon the effort we put into our own work, and our willingness to respond adequately to the criticism that has been levelled against cartography in general. Maps are socially more important than ever before, but that does not mean that our work is done, or that we have finally arrived as a discipline. What it means is that our work is just beginning and that we might really be going somewhere now.

REFERENCES

Bell, D. (1973). *The coming of post-industrial society*. New York: Basic Books.

Belyea, B. (1992). Images of power: Derrida/Foucault/Harley. *Cartographica*, 29(2), 1-9.

Berman, M. (1982). *All that is solid melts into air*. New York: Simon and Schuster.

Berman, M. (1988). The experience of modernity. In J. Thackara (Ed.), *Design after modernism: Beyond the object* (pp. 35-48). New York: Thames and Hudson.

Bertin, J. (1983). A new look at cartography. In D.R. Fraser Taylor (Ed.), *Graphic communication and design in contemporary cartography* (pp. 69-86). II. New York: John Wiley & Sons.

Best, S., and Kellner, D. (1991). *Postmodern theory: Critical interrogations*. New York: The Guilford Press.

Bolton, R. (1988). Architecture and cognac. In J. Thackara (Ed.), *Design after modernism: Beyond the object* (pp. 85-94). New York: Thames and Hudson.

Branzi, A. (1989). We are the primitives. In V. Margolin (Ed.), *Design discourse: History, theory, criticsm* (pp. 37-41). Chicago: The University of Chicago Press.

Buchanan, R. (1989). Declaration by design: Rhetoric, argument, and demonstration in design practice. In V. Margolin (Ed.), *Design discourse: History, theory, criticsm* (pp. 91-110). Chicago: The University of Chicago Press.

Buckley, C. (1989). Made in patriarchy: Towards a feminist analysis of women and design. In V. Margolin (Ed.), *Design discourse: History, theory, criticsm* (pp. 251-262). Chicago: The University of Chicago Press.

Burkhardt, F. (1988). Design and 'Avantpostmodernism'. In J. Thackara (Ed.), *Design after modernism: Beyond the object* (pp. 145-151). New York: Thames and Hudson.

Caws, P. (1988). *Structuralism: The art of the intelligible*. Atlantic Highlands, NJ: Humanities Press International, Inc.

Derrida, J. (1967). *De la grammatologie*. Paris: Minuit.

Diani, M. (1989). The social design of office automation. In V. Margolin (Ed.), *Design discourse: History, theory, criticsm* (pp. 67-76). Chicago: The University of Chicago Press.

Dorling, D. (1993). Map design for census mapping. *The Cartographic Journal*, 30(2), 167-183.

Foucault, M. (1969). *L'Archeologie du savoir*. Paris: Gallimard.

Graham, J. (1992). Theory and essentialism in Marxist geography. *Antipode*, 22(1), 53-66.

Habermas, J. (1983). Modernism—An incomplete project. In H. Foster (Ed.), *The anti-aesthetic: Essays on postmodern culture* (pp. 3-15). Seattle: Bay Press.

Haraway, D. (1991). *Simians, cyborgs, and women: The reinvention of nature*. New York: Routledge.

Harley, J.B. (1988a). Maps, knowledge, and power. In D. Cosgrove and S. Daniels (Eds.), *The iconography of landscape* (pp. 277-312). Cambridge: Cambridge University Press.

Harley, J.B. (1988b). Silences and secrecy: The hidden agenda of cartography in early modern Europe. *Imago Mundi*, 40, 57-76.

Harley, J.B. (1989). Deconstructing the map. *Cartographica*, 26(2), 1-20.

Harley, J.B. (1991). Can there be a cartographic ethics? *Cartographic Perspectives*, 10,

Harley, J.B. (1992). Cartography, ethics and social theory. *Cartographica*, 27(2), 1-23.

Harley, J.B., and Zandvliet, K. (1992). Art, science, and power in Sixteenth-century Dutch cartography. *Cartographica*, 29(2), 10-19.

Harvey, D. (1989). *The condition of postmodernity*. London: Basil Blackwell.

Helgerson, R. (1986). The land speaks: Cartography, chorography, and subversion in Renaissance England. *Representations*, 16, 51-85.

Hutcheon, L. (1989). *The politics of postmodernism*. London: Routledge.

Jameson, F. (1984). Postmodernism, or the cultural logic of late capitalism. *New Left Review*, 146, 53-93.

Jones, J.C. (1988). Softechnica. In J. Thackara (Ed.), *Design after modernism: Beyond the object* (pp. 216-225). New York: Thames and Hudson.

Keates, J.S. (1993). Some reflections on cartographic design. *Cartographic Journal*, 30(2), 199-201.

Kinross, R. (1989). The rhetoric of neutrality. In V. Margolin (Ed.), *Design discourse: History, theory, criticsm* (pp. 131-143). Chicago: The University of Chicago Press.

Latour, B. (1987). *Science in action*. Cambridge, MA: Harvard University Press.

Latour, B. (1990). Drawing things together. In M. Lynch and S. Woolgar (Eds.), *Representation in scientific practice* (pp. 19–68). Cambridge, MA: MIT Press.

Lemoine, P. (1988). The demise of classical rationality. In J. Thackara (Ed.), *Design after modernism: Beyond the object* (pp. 187-196). New York: Thames and Hudson.

Lynch, M., and Woolgar, S. (Eds.) (1990). *Representation in scientific practice*. Cambridge, MA: MIT Press.

Lyon, D. (1994). *The electronic eye: The rise of surveillance society*. Minneapolis, Minnesota: University of Minnesota Press.

Margolin, V. (1989). Introduction. In V. Margolin (Ed.), *Design discourse: History, theory, criticsm* (pp. 3-28). Chicago: The University of Chicago Press.

Mitchell, T. (1988). The product as illusion. In J. Thackara (Ed.), *Design after modernism: Beyond the object* (pp. 209-215). New York: Thames and Hudson.

Moles, A.A. (1989). The legibility of the world: A project in graphic design. In V. Margolin (Ed.), *Design discourse: History, theory, criticsm* (pp. 119-129). Chicago: The University of Chicago Press.

Monmonier, M. (1991). *How to lie with maps*. Chicago: The University of Chicago Press.

Muller, J–C. (1981). Bertin's theory of graphics/A challenge to North American thematic cartography. *Cartographica*, 18(3), 1-8.

Nicholson, L.J. (Ed.) (1990). *Feminism/postmodernism*. New York: Routledge.

Norris, C. (1982). *Deconstruction: Theory and practice*. New York: Methuen.

Pickles, J. (1992). Texts, hermeneutics and propaganda maps. In T.J. Barnes and J.S. Duncan (Eds.), *Writing worlds: Discourse, text, and metaphor in the representation of landscape* (pp. 193-230. London: Routledge.

Review, Columbia Human Rights Law (Ed.) (1972). *Surveillance, dataveillance and personal freedoms*. Fair Lawn, NJ: R.E. Burdick, Inc.

Rule, J.B. (1974). *Private lives and public surveillance: Social control in the computer age*. New York: Schocken Books.

Rundstrom, R.A. (1991). Mapping, postmodernism, indigenous people and the changing direction of North American cartography. *Cartographica*, 28(2), 1-12.

Said, E.W. (1978). *Orientalism*. New York: Pantheon.

Said, E.W. (1994). Rally and resist: For Palestinian independence. *The Nation*, February 14, 190-193.

Simon, H. (1969). *The sciences of the artificial*. Cambridge, MA: MIT Press.

Strohmayer, U., and Hannah, M. (1992). Domesticating postmodernism. *Antipode*, 24(1), 29-55.

Vitta, M. (1989). The meaning of design. In V. Margolin (Ed.), *Design discourse: History, theory, criticsm* (pp. 31-36). Chicago: The University of Chicago Press.

Wood, D. (1984). Cultured symbols/thoughts on the cultural context of cartographic symbols. *Cartographica*, 21(4), 9-37.

Wood, D. (1985). Review article/graphics and graphic information-processing. *Cartographica*, 22(2) 118–121.

Wood, D. (1987). Pleasure in the idea/The atlas as narrative form. *Cartographica*, 24(1), 24-45.

Wood, D. (1992). How maps work. *Cartographica*, 29(3 & 4), 66-74.

Wood, D. (1993a). Are maps sending society in a wrong direction? *Boston Globe*, January 24, 69-70.

Wood, D. (1993b). Maps and mapmaking. *Cartographica*, 30(1), 1-9.

Wood, D., and Fellowes, L. (1994). *The power of maps*. Washington, DC: The Smithsonian Institute.

Wood, D., and Fels, M. (1986). Designs on signs/Myth and meaning in maps. *Cartographica*, 23(3), 54–103.

Wood, D., and with Fels, J. (1992). *The power of maps*. New York: The Guilford Press.

Wood, M. (1984). An historical view of ideas and techniques in cartographic visualization. In A. MacEachren and D.R.F. Taylor (Eds.), *Visualization in modern cartography* (pp. 13-26). Oxford: Pergamon.

5

Automated Cartography and the Human Paradigm

William Mackaness

Department of Geography, University of Edinburgh

INTRODUCTION

This chapter challenges the worth of investigating traditional cartographic techniques in the hope of devising automated (intelligent) visualization techniques for Geographic Information Systems (GIS), and suggests that because the technology itself alters the methodology of design that it needs to be more considered in the design of automated mapping systems. Furthermore, it proposes that, with respect to automated design, there are a number of critical considerations to be addressed. Beyond the acknowledged nebulous nature of design, for the study of traditional cartography to be both relevant and fruitful in the development of a cartographic expert system, specific impediments and considerations need to be overcome:

- though currently the understanding of the generalization process is very incomplete (Rieger and Coulson, 1993), there is a need for a formalized, overall conceptual framework for automated cartography.
- there is a need for a more user-centred perspective in the design of such an automated cartographic system, one that acknowledges the impact of the cognitive artifact and how it has fundamentally changed the process of design.
- there is a widening gulf between analog cartographic products and interactive visualization (animation/sound/data linking between spatial and lexical information).

For the study of the human cartographer to be of value in the development of knowledge-based cartographic expert systems, the above considerations need to be taken into account. This chapter begins with a discussion of the human paradigm and follows with a description of the cognitive artifact and how it has changed the fundamental way in which we go about designing maps.

THE HUMAN PARADIGM

The simplest view of a paradigm is that it is a 'good example' and provides a methodology by which something is done. Much effort continues to be devoted towards elicitation of human cartographic knowledge in the hope of emulation in an automated environment (for example Buttenfield et al., 1991; Muller, 1990; Muller and Mouwes, 1990). The idea that the human is a good model is implicit in much of this research. In studying the cartographer at work (or indeed through the analysis of human cartographic product), we aim to model the human paradigm, and encapsulate (through formalization) the design process in an automated environment.

But in this software engineering process of translating human process into automated systems, there are potential pitfalls. Boehm (1981), for example, illustrates by case study the pitfalls in software engineering when attempting to design systems that automate some human task. In one case study, a classic top-down, stepwise refinement approach was adopted—refining the top level functions into subsequent lower level functions, and continuing this refinement to lower and lower levels as required. The implemented system failed at a number of levels, primarily because some key operational elements were overlooked. A different consultant was hired who viewed the problem from a different, more user-centred perspective, and though asked in a different context from that of map design, addressed the problem as a set of five questions (Table 5.1).

Table 5.1 A more user-oriented approach (after Boehm, 1981: 6).

1. What objectives is the user trying to satisfy?
2. What decisions do we control which affect these objectives?
3. What items dictate constraints on our range of choices?
4. What criteria should we use to evaluate candidate solutions? How are the criterion values related to the decision variables involved in the candidate solutions?
5. What decision provides us with the most satisfactory outcome with respect to the criteria we have established?

From this perspective a system was successfully designed and implemented. With respect to the initial failed approach, Boehm (1981) quotes the proverb: 'Give a man a hammer and he will begin to see the world as a collection of nails'; in other words, the programmers identified a portion of the problem—'a nail', and proceeded to hammer it in with a programming solution without considering the broader context.

With respect to automated cartography, it is suggested that a similar top-down stepwise refinement approach is being adopted, focusing on parts of the automated process (such as line generalization and text placement), and that these are being hammered with insufficient concern for broader objectives and issues raised by the questions of Table 5.1. This is not to say that the extensive research carried out on human approaches to map design have not borne fruit, especially since 'design and development proceed chiefly by emulation of prior art' (Carroll et al., 1991: 75). A large knowledge of cartographic design has been transposed from the literature, compilation manuals, and the study of humans (for example, through protocol analysis). But it would be equally foolish not to acknowledge the fundamental changes that have taken place in the way we explore and visualize spatial information through the use of GIS. Beyond the relatively simple task of automated map production, the computer has changed the means by which we reach the end product and has changed the role of the map in the way that we analyse, interpret, and explore geographical phenomenon. Any study of the human paradigm needs to be made in the context of this changing methodology, and new paradigms need to be considered.

Focus within the Human Paradigm

Nickerson (1987), in describing manual generalization, suggests that traditionally features were selected from a source map for inclusion in a target map; a preliminary version of the target map was made by tracing over the source map using a pen of large width sufficient for the intended scale reduction. During this tracing, features are smoothed, exaggerated, combined, simplified, and eliminated. Finally the map is photographically reduced to the now generalized target scale. Thus, the process is a sequential one (select content, create preliminary version using a pen of high width, and photographically reduce to the target scale). The generalization process takes place in the presence of all the selected features (the cartographer is generalizing individual objects while knowing what other objects are in the immediate vicinity and included on the map as a whole). It is anticipated that this is one version of the map (general in nature, such as the topographic series). The choice of generalization operators would, in part, be governed by the cartographers knowledge of the intended map use. The photographic medium and scribing tools act to constrain the choice and sequence of generalization operators. Time/cost are important constraints. Once scribed, the process is not easy to reverse (in other words, there is little room for experimentation) and the human carries out their own actions based on their decision. Some tasks are performed effortlessly, while others reflect our poor memory, boredom threshold and poor use of analytical techniques.

Conversely, in the automated environment the arduous task of thick pen tracing does not occur, the medium of operation (dark room vs. computer screen) is entirely different, the process need not be sequential, decisions are reversible, multiple representations of phenomenon may be generated in the hope of 'lying less by telling more of the truth' (Monmonier, 1991a). Current automated generalization techniques do not consider the presence of other features (Mackaness, 1994), and the decision making is shared between the user and the system (Turk and Mackaness, 1994). These differences and others (McMaster and Shea, 1992) suggest that the human is not a good paradigm for an automated solution.

Additionally, the study of the human paradigm has tended to occur at a rather simplistic level, with much effort focusing on particular generalization operators (for example, line simplification and text placement), and evaluation techniques that take place at a geometric level. Furthermore, research has focused on mimicry rather than advancement (Chrisman, 1983). This tendency to focus upon geometric elements and operators based upon the dimensionality of space is perhaps due to 'the straightforward manipulations of these elements on a conceptual level' (Claire and Guptill, 1982: 189) and which, of course, are observable during the human map compilation process. Study and re-evaluation of these operators has tended to stymie or occlude research on the underlying process. What is apparently lacking (and not well understood) is the framework of decision making that is going on *behind* the movement of the hand. It might be suggested that this implicit 'essentially creative process' (Robinson, 1960: 132) is where new focus should be, but even here caution is required. It should be remembered that the creative process of design is, itself, context dependent and that technology alters the fundamental process of design. Thus, the opportunity exists to re-engineer cartography through the use of new technology and avoid the capturing of labour intensive tasks that were constrained by the tools (such as scribing), and inadequate graphing techniques through the misplaced observation of the human cartographer. An example of misplaced mimicry in cartographic design is seen in the widespread use of choropleth maps in GIS. Despite the pitfalls of choropleth maps (for example, see Monmonier, 1991b), they remain a predominant form of representation.

There are those who suggest a distinction in classification exists between scientific visualization and traditional cartography. To wit, that emphasis in scientific visualization is on the *development* of ideas, and traditional cartography emphasizes the *presentation* of ideas (Fisher et al., 1993). Somewhat in parallel, Muller (1989) identified two schools of thought with respect to map generalization: one stresses the search for geographic meaning and exploration of geographic process at various scales and extents (Stuetzle, 1988; MacDougall, 1992), and another emphasizes reduction of data imposed by scale reduction and legibility (epitomized by the Topfer and Pillewizer (1966) rule. But these distinctions appear rather academic since presentation is intrinsic to visualization. Furthermore, the strength and growth in the utilization of GIS will depend on the exploratory functionality of visualization techniques such as animation (DiBiase et al., 1992; Monmonier, 1992), sound (Cassettari and Parsons, 1993; Fisher, 1994), and cartograms (Dorling, 1994) and other graphical techniques. An example of an alternative graphing technique is the interactive query of map objects. For example, an object-oriented approach facilitates interactive query of map objects and allows objects to display themselves in finer geometric or attribute detail (Figure 5.1a and b), or to describe how they have been generalized (Figure 5.1c). Such a facility increases the flexibility with which a feature can be represented and provides potential for more radical or abstracted designs.

The distinction between the development and presentation of ideas has been further blurred by the advent of exploratory data analysis (EDA) techniques. EDA

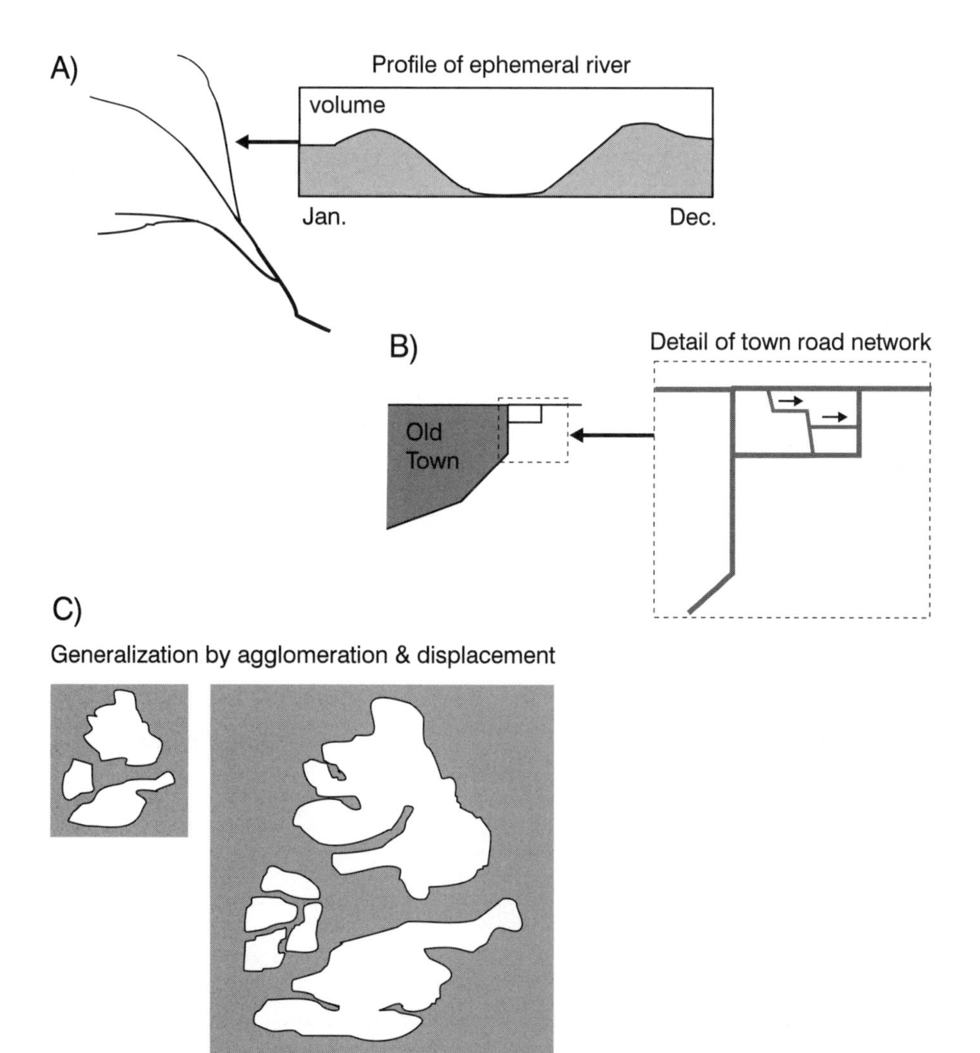

Figure 5.1 Examples of 'intelligent' representations.

is intrinsically linked with the visualization of data. It might be argued that the nature of optimal design requires exploration of representation of interdependent variables. Thus map design is, itself, a form of EDA.

Thus the elicitation of cartographic knowledge, by whatever method, needs to be assessed for its relevancy in an automated environment—an automated environment that has 1) little in common with human constraints, and 2) brings to bear an ever broadening range of visualization techniques that were never available to the human hand. The questions of Table 5.1 are, therefore, pertinent to the successful design of an automated cartographic system.

COGNITIVE ARTIFACTS

A cognitive artifact is 'an artificial device designed to maintain, display, or operate upon information in order to serve a representational function' (Norman, 1991: 71). Thus, they affect human cognitive performance—hence the term *cognitive artifact*. Cognitive artifacts have the potential to change the *nature* of the task and so enhance performance. They can also be used as tools for understanding human cognition. As Carroll et al. (1991) succinctly state, a task (such as designing a map) implicitly sets requirements for the development of artifacts to support it, and it is these requirements that have fashioned GIS hardware and software. In turn, 'a cognitive artifact suggests possibilities and introduces constraints that often radically redefine the task for which the artifact was originally developed' (Carroll et al., 1991: 79). This process by which the task changes in response to evolving technology (which in turn structures the technological evolution in artifact design) is called the task-artifact cycle (Carroll et al., 1991), and is illustrated in a simplified manner in Figure 5.2.

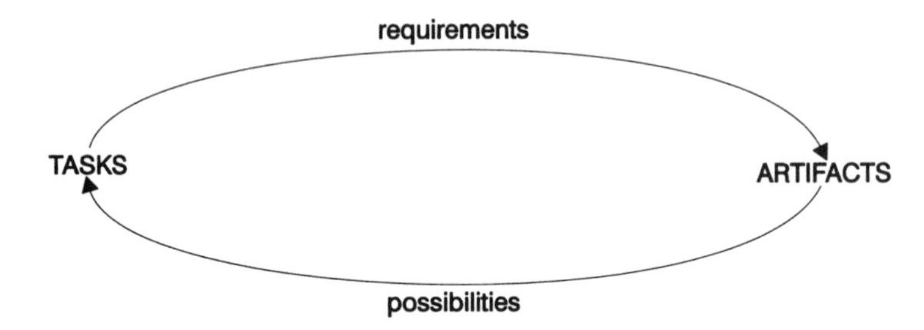

Figure 5.2 The task-artifact cycle for GIS (after Carroll et al., 1991: 80).

An example of where the artifact has changed the task is found in office automation: the electric typewriter altered the ease of office tasks. Subsequently, word processors facilitated editing, and desktop publishing coupled with electronic networking made international authorship and review feasible. Given this task-artifact cycle, it is important that we study the artifact as well as the task in order to reflect the synergistic evolution of task and artifact—the mechanics of design emerging from consideration of particular contexts or tasks.

An artifact is seen as having psychological consequences for users involved in a particular task. An example, with respect to GIS, is the empowerment given to users to create their own maps, including the capacity to 'learn as you go', and to create, reverse, and experiment in the very process of design (an empowerment, incidentally, that apparently obviates the need for cartographers in GIS). Radical redefinition of the task may manifest itself in constraint changes. For example, GIS provides the ability to select, transpose/project, scale, colour, explore, and experiment with spatial data in endless permutations. Ironically, in a highly inefficient manner the user becomes constrained not by design parameters, but by the learning curve of endless design permutations. The user becomes lost among the competing design objectives of context, abstraction and aesthetics.

With respect to GIS and automated cartography, task changes have been dramatic and have apparently occurred at several levels:

- Specialized, laborious and time consuming analog map production techniques have been replaced by digital representations and computational equivalents of the 'overlay';
- The human cartographer has supposedly been made redundant by the 'GIS artifact';
- The first abstraction (field collection) which was circumscribed by the anticipated map product is now replaced by the collection of rich/temporal multi-shared data sets of varying scale and theme;
- Visualizations (digital elevation models/cartograms/sound/animated sequences) can be created that were not previously feasible by hand;
- Linking between spatial and non-spatial data can now take place (Steutzle's (1988) grand tour);
- Representation of abstract phenomenon (such as quality components) is now feasible with the inclusion of spatial metadata (data about data).

Therefore, the demands on graphic design in automated environments go far beyond traditional techniques and alternative paradigms need to be considered.

A CONSTRAINT BASED PARADIGM

Given the amorphous process of map design and the complexity and subjective nature of generalization, what alternative paradigm might we consider that encapsulates both the nature of the task (map design) and the influence of the artifact? How might we view design in an automated environment? Map design and generalization involve a decision making process which is iterative and evolutionary (developing from a hazy thumbnail sketch through to a final product). Its basic *modus operandi* is one of analysis, synthesis and evaluation (Figure 5.3).

Most importantly, it is not a single linear sequence. It occurs iteratively, at many different conceptual and physical levels. Design is not a logical or deterministic activity (Cathaine, 1982), and the user may not have complete knowledge of how the final map will look. Indeed, because design problems are complex and inevitably under-specified, there is likely to be 'a range of acceptable solutions rather than a single optimum solution' (Whitefield, 1990: 14). Robinson et al. (1984) have suggested that the complexity and interdependence of geographical data ensure a considerable number of solutions are potentially meaningful. It is apparent that the utilization of cognitive artifacts has 'unconstrained' the amorphous process of design. For example, users can choose from data of varying quality and resolution, from differing sources, and can transpose, scale, colour, explore, and experiment in

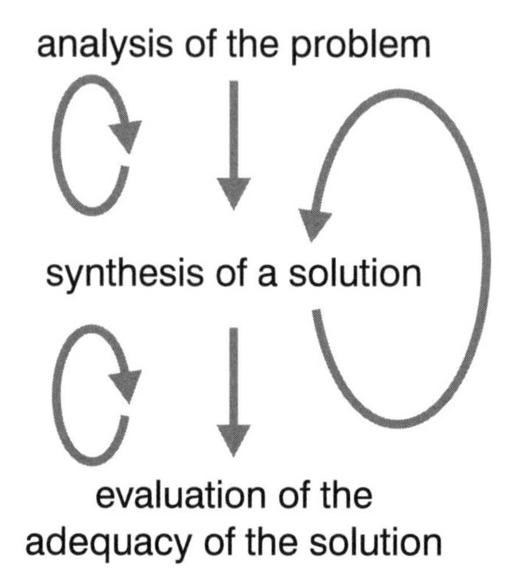

Figure 5.3 Map design as an iterative, non-linear process with an ill-defined goal.

endless permutations. What is needed in the human/machine interaction is a method of informing the user of changing constraints based on the choices made during the design phase. Such constraints variously arise as a result of scale, purpose, and the cognitive artifact. Accuracy is also constrained by resolution, complexity of data constrains content. The system design of the cognitive artifact imposes constraints in terms of functionality and route by which the map is designed.

Automated design can be described as a search problem characterized by an initial state which evolves into a goal state via a suitable sequence of operators (Newell and Simon, 1972). This view mirrors Caplan's definition of design as being 'the process of making things right' (Caplan, 1982: 9)—the interplay of two entities—'the process' and 'the getting it right' requiring interaction and experimentation. As an initial step in an automated environment, heuristics (rules of thumb) can be used to constrain the search space in map design, and during the compilation phase, the choice of generalization operators can also constrain subsequent choices. The idea that design constraints are needed in the design process is not a new one (Gross et al., 1978); indeed, design constraints play a crucial role in any design process principally because 'constraint knowledge is the main method of detecting design failures' (Horner and Brown, 1990: 165). In map design we can say that from an initial, almost speculative state, the cartographer moves through a series of stages that are distinguished not so much by a particular operation, but by the degree and extent of application of a particular technique. Therefore, operations that produce wholesale effects tend to be applied early in design (for example, scaling or selection of whole classes of objects for inclusion on a map), while fine tuning occurs towards the end (for example, selectively omitting, or marginally displacing objects). Figure 5.4 graphically conveys the conceptual nature of this 'honing' effect. The initial state starts with a large (potentially infinite) number of options and operations with global, or large scale effects. With progression along a design process line, options are constrained and operators become progressively more subtle, but nevertheless essential to the final outcome.

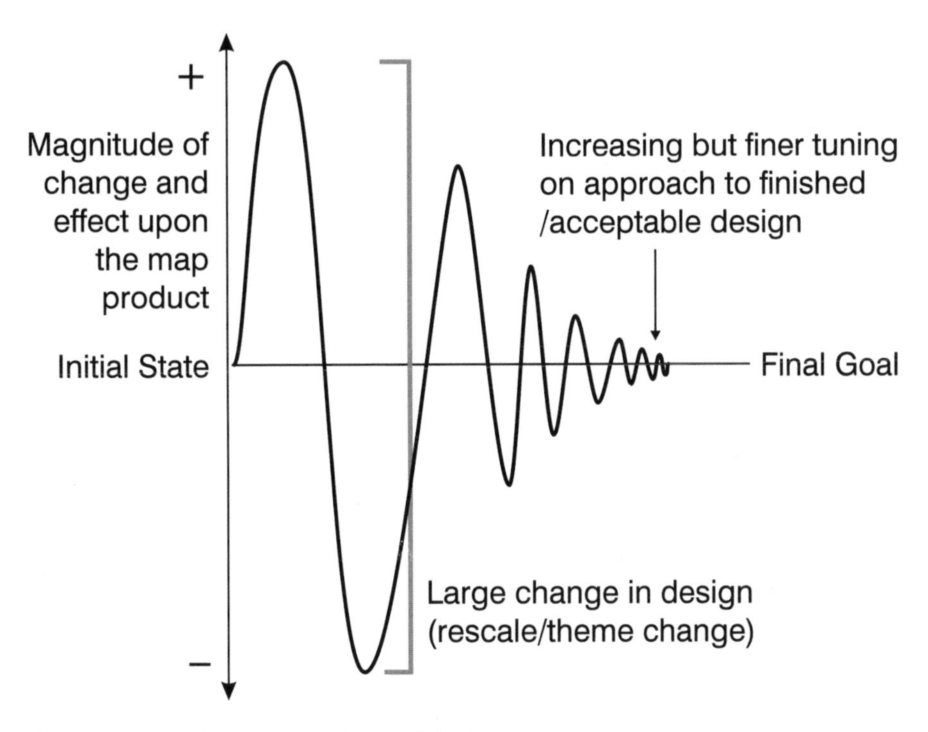

Figure 5.4 Honing in on the final design.

While it is tempting to position particular operators within the design process space, the pertinent message of Figure 5.4 is twofold. First, large effect operators tend to be used early in the map creation process (such as selection, classification), while small effect operators (such as displacement and individual object omission) become the potential choices in the closing stages. In any given situation, any number of operators may resolve a design problem; however, an operator which has a global effect on the map (involving wholesale map changes) and wide implications for subsequent design alterations may not be desirable later in the design process. Small effect operators (such as displacement or individual object inclusion) tend to group together as potential choices. For example, the choices for resolving a single symphysis (some objects sharing very similar space) would **not** be: displace or rescale map. While both would resolve the problem, rescaling would have a major wholesale effect over the entire map.

The second point is that, as the user approaches an acceptable solution, one can envisage micro operations being applied. Thus, a class of features may be resymbolized or reclassified, but their new 'values' might only be marginally different or subtle in colour tone in order to achieve a more aesthetic solution. Choices become more limiting both in degree and extent, as the user approaches a solution; thus users are in essence constrained by their previous actions. If, for some reason, at a later stage the user wishes to rescale the whole map, this effectively moves the design back along the design process line towards the initial state, and once again to a broader choice of options.

By advising the user of a set of nonbinding constraints, it is possible for the system to funnel the user towards a solution. Additionally, by providing the user with a 'map' of progress, the system can provide 1) the flexibility required to explore; 2) an ability to navigate through the decision space; 3) guidance on task sequence and design task parameters. This provides a model for organizing design tasks that support logical steps toward a solution, as well as reversal if required.

CONCLUSION

In general, it would appear that researchers have been too optimistic in their attempts to elicit the nature of design from cartographers. Literature on the nature of cartography has not proved as useful in the formulation of cartographic design knowledge as was once hoped. Indeed, generalization is rarely mentioned in early texts (Rieger and Coulson, 1993), and in some instances, does not reflect reality (Wu and Buttenfield, 1991). The elicitation and subsequent formalization of cartographic knowledge has proved elusive (Frank and Egenhofer, 1992), but in any case, rules for map design should be structured to handle data in a manner that is compatible with the objectives of GIS, EDA and spatial decision making.

In considering the human paradigm and the impact of cognitive artifacts on the process of design, the following three observations are made:

- GIS as an artifact has changed the way in which we explore, design, and ultimately represent maps. While study of the human cartographer has aided and initiated research in automated cartography, its study will become increasingly less relevant in our attempts to model and optimize the process of automated design in cartography. Study of traditional cartography has revealed methods by which maps are generalized, but the overall/collective approach is not applicable in the automated environment, and alternative paradigms need to be considered in the context of automated map design (for example, Mark, 1990).
- Highlighting the inadequacies/differences of the human paradigm does not imply that the human should be removed from the design process. On the contrary, with respect to human computer interaction, there is a real need to maintain a user-centred perspective (as described by Karat and Bennett, 1991).
- More specifically, the design of any knowledge-based system, or cartographic expert system should implicitly support the user's opportunistic sense-making efforts. With respect to map design, this can be achieved in a number of ways—through navigation of design choices, conveying of outcomes, and interface design.

ACKNOWLEDGEMENT

The author is grateful for support through the National Center for Geographic Information and Analysis from the National Science Foundation (SES 88-10917) and reviewer's comments. This research is part of Initiative 8—Formalization of Cartographic Knowledge.

REFERENCES

Boehm, B.W. (1981). *Software engineering economics.* Englewood Cliffs: Prentice Hall.

Buttenfield, B.P., Weber, C.R., Leitner, M., Phelan, J.J., Rasmussen, D.M., and Wright, G.R. (1991). How does a cartographic object behave? Computer inventory of topographic maps. *Proceedings,* GIS/LIS, 91(2), 891-900.

Caplan, R. (1982). *By design: Why there are no locks on the bathroom doors in the Hotel Louis XIV and other object lessons.* New York: McGraw Hill.

Carroll, J.M., Kellogg, W.A., and Rosson, M.B. (1991). The task artifact cycle in designing interaction. In J.M. Carroll (Eds.), *Psychology at the human computer interface* (pp. 74-102). Boston, MA: Cambridge University Press.

Cassettari, S., and Parsons, E. (1993). Sound as a data type in a spatial information system. *Proceedings,* European Geographical Information System Conference EGIS, 93(1), 194-202.

Cathaine, C.S. (1982). Why is design logically impossible? *Design Studies,* 3(3), 123-5.

Chrisman, N.R. (1983). The role of quality information in long term functioning of a GIS. *Proceedings,* Auto Carto, 6(1), 303-312.

Claire, R.W., and Guptill, S.C. (1982). Spatial operators for selected data structures *Proceedings,* Auto Carto, 5, 189-200.

DiBiase, D., MacEachren, A.M., Krygier, J.B., and Reeves, C. (1992). Animation and the role of map design in scientific visualization. *Cartography and Geographic Information Systems,* 19(4), 201-214, 265-266.

Dorling, D. (1994). Cartograms for visualizing human geography. In H.M. Hearnshaw and D.J. Unwin (Eds.), *Visualization in geographic information systems* (pp. 85-102). Chichester: John Wiley and Sons.

Fisher, P.F., Dykes, J., and Wood, J. (1993). Map design and visualization. *The Cartographic Journal,* 30, 136-142.

Fisher, P.F. (1994). Hearing the reliability in classified remotely sensed images. *Cartography and Geographic Information Systems,* 21(1), 31-36.

Frank, A.U., and Egenhofer, M.J. (1992). Computer cartography for GIS: An object oriented view of the display transformation. *Computers and Geoscience,* 18(8), 975-987.

Gross, M., Ervin, S., Anderson, J., and Fleisher, A. (1987). Designing with constraints. In Y.E. Kalay (Ed.), *Computability of design* (pp. 53-84). New York: John Wiley and Sons.

Horner, R., and Brown, D.C. (1990). Knowledge compilation using constraint inheritance. In J.S. Gero (Ed.), *Applications of artificial intelligence in engineering, Volume 1: Design* (pp. 161-174). Berlin: Springer Verlag.

Karat, J., and Bennett, J.L. (1991). Working within the design process: Supporting effective and efficient design. In J.M. Carroll (Eds.), *Psychology at the human computer interface* (pp. 269-285). Boston, MA: Cambridge University Press.

MacDougall, E.B. (1992). Exploratory analysis, dynamic statistical visualization, and geographic information systems. *Cartography and Geographic Information Systems,* 19(4), 237-246.

Mackaness, W.A. (1994). Knowledge of the synergy of generalization operators in automated map design. *Proceedings*, Canadian Conference on GIS-1994, 1, 525-536.

Mark, D.M. (1990). Competition for map space as a paradigm for automated map design. *Proceedings*, GIS/LIS 90 Anaheim California, 1, 97-106.

McMaster, R.B., and Shea, K.S. (1992). *Generalization in digital cartography*. Resource Publications in Geography, Association of American Geographers.

Monmonier, M. (1991a). Ethics and map design—Six strategies for confronting the traditional one map solution. *Cartographic Perspectives*, 10, 3-8.

Monmonier, M. (1991b). *How to lie with maps*. Chicago: University of Chicago Press.

Monmonier, M. (1992). Summary graphics for integrated visualization in dynamic cartography, *Cartography and Geographic Information Systems*, 19(4), 247-260, 272.

Muller, J.C. (1989). Theoretical considerations for automated map generalization. *ITC Journal*, 3(4), 200-204.

Muller, J.C. (1990). Rule based generalization: Potential and impediments. *Proceedings*, 4th International Symposium on Spatial Data Handling Zurich, Switzerland, 1, 317 - 334.

Muller, J.C., and Mouwes, P.J. (1990). Knowledge acquisition and representation for rule based map generalization: An example from the Netherlands. *Proceedings*, GIS/LIS 90, Anaheim, California, 1, 58-68.

Newell, A., and Simon, H.A. (1972). *Human problem solving*. Englewood Cliffs, NJ: Prentice Hall.

Nickerson, B.G. (1987). *Automated cartographic generalization for linear map features*. PhD Thesis, Rensselaer Polytechnic Institute, Troy, New York.

Norman, D.A. (1991). Cognitive artifacts. In J.M. Carroll (Eds.), *Psychology at the human computer interface* (pp. 17-38). Boston, MA: Cambridge University Press.

Rieger, M.K., and Coulson, M.R. (1993). Consensus or confusion: Cartographer's knowledge of generalization. *Cartographica*, 30(2 & 3), 69-80.

Robinson, A.H. (1960). *Elements of cartography*, 2nd ed. New York: John Wiley and Sons.

Robinson, A.H., Sale, R.D., Morrison, J.L., and Muehrcke, P.C. (1984). *Elements of cartography*, 5th ed. Chichester: John Wiley and Sons.

Roth, K.S. (1991). Practical problems in developing rules for automated mapping. *Proceedings*, ACSM/ASPRS, vol. 2: Cartography and GIS, 287-292.

Stuetzle, W. (1988). Plot windows. In W.S. Cleveland and R. McGill (Eds.), *Dynamic graphics for statistics* (pp. 225-245). Pacific Grove, CA: Wadsworth and Brooks.

Topfer, F., and Pillewizer, W. (1966). The principles of selection. *Cartographic Journal*, **3**(1), 10-16.

Turk, A.G., and Mackaness, W.A. (1994). Computer-assisted visualization design and the modeling of cognitive ergonomics, *SIRC Colloquium*, University of Otago, Dunedin, New Zealand, 17/18 May 1994, 281-288.

Whitefield, A. (1990). Human computer interaction models and their roles in the design of interactive systems in cognitive ergonomics. In P. Falzon (Ed.), *Understanding, learning, designing human computer interaction*, (pp. 7-25). London: Academic Press.

Wu, C.V., and Buttenfield, B.P. (1991). Reconsidering rules for point feature name placement. *Cartographica*, 28(1), 10-27.

The Practitioner's View? A Pilot Study into Empirical Knowledge About Cartographic Design

6

Michael Wood
Centre for Remote Sensing and Mapping Science, Department of Geography, University of Aberdeen

Kenneth J. Gilhooly
Department of Psychology, University of Aberdeen

INTRODUCTION

"There is no such thing as cartographic design . . . the stereotyped nature of extant map work is evidence of the absence of design as a force in cartography" (Collinson, 1993). These words come from a challenging paper by a practising professional cartographer whose livelihood depends largely on his ability to design maps that do more than merely satisfy the information needs of their users, but also stimulate them visually and give pleasure. His are controversial statements but even without reading more of Collinson's words some readers may already appreciate what he is driving at: What has happened to map design? Despite centuries of cartographic productivity "most maps" as he concludes "are like pyramids, they are all the same . . .". The implication is that while maps may have increased in number over the decades perceived constraints of their utility seem to have blinkered cartographers and limited the variety of graphic solutions which they employ.

Standard cartographic texts do not concur exactly on a definition of map design. Dent (1993) explains it as "the aggregate of all the thought processes that cartographers go through during the abstraction phase of the cartographic process." Other books take slightly different approaches, such as the 6th Edition of *Elements of Cartography* (Robinson et al., 1995) in which the authors consider the objectives, the function and the scope of map design. One of the most recent discourses on maps sees cartography's function as "creating interpretable graphic summaries of spatial information (i.e. representations) and the goal of producing more consistently functional maps" (MacEachren, 1995). Traditionally, some maps, such as the necessarily standardized topographic map, have been designed to fulfil just such a set of functions or goals. Design has been controlled by geographic constraints of content as well as the need to achieve a balanced representation that avoids over-emphasis on any one feature. The designers of other categories of maps, however, should be able to ignore many constraints and adopt more creative and innovative approaches to their problems. Collinson's frustrations, therefore, are that too many cartographers have failed to escape from the graphic traditions of cartography even when the constraints are no longer present. At a time when the flexibility and potential of new computer graphics packages are so great, this reticence is even more regrettable. What do other contemporary cartographers think?

The present chapter introduces a pilot project, currently under way, to find out more about map designing, in particular, through investigation of the ideas and opinions of practising professional (bench) cartographers. Preliminary analysis of the responses received to date supports the importance of the practitioners' contribution to design. It also helps reveal the implicit nature of their skills and knowledge and how feelings and emotion may play a vital part in the design decision-making processes.

BACKGROUND AND APPROACH

Much scientific research has been carried out (Wood, 1993), and there are established academic texts which deal effectively with map design (e.g. Robinson, 1952; Keates, 1989; Dent, 1990; MacEachren, 1994, 1995). But although research problems have been addressed and the texts may have distilled much of the essence of the subject, there is often a sensation, after reading them, that something is missing . . . a clearer insight into the actual knowledge and experience of real cartographic practitioners—those who spend their lives making maps.

Although there has been a considerable volume of research on the cognitive psychology of expertise over a wide range of domains (see Ericsson and Smith, 1991), the area of expertise in map design does not appear to have been given much attention. However, since map making is a communicative activity, it may be fruitful to draw analogies between skill in map design and skill in expository writing, which has been studied (e.g. Scardamalia and Bereiter, 1991). Both domains involve communication of complex information in a form intended to be easy to use in multiple ways by a target audience. In writing, experts are found to engage in extensive planning with frequent revisions being the norm. Less expert writers are more likely to write quickly, without planning and without revision. In a range of other domains (e.g. physics, mathematics, political science) it has been found that experts spend longer constructing an initial representation of what the current problem actually involves in terms of materials, goals, constraints of time and costs, and so on (Glaser and Chi, 1988). Thus, we would expect that experts in map design, when faced with problems that are not completely routine, would engage in extensive task analysis, planning, goal definition, exploration of possible solutions, and revisions. Evaluation of possible designs is clearly an important part of the design process and is one which is frequently regarded as difficult to verbalize. It may well be that with experience and exposure to examples generally agreed to be 'good' and 'bad' designs, map designers implicitly learn an assemblage of cues for goodness of design; such implicitly learned patterns of cues are typically not accessible to conscious report. Medical experts generally become skilled in using a multiplicity of subtle cues to recognize diseases rapidly without explicit reasoning and may not be able accurately to report the cues that they used (Harries, Dennis, and Evans, 1994). Thus, we may expect that map designers will similarly be unable to report design criteria fully or explicitly.

Damasio (1995) has recently presented a hypothesis which might help explain the apparently implicit, but undoubtedly effective, processing methods employed by experts in many fields. The links, he believes, are feelings and emotions. His ideas have grown from extensive research carried out with patients who have suffered damage to parts of the brain, notably the pre-frontal area. Despite the severity of such damage, many patients recover and appear to retain all the characteristics of fully rational behaviour. What they often lose, however, are emotions or feelings and this deficiency seems to correlate closely with a pathological curtailment of the ability to reach effective decisions or, sometimes, any decisions at all! In his so-called 'somatic marker' hypothesis, Damasio believes that certain parts of the normal brain contain myriads of accumulated somatic markers which are 'a special instance of feelings generated from secondary emotions. Thus emotion and feelings have been connected, by learning, to predicted future outcomes of certain scenarios'. A negative marker will sound an alarm bell in the brain while a positive one can become 'a beacon of incentive'. He suggests that the absence of somatic markers reduces the accuracy and efficiency of the decision-making process and it can help explain why an experienced expert can apparently skip through a complex problem and quickly offer a series of workable solutions. The hypothesis could also explain the emotional reaction an experienced cartographer may have to a map whose satisfying design triggers off a whole series of positive sensations.

Returning to the conclusion of the previous paragraph, this hypothesis could also contribute to the puzzle as to why map designers often cannot report design criteria in any detail. They just 'feel' that things are 'right'. This also tallies with the authors' experience when talking to practising cartographers. Feelings and emotions are seldom far from any discussion about design.

Discussion with cartographic professionals reveals a clear overlap between their knowledge and what has already been encapsulated in academic texts. However, there is more. Imhof, one of the most verbally and graphically skilled cartographers of recent times, identifies the need for "... balanced expression which emphasises the significant and subdues the insignificant; and . . . a well-balanced harmonious interplay of all the elements concerned" (Imhof, 1982). You will see similar phrases in the cartographic instruction manuals, but what do these words really mean? It is almost like telling a complete beginner that riding a bicycle involves a subtle interplay of balance, motion and confidence! The words mean nothing until the process has been experienced physically, again and again, and the skill is acquired. The words of the textbook are wise but, as Keates has observed, ". . . there is a great body of empirical knowledge about 'everyday' cartographic design problems, and a considerable understanding of the principles by which they are resolved" (Keates, 1989). The authors were further motivated by reading interviews with four outstanding cartographic practitioners (Dunlavey, DiBiase, Parsons and Furno), reported in *Aldus Magazine* (Brown, 1992). Although it has been said that "especially creative people may tend to be so thoroughly non-verbal . . ." that it is impossible for them to explain their spatially-intuitive skills and achievements in words (Shepard, 1978), some cartographers, like Imhof, Keates, and the four mentioned above, are able to translate complex graphic ideas into words and go some way towards contradicting that statement.

With the current growth in sophisticated computer software such as geographic information systems (GIS), which offer mapping tools to users who may have no background in cartography, the need to analyse the nature and structure of map design expertise is paramount and is being addressed (Weibel and Buttenfield, 1992). While this current project may eventually offer some input for the building of cartographic expert systems to support the use of such software, that is not its primary aim at present. Its purpose may seem simpler, although, perhaps, no less ambitious: To try to tap the 'body of empirical knowledge' mentioned by Keates and ascertain the status of 'design' within 'everyday' map production as perceived by current cartographic practitioners.

The chosen methods of investigation are familiar. The initial stage involved informal discussions with individual cartographers (that is, those who are intimately involved, directly or indirectly, in the creation of maps). This was followed by questionnaire surveys, and more detailed studies with selected practitioners will follow. That stage will focus on a smaller number of individuals, identified by the interest and motivation they have shown and by their special fields of experience. These methods have been selected for their simplicity and, initially at least, for their low cost.

The preliminary discussion phase and a pilot questionnaire survey have already taken place. Some analysis has been carried out and early results are encouraging. The respondents have shown an eagerness to discuss what they do, although often down-playing some of the details as too mundane: thoughts and techniques which they regard as just part of their daily experiences of finding cartographic solutions! This may be one of the defence mechanisms used to rationalize their inability to verbalize aspects of their practical knowledge. Analysis of the results of the pilot questionnaire is still incomplete, but results look promising and the process of interacting with cartographers continues to be rewarding for the investigators. While not necessarily promising lists of previously hidden 'rules-of-thumb,' it is focusing attention on aspects of what continues to be a unique profession.

In the introductory paragraph of this chapter the word 'design' seems to refer to much more than the mere manipulation of the elements of the map image to help achieve clarity and legibility. The paper quoted seems to criticize some maps at least for an absence of truly creative thinking. But is this criticism valid? How much creativity is possible when designing maps? One of the most compact summaries of the concept of design in a cartographic context is offered by Keates in which he states that "the purpose of (all) design is to create something which did not exist before" (Keates, 1989). In that sense designing has always been part of the process of making functional artefacts, but only over the past 50 years has it been recognized as a process in its own right. Today there are architectural designers, product designers, fashion designers, but are there any true cartographic designers? Collinson highlights the domination, until quite recently, of the craft element in cartography, that is, the methods and techniques of creating the images (such as pen drawing and scribing). This has protected the profession in the past from becoming subsumed within other disciplines, but it has also diverted the attention of practitioners themselves away from more basic topics, such as the formulation of design principles on graphic image design and composition. Few practising cartographers have written about such detailed design issues. One reason may be that the procedures are so extensively nonverbal and spatially intuitive in character, for example, the subtle visual analysis and solution of spatially-conflicting clusters of symbols and names or the often mandatory employment of imagination and mental visualization during the preliminary stages of map design. It has been left to latter-day geographer-cartographers to analyse cartographic products and try to distil the elements and principles of the subject for academic teaching purposes. Design is referred to in various texts on cartography, but these sections tend to concentrate, in a fairly constrained way, on aspects such as generalization, the creation of appropriate symbols, and the *need* to resolve the complex assembly of image elements: symbols and names. The readers are left to investigate possible solutions themselves. More recent publications have further highlighted the additional importance of overall strategies such as contrast and visual hierarchy to the achievement of acceptably legible images (e.g. Dent, 1990). However while this interpretation of 'design' is important for the production of legible and comprehensible maps, only concepts such as visual hierarchy transcend the spatially localized issues of merely resolving clutter. Perhaps, having been restricted for so long by convention and standardization, cartography has never given 'real' design a chance. Collinson, again from personal experience, observes that publishers normally allocate 'the lion's share' of map development to compilation processes. The cartographic designer effectively does not exist.

One of the most relevant areas of map design research today relates to the increasingly widespread use of both GIS and digital mapping software. Development of these tools has meant that "many individuals with little or no formal experience in cartographic design are now engaged in map making" (MacEachren, 1994). Increasingly, researchers or managers who need to examine and present geographic information will not have to turn to professional cartographers, but will be able to produce perfectly adequate maps themselves from current software. Support for such an essentially 'utilitarian' approach to design will increase, but is never likely to reach the plane of truly innovative creativity. Indeed, using his own definition of cartographic design as an innovative creative process, Collinson observes that "expert systems for design are merely figments of crazed imaginations." MacEachren, in his recent publication, *Some truth with maps: A primer on symbolization and design,* expresses the now common view of the need for an acceptable level of utilitarian design knowledge. His text is directed, primarily, at maps being used "in the context of environmental research, policy formation, monitoring and management" and "from data exploration . . . to presentations" (MacEachren, 1994). Although maps, even produced in this way, will also be judged "in terms of their appearance as well as their utility" (Keates, 1989) most 'makers' (e.g. GIS users) are likely to be relatively unconcerned with such aesthetic issues. On the other hand, when you turn to the world of the professional cartographer (responsible for the production of special-purpose maps for textbooks, academic papers, atlases and the media), their frequently deep involvement with the more creative aspects of cartographic design reveals a different viewpoint. While they might use the functionality of a GIS to help them assemble and customize their spatial data (as in commercial organizations such as Bartholomew), they are normally faced with finding new cartographic solutions to specific needs. The concepts which emerge should then control the whole operation of design and compilation. You seldom meet a professional cartographer in such circumstances who is not concerned with aesthetic issues, even if functionality is still uppermost in the design brief. This could be contrasted with an urban planner using a GIS to assemble and manipulate all the spatial (and aspatial) data required to guide his decisions. Involvement with, and understanding of, such specialized data will reduce the need for highly resolved design solutions for maps for private contemplation or discussion between experts in that subject area. Clarity and legibility will often suffice.

So, as we always suspected, map design exists on at least two levels and professional cartographers (in particular the target group of this investigation) always seem concerned with more than the purely functional design of the images they create.

METHODS AND PRELIMINARY RESULTS

The decision to instigate this project followed discussions at the 1993 Annual Summer School of the (British) Society of Cartographers (SoC) in Lampeter, Wales. Permission was granted to employ the SoC membership list from which those most directly involved in map production could be identified. Discussions with members had revealed enthusiasm for the proposal and a questionnaire was eventually composed and distributed to a sample group. The wording of the first questionnaire is as follows and was intended to be non-directive and to evoke the respondents' most salient beliefs:

A: GENERAL THOUGHTS ON CARTOGRAPHIC DESIGN

This is what I believe about good cartographic design: *(Please just put down your thoughts, in sentences, phrases or even as a list, in the space below.)*

B: PRACTICAL KNOWLEDGE/EXPERTISE ABOUT MAP DESIGN

1. Please name (with explanatory note, if necessary) UP TO ten (10) practical techniques which you use in (and believe to be essential to) map construction/design, but are **NOT GENERALLY DESCRIBED IN CARTOGRAPHY TEXT BOOKS** IN SUFFICIENT DETAIL.

---

2. Choose ***at least ONE of your selected techniques*** and attempt to explain the detailed procedures involved (e.g. by referring to an example). *Although it might be difficult please try to avoid using generalizations.*

---

3. ***(OPTIONAL)*** **Please** send me a photocopy of one map whose design you are pleased with, explaining why (on an accompanying sheet).

   ❑ *map enclosed* ❑ *map not enclosed* (please tick)

---

4. Would you be willing to **help me further** in this study? ❑ *yes* ❑ *no*

An accompanying letter outlined the aims and background to the project and subjects were also asked to provide information about their job description/duties, number of years of experience and computer/software, if used. As shown above, respondents were also invited to participate further in the study and the majority have agreed to do so.

Analysis of non-directed free text responses is difficult. One of the first methods applied to a sample of 30 responses to Section A of the questionnaire (General thoughts on cartographic design) was a frequency-of-mention count of features related to good map design. Those most often mentioned were as follows:

Clarity/Not cluttered; Look of map—Attractive/Joy/Beautiful to look at and use	52%
Balance	48%
Consideration of purpose of map/Fulfilling client's stated purpose	24%
Communicate effectively/Understandable	21%
Emphasis (Linework, lettering, colour tints); Accuracy; No ambiguity/Uncomplicated symbols; Lettering style; Correct weight of line information	17%
Size of lettering; Consideration of final user; Simplicity; Innovative/Choice of colour use	14%

More careful analysis is under way and the results are becoming more useful as the number of responses increases. The detailed structure of the follow-up stage has not yet been finalized, but will include in-depth interviews and more formal studies of expertise.

There are, as yet, no real results in this study, only data, in the form of pages of written notes. The extent and variety of the ideas contained are already too great for justice to be done to them without formal analyses but, to add some 'colour' to this report and to provide a flavour of the responses received so far, a **very** small (and random) selection of the more original quotes is listed below, without comment. They are unattributed and both thanks and apologies are offered to their authors.

A. GENERAL THOUGHTS ON MAP DESIGN:

- A growing number of other computer users can now produce maps with apparent ease but not necessarily 'expertise'.
- Good design makes the product a joy to look at *and* use.
- Good design has to allow some breaking of traditional rules.
- Good design can be messed up by too many opinions and is difficult to achieve by consensus.
- Intuitive skills are (perhaps) more important than any skills that can be taught. This diminishes the role of the instructor and relegates his/her role to one of enforcing the *golden rules of convention*. But it also suggests that good natural cartographic designers will emerge from the pack despite computers and despite the lack of good design instruction that has always existed in our profession.
- Good design is simple but effective. (Most academic authors insist on including far too much information.)
- It takes time to know what 'feels right'.
- The cartographer's use of intuition defies attempts to formulate rules and guidelines.
- If it looks right it is right.
- An illusion of reality is better than reality itself
- Maps must have an emotive as well as a rational content.
- Creativity in design enhances comprehension.
- Add to the harmony not the discord.
- Concept before compilation.
- Good design moves in the direction of simplicity.
- Why has cartography become totally dominated by 'A' paper sizes?
- Style is the hint of uniqueness in the way all the graphic elements of a map are brought together.
- There are more people talking about cartography than actually doing it.

B. SOME NOTES ON TECHNIQUES:
(Summary topics only—details were provided in the questionnaires.)

- Locating lettering on maps.
- Use of artistic licence.
- Practice is the best way and *time* to experiment...of which I never seem to have enough!
- Saving templates...and also files with effective hatches, tints and line symbols.
- Design for less than optimum eyesight.
- Play around with different roughs/layouts.
- The only 'scientific' procedure I know for (the techniques listed) is *experience*.

CONCLUDING THOUGHTS

At this stage in the project the investigators feel as if they have reached into a treasure chest of knowledge but have, as yet, been unable to grasp, fully, any of the jewels. Unlike objective/neutral questionnaire surveys, there is an enthusiasm here amongst the subjects, a desire to know more about their own expertise and, for many, a desire to participate much more fully in the investigation. Even without a full set of formal results, the principles behind the project seem to hold good. Practitioners do have something distinctive to offer, although the initial technique, the questionnaire, may be too blunt an instrument to tease out the detail which is there. The early batch of responses are mainly from professionals with long working experience (up to 30 years, or more). This group is most likely to contain the people who have not had the opportunity for formal education in their subject. Inevitably when the final tally of knowledge is made, deriving from respondents from all ranges of education and experience, there will be a significant overlap between 'textbook-based' ideas and those gained from working experience. This and other issues will have to be addressed.

It is hard to predict the likely effects, if any, of this study on the target population. If the investigation receives full participation it might help to strengthen the professional bonding already in existence among cartographers. The sharing of practical ideas is already facilitated through professional journals such as the *Bulletin of the Society of Cartographers* and *Cartographic Perspectives* (the bulletin of the North American Cartographic Information Society). The historical craft influence on cartography is fading, although today's novice professional may still regard the acquisition of a full working knowledge of a software package such as Aldus Freehand (or even Arc/Info!) in much the same way as draughtsmen of the past regarded perfect skills in pen drawing. It is clear, however, that the cartographers of today need not work in isolation. Not only are there numerous means of rapid communication, but there is an enthusiasm among knowledgeable practitioners, for example of new software packages, to share their knowledge freely. If this project can contribute to this networking of ideas, even if it seems only to have extended the knowledge-sharing characteristics of the group, it will have succeeded.

The formal literature of a subject contains much of value to students, particularly for degree course assessment. However, equally valuable is knowledge obtained through experience. It is self-evident that such knowledge must be personally acquired, but a fuller awareness of its existence and authentic nature might engender greater respect for the subject and contribute to greater maturity of approach amongst young professionals. Cartographic design may not have been part of the wider field of design in the past, but a fuller awareness of the practitioners' viewpoint may help clarify uncertainties and complement the more academic writings which, not infrequently, have been criticized for their overly theoretical nature. This study may encourage others to carry out more effective investigations and help expand this 'database' of empirical knowledge.

REFERENCES

Collinson, A. (1993). Cartographic design does not exist (and never has)! *SUC Bulletin*, 27, 3-6.

Brown, C. (1992). All over the map. *Aldus Magazine*, May/June, 14-24.

Damasio, A. (1995). *Descartes' error: Emotion, reason and the human brain*. London: Picador.

Dent, B.D. (1990). *Cartography: Thematic map design*. 2nd ed. Dubuque: WCB Publishers.

Dent, B.D. (1993). *Cartography: Thematic map design*. 3rd ed. Dubuque: WCB Publishers.

Ericsson, K.A., and Smith, J. (Eds.) (1991). *Toward a general theory of expertise: Prospects and limits*. Cambridge, MA: Cambridge University Press.

Glaser, R., and Chi, M.T.H. (1988). Overview. In M.T.H. Chi, R. Glaser and M.J. Farr (Eds.), *The nature of expertise*. Hillsdale, NJ: Lawrence Erlbaum Associates.

Harries, C., Dennis, I., and Evans, J. (1994). General practitioners' insight into their own decision-making policies: What are they telling us? Paper presented at the 11th Annual Conference of the British Psychological Society, Cognitive Psychology Section, September, 1994. Cambridge.

Imhof, E. (1982). *Cartographic relief presentation*. Berlin: de Gruyter.

Keates, J.S. (1989). *Cartographic design and production*. London: Longman Scientific and Technical.

MacEachren, A.M. (1994). *Some truth with maps: A primer on symbolization and design*. Washington, D.C.: Association of American Geographers.

MachEachren, A.M. (1995). *How maps work: Representation, visualization and design*. London: The Guilford Press.

Robinson, A.H. (1952). *The look of maps*. London: University of Wisconsin Press.

Robinson, A.H., Morrison, J.L., Muehrcke, P.C., Kimerling, A.J., and Guptill, S.C. (1995). *Elements of cartography*. 6th ed. Chichester: John Wiley and Sons.

Scardamalia, M., and Bereiter, C. (1991). Literate expertise. In K.A. Ericsson and J. Smith (Eds.), *Toward a general theory of expertise: Prospects and limits*. Cambridge, MA: Cambridge University Press.

Shepard, R.N. (1978). Externalization of mental images and the act of creation. In B.S. Randhawa and W.E. Coffman (Eds.), Visual learning, thinking and communication (pp. 133-189). London: Academic Press.

Weibel, R., and Buttenfield, B.P. (1992). Improvement of GIS graphics for analysis and decision-making. *International Journal of Geographical Information Systems*, 6, 223-245

Wood, M. (1993). The map-users' response to map design. *The Cartographic Journal*, 30, 149-153.

7

Cartographic Complementarity: Objectives, Strategies, and Examples

Mark Monmonier

Department of Geography, Syracuse University

INTRODUCTION

'Cartographic complementarity' addresses the inherent interrelationship of map content and graphic symbols within a series of maps. It seeks efficiently informative presentations by minimizing needlessly redundant information as well as by juxtaposing or integrating multiple measurements or themes needed for insightful interpretation. As a coherent approach to map design, cartographic complementarity also attempts, within or between maps, to avoid pointless and distracting inconsistencies in symbols and text. An essential element in dynamic narrative cartographic presentations (including animated maps), cartographic complementarity is also important in the design of atlases, narrative sequences of static maps, and expository cartography in general.

Less formal than most attempts at a conceptual framework, this essay examines strategies for promoting cartographic complementarity in diverse contexts. The first part of the chapter not only advocates a fuller awareness of the need for complementarity but illustrates with examples the broad range of situations in which the concept is relevant. The second section explores briefly three design problems in dynamic cartography: graphic narratives, geographic correlation, and spatial-temporal displays.

THE NEED FOR COMPLEMENTARITY

Topographic maps with standardized symbols and format reflect a cartographic complementarity in which one quadrangle map complements all other maps in the series. Experienced users benefit from similarities in scale, projection, content, feature categories, and map symbols. Inexperienced users quickly recognize that a standard design obviates having to verify whether a symbol has the same meaning on different sheets. A uniform set of cartographic decisions applied to multiple maps sharing a common goal simplifies the work of map user and map maker.

As I argue here, cartographic complementarity is more than standard symbols, definitions, and content; it also relies on the graphic equivalent of two key concepts in expository writing: coherence and flow. To describe accurately an event, process or idea, an author must provide a comprehensible succession of examples and explanations as well as avoid distracting readers with extraneous material. Like expository writing, expository cartography depends on the author's skill in linking relevant facts in a coherent sequence, in which new information builds on old information. Unfortunately, map authors all too frequently ignore complementary data, features, or graphic representations that could enlighten their audience.

Words are particularly important in cartographic complementarity because words provide an efficient link between the graphic vocabulary of the map author and the natural language of everyday speech and intellectual discourse. Although verbal ties are crucial, most textbooks on map design say little about words on maps —an unfortunate consequence, perhaps, of the scientist-technologist's preoccupation with geometric accuracy, legibility, and conceptual frameworks. Moreover, maps in atlases, academic journals, and government reports demonstrate widespread ignorance of how a concise, carefully crafted map title can announce the map's objectives and establish a context for reliable interpretation; how subtitles and other descriptive text can warn against misinterpretation; and how place names can tie mapped features to the reader's mental map. Because written text and named features provide links between graphic symbols and our understanding of reality, the map author must reflect carefully on how the report, book, or atlas discusses its maps, how each map relates to other maps in the same publication, and how the reader's knowledge of the area or subject matter provides a foundation for further understanding. By referencing the same places and features, for instance, a map and its accompanying text not only reinforce each other but avoid frustrating readers attentive to inconsistencies. An editor reviewing a manuscript for cartographic complementarity must look closely at the integration of maps and text, either (or both) of which might require revision.

Cartographic complementarity is especially important in dynamic cartography, in which complex sequences and moving images can readily overwhelm the senses. Complementary explanations that precede or accompany maps are indispensable because understanding a rapidly paced graphic sequence is difficult without first comprehending its point, or purpose. Although interactive software can avoid many difficulties by letting the user control the pace and direction of the presentation, a 'closed' narrative designed for multi-person audiences typically requires a preliminary announcement of its objective, a running commentary to point out key spatial-temporal patterns, and an interpretative summary. Because an audio channel can be valuable for interactive cartographic presentations, cartographic complementarity takes on the fuller burden of linking dynamic graphic symbols with (a) visual text directly on the graphic or in a separate text window, (b) spoken text in a voice-over or a voice annotation triggered by a sound button, and (c) sound variables such as pitch.

Among the more radical consequences of cartographic complementarity is a replacement of traditional one-map solutions with complementary sets of maps, graphs, and diagrams. Multi-view solutions, of course, are by no means new; geologic maps, for instance, have long included structural profiles to help viewers understand planimetric three-dimensional representations of the earth's crust. In most cases, though, the cartographer who needs to describe a spatial variable or phenomenon is satisfied with just one more or less optimal map. George Jenks and other cartographic theorists, myself included, have privileged the single-map approach with optimization algorithms addressing a variety of cartographic decisions (Jenks and Caspall, 1971; Monmonier, 1982). But the problem is not optimization per se—computational strategies might, after all, seek out optimally informative multiple views.

Optimization in pursuit of cartographic complementarity is an intriguing possibility. Although any attempt at optimization (algorithmic or otherwise) might seem hopelessly positivist, even postmodern critics should recognize the value of complementary graphics offering diverse interpretations. By encouraging viewers to consider best-case and worst-case scenarios and other "alternative" interpretations, cartographic complementarity can avoid at least some of the inherent bias of single-map solutions (Monmonier, 1991).

A few examples in statistical mapping illustrate nicely the need for cartographic complementarity. These examples are based on unemployment data for the United States.

By tradition if not instinct, labour economists and the news media usually work with rates, not counts, when dealing with employment data. The unemployment rate measures the number of unemployed persons as a percentage of the total labour force, which includes persons actively seeking work as well as those holding jobs. (Ironically, the unemployment rate can fall when chronically unemployed persons lose hope, stop looking, and deflate the denominator.) Widespread use of percentage rates spares professors and other cartographic purists the pain of conceptually incorrect choropleth maps showing that—Surprise! Surprise!—most unemployed workers live in places with large populations. For other geographic data, though, intellectual tourists on the map maker's turf often ignore the folly of mapping counts with intensity symbols as well as the limited value of conceptually correct proportional-point-symbol maps.

Even so, the full story typically requires more than one map. Although the choropleth map of intensity symbols portraying percentage rates is, in theory, the single most informative presentation of unemployment data, a companion display portraying counts can be a fittingly informative complement. In Figure 7.1, for instance, a choropleth map (above) shows that in 1990 California and New York had comparatively moderate rates of unemployment. But policy makers need to know that even modest rates in states with relatively large populations mean huge numbers of unemployed persons, as the map of counts (below) reveals. Because a map of magnitude informs the interpretation of a map portraying intensity, the conscientious map author might juxtapose the two complementary graphics within the same neat line or on the same slide—no point in letting an insensitive editor separate them or inattentive viewers ignore their joint meaning.

For some audiences there is more to the story than two maps can easily accommodate. Local viewers, for instance, might appreciate a pair of maps that not only

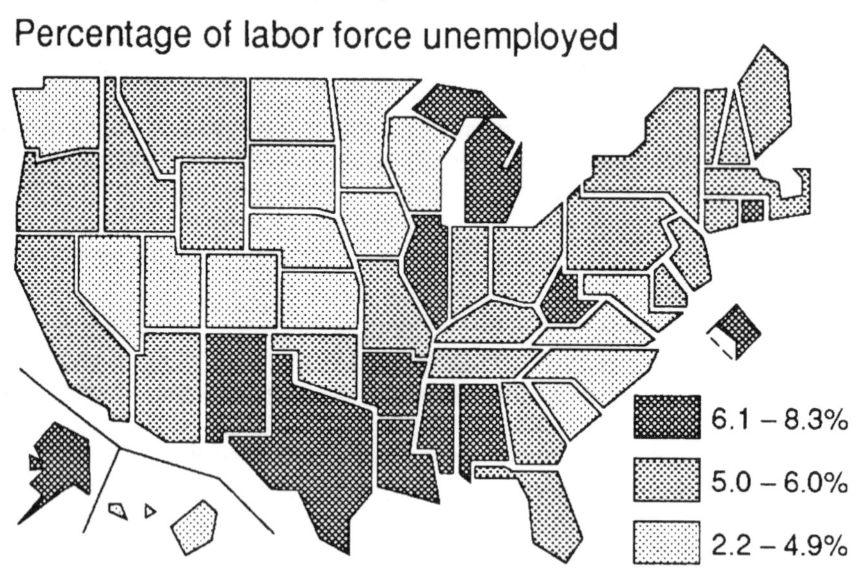

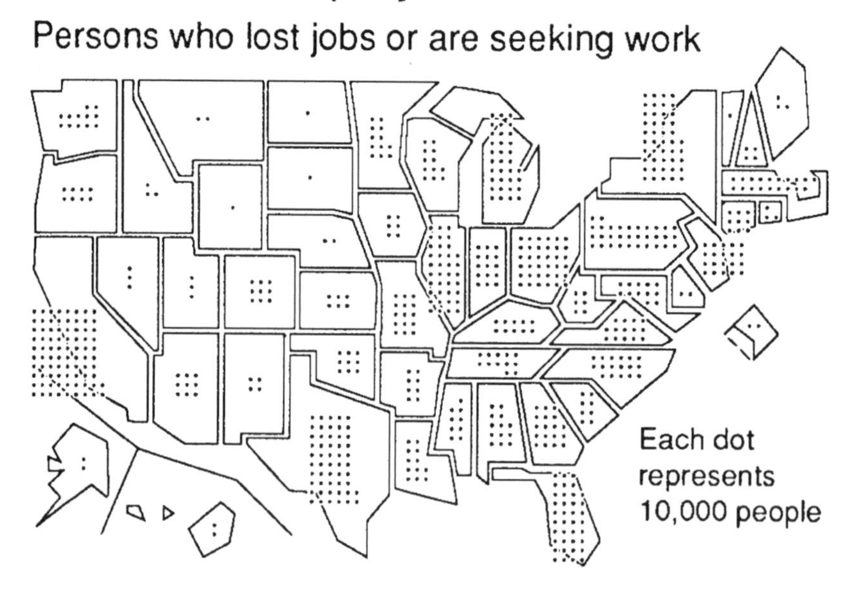

Figure 7.1 Complementary pair of maps showing intensity (above) and magnitude (below) of civilian unemployment in the United States, 1990.

relates their own state to the other 49, but compares counties within their state to the country as a whole. For residents of New York, Figure 7.2 complements the preceding maps of rates and counts by relating the regional pattern of unemployment to the national picture. On these and other two-class maps, a straightforward title obviates a complex key. The upper map puts the state's predicament in perspective by showing higher unemployment rates throughout the "Sun Belt," whereas the lower map points out that unemployment is higher than the national rate in New York City and northern New York State. Other useful cartographic complements include a multi-category county-level choropleth map showing geographic variations within the state and a municipal-level choropleth map showing variation within Onondaga County or central New York.

Thorough cartographic treatment of unemployment data demands disaggregation by gender, race, and ethnicity. Additional views might address unemployment among the young, the aged, high-school dropouts, and college graduates. Where the data permit, a more thorough treatment might describe geographic variations in the population of unemployed workers and present such themes as the proportions of unemployed who are women, minorities, over 50, under 25, and recent college graduates. Also worth considering are maps that disaggregate unemployment by occupational category and industrial sector.

EXAMPLES IN DYNAMIC CARTOGRAPHY

This section discusses the value of cartographic complementarity in addressing three design problems in dynamic cartography. Despite the emphasis here on new technology, many of the principles are relevant, as well, to traditional, static maps.

Graphic Narratives

Narrative maps typically have a logical, natural sequence reflecting the order in which events occur. Cartographic sequences are often based on chronological or historical time, for example, when a series of maps describes the course of a war or battle. An important variation of the time-based narrative is the use of maps to describe a process such as the evolution of a city, or the birth and death of a major storm. Another kind of narrative theme is the causal relationship, which requires maps or other graphics organized to reflect a coherent exposition of the link between cause and effect. Devising an effective organization can be challenging when a graphic narrative must integrate multiple causes or describe a complex chain of events. Other challenges include the need to explain unfamiliar terms, introduce new data, or examine secondary consequences.

Static cartography can cope effectively with simple narratives, such as a travelogue, a scientific expedition, or the flight of an elusive criminal. For these examples, a single map might describe the route as well as point out significant intermediate events. Traditional cartography is equally adept in presenting a narrative

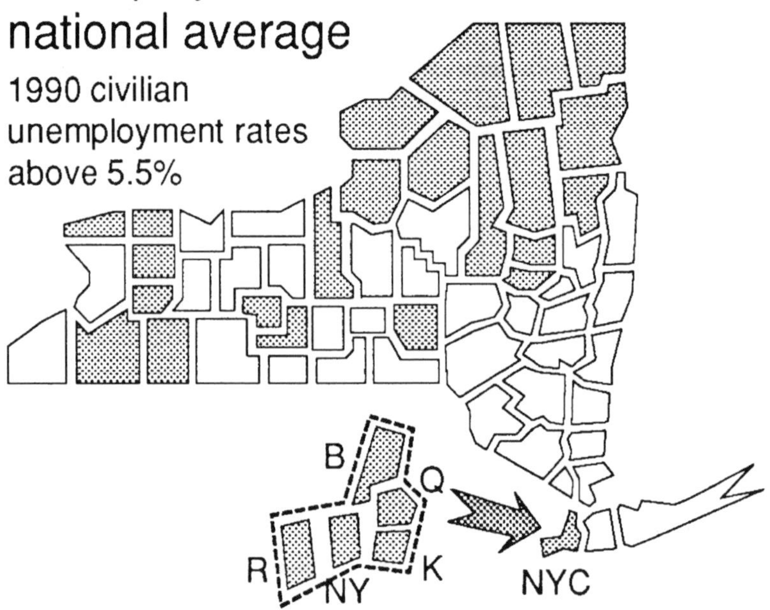

Figure 7.2 Complementary two-class maps of unemployment compare New York to other states (above) and New York's counties to the nation as a whole (below).

series of geographic snapshots showing change over time for a single distribution, such as population density. Time is treated by the orderly juxtaposition of numerous small maps with year-date titles. To be reliable, though, a narrative cartographic tableau requires the complementarity of consistent data and, if based on choropleth maps, a common set of categories. Similarly, for a temporal series of dot-distribution maps, a single dot should represent the same amount throughout the display so that densities are comparable.

Complementarity is especially important when a graphic narrative needs to link different kinds of maps. For example, to describe the progressive advance of settlement from the coast to the interior, a series of single-year dot-distribution maps of population might be linked to a multi-year summary map by lines describing the settlement frontier. Each map in a tableau of cartographic snapshots can describe the settlement frontier for a single year—and illustrate as well the meaning of a 'frontier' far removed from areas of dense settlement. Collecting exact duplicates of all frontier lines on a single isochronic map can then highlight geographic obstacles and periods of rapid expansion. By focusing attention on these relationships, the isochronic map is a meaningful complement to the temporal series of dot maps.

A dynamic presentation of the advance of settlement might use two cartographic windows (Figure 7.3), one displaying single-year snapshots of population density and the settlement frontier and the other showing a cumulative set of frontiers. To complement these maps, I would add a dynamic time line as well as a dynamic time-series graph to describe the growth of the national population. To avoid overwhelming the viewer, my presentation might introduce these elements gradually, at first running through the entire historic sequence with only the time line and the cumulative population graph. I would then repeat the sequence four more times: adding the dot map alone, then adding a frontier line to the dot, then displaying an isochronic frontier map next to the dot map, and finally by showing the dynamic frontier map by itself. To make certain the viewer appreciates these enhancements, spoken or visual text would announce each addition and its intent.

Multiple graphic narratives can be informatively complementary. For example, a dynamic summary of the advance of European settlement might be followed by a narrative addressing the forced removal and decline of native peoples. Cartographic complementarity involves not only the synergy of graphs with conceptually different points of view, but also the meaningful integration of related data.

A Graphic Script for Geographic Correlation

Three years ago, I explored the potential for dynamic graphic narratives by authoring two 'graphic scripts' (my term for a narrative sequence of maps, graphs, and texts). One script examines the correlation of two variables, and the other addresses a spatial-temporal distribution (Monmonier, 1992). Although both scripts demonstrate the usefulness of complementary graphics, the bivariate-correlation script is a particularly cogent example of cartographic complementarity because it integrates one or two maps with a bivariate scatterplot.

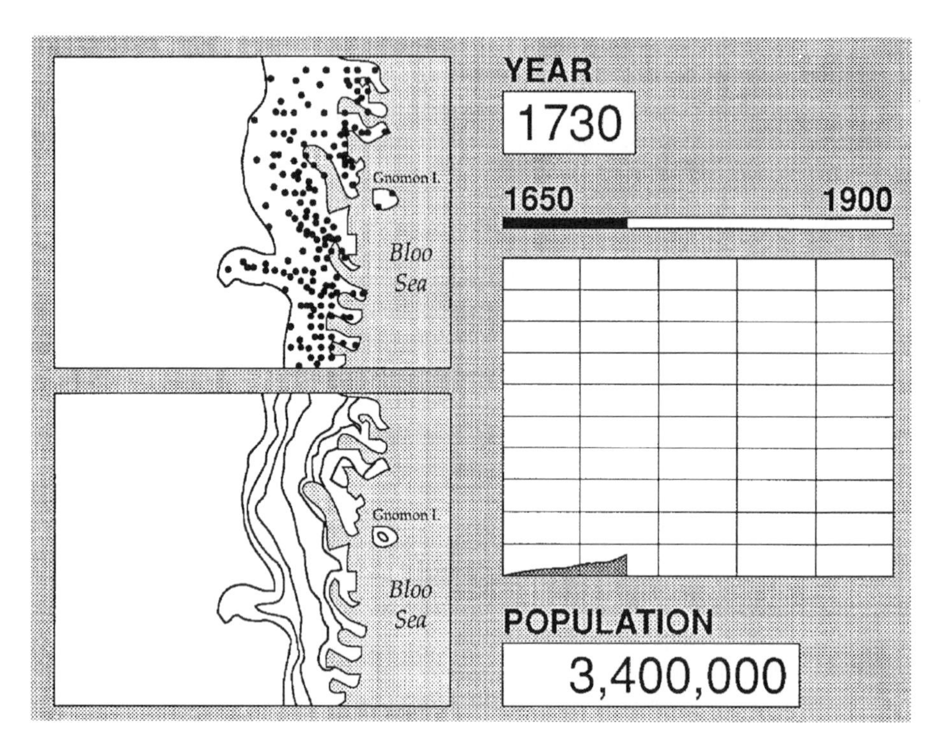

Figure 7.3 Concept diagram showing the juxtaposition of a time bar, a dynamic time-series graph of population growth, and dynamic maps of population density and the settlement frontier.

I wrote this script for two specific variables measured for the 50 states of the U.S. The dependent variable was the female percentage of elected officials in local government; "female officials" for short. The independent variable was female workers (employed or seeking employment) as a percentage of the labour force; "females working" for short. My underlying hypothesis was that states where females comprised a higher-than-average share of the labour force would tend, on average, to elect proportionately more women.

I developed the "correlation" script like a play with three acts. The first act introduces the variables separately and explores geographic variation, the second act examines geographic and statistical covariation, and the final act imposes a linear regression model and examines the unexplained, residual variation. Act one's first scene begins with a text window that defines the dependent variable and states its full and short names. Several short graphic sequences called 'graphic phrases' then examine the frequency distribution and geographic pattern of data values; these graphic phrases relate the map to a frequency histogram, demonstrate the effect of different classifications on a choropleth map, highlight the locations of extremely high and extremely low values, and force the viewer to examine the relative homogeneity (or lack thereof) of the country's nine census divisions. The second scene provides a similar overview of the independent variable.

In the 10 minutes required to run through all three acts, the script presents a variety of statistical and cartographic views linked dynamically to each other. So as not to confuse viewers, I avoided graphic noise by restricting the variety of labels, colours, and locations on the screen. Each type of map and graph has a fixed position, as do specialized text windows. Because viewers are more likely to understand a dynamic display if they comprehend its point, a special text window announces the purpose of the scene or graphic phrase to follow. Each variable has a unique fill pattern with a contrasting 'signature hue': red for the dependent variable, blue for independent variable, and magenta for the residuals. Act one's first two scenes introduce the signature hues and relate them to the brief titles, "female officials" and "females working."

Signature hues are especially useful in act two, which relies heavily on the juxtaposition of two maps and a scatterplot (Figure 7.4). The scatterplot is in the right half of the screen, and the maps occupy the left. The upper map and vertical axis of the scatterplot represent the dependent variable, and the lower map and horizontal axis represent the independent variable. The second act's first scene uses a dynamic 'conditioning brush' to relate each map individually to the scatterplot. The long side of the brush spans the complete range of one variable, while the short, movable side defines a comparatively narrow region (a 'condition' with an instantaneous minimum and maximum) for the other variable. As the brush sweeps back and forth along the axis for the 'conditioned' variable, states instantaneously within the brush (that is, states satisfying the condition) are simultaneously highlighted on the corresponding map. Figure 7.4, a snapshot captured during brushing of the dependent variable, reveals slightly-above-average proportions of female officeholders in Alaska, Michigan, and three northeastern states. Signature hue identifies the variable being brushed and reinforces the link between map and scatterplot. Fill patterns for the horizontal brush and highlighted polygons on the upper map in Figure 7.4 are in red, but later in the scene, a blue vertical brush sweeps across the horizontal axis while blue interiors highlight polygons in the lower map.

In act three, dynamic links between a map and scatterplot explore the significance of a linear regression model and the residuals therefrom. Figure 7.5 is one of nine unique views displayed during a 'regional canvass' based on the nine census divisions of the U.S. The scatterplot and map are filled with black, the regression line is magenta, and the polygons and points for the five East North Central states are highlighted in yellow. This view reveals considerable consistency in this part of the country, where females are moderately prominent in the work force and slightly more conspicuous as elected officials. The script's final act uses a somewhat similar sequence of linked views to relate a four-category choropleth map of residuals to the scatterplot. In this 'canvass-by-category' sequence, a label above the map provides a verbal statement of each category, while the scatterplot reveals

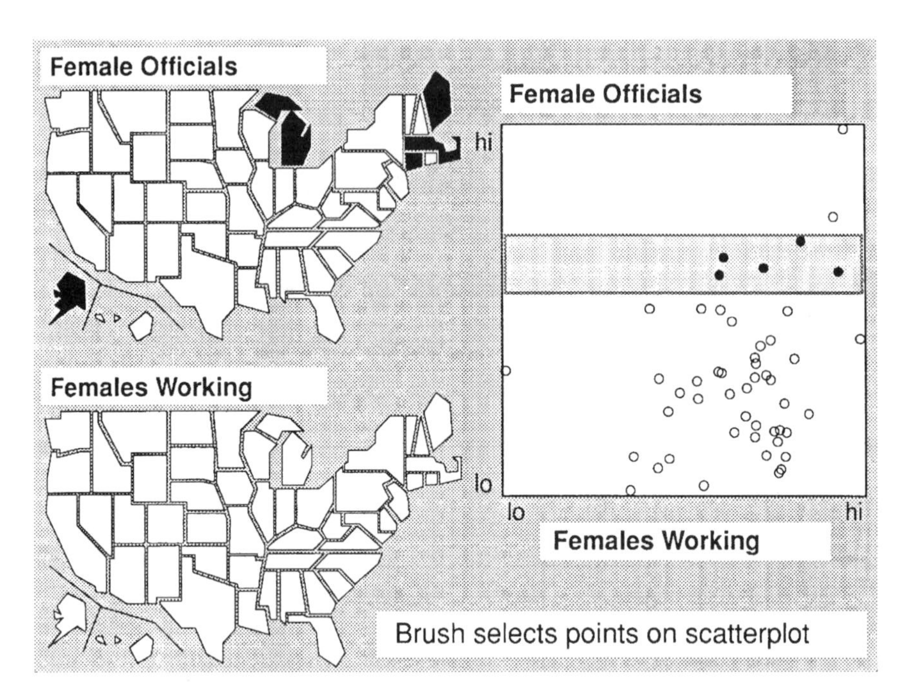

Figure 7.4 Screen snapshot captured during brushing of the dependent variable.

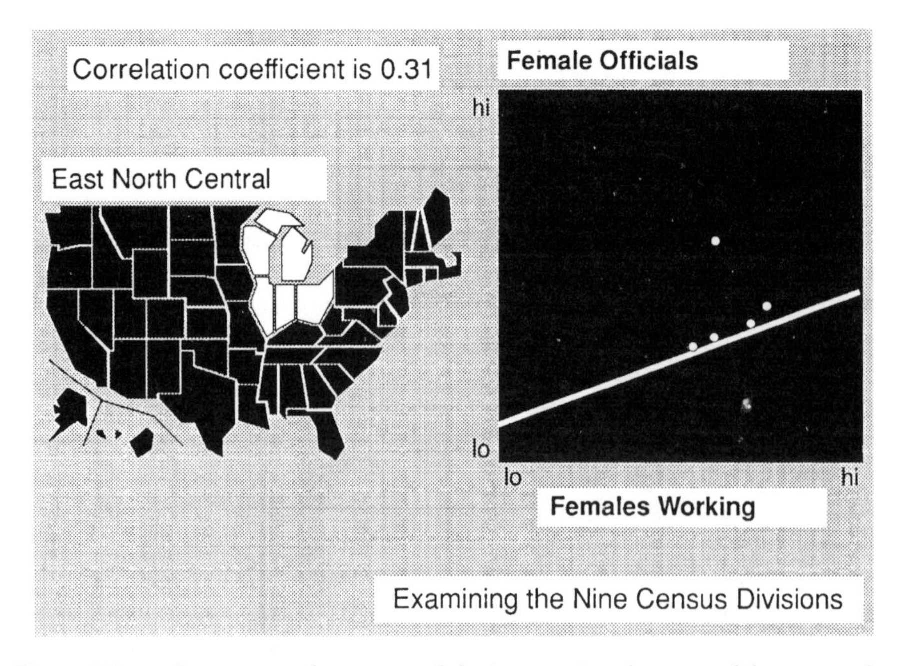

Figure 7.5 Screen snapshot captured during a regional canvas of the scatterplot shows positive residuals for the East North Central states.

only the regression line and data points of states in the category and the dynamic choropleth map momentarily substitutes black for polygons in the other three categories.

My graphic scripts can bewilder and frustrate new viewers unfamiliar with choropleth maps and statistical correlation. Nonetheless, focus groups with four sets of experienced analysts suggest that the dynamic integration of complementary maps and graphs affords an informative examination of a bivariate geographic correlation (Monmonier and Gluck, 1994). Participants seem able to grasp and apply the graphic logic of signature hues and linked windows. Their single most persistent complaint is a compelling urge to interact with the display—a need easily addressed by an 'open script' that introduces the technique or relationship to viewers and then allows them to explore the data at their own pace.

Class Breaks for Dynamic Temporal Choropleth Maps

My final example involves an animated series of choropleth maps that describe a single variable for different instants or periods of time. So that each area symbol has a consistent meaning throughout the presentation, a temporal sequence of choropleth maps needs a single, constant set of class breaks. But to be effectively and efficiently informative, the classification should minimize meretricious category changes.

The composite time-series graph in Figure 7.6 explains the problem. The vertical axis represents time, and the horizontal axis represents value. Fifty narrow trend lines describe state-level patterns in the average yearly unemployment rate for the 21-year period 1970 through 1990. A thick line representing the national average reflects a sharp rise in unemployment in the early 1980s. Most states reflect the national trend, but some experienced much higher rates of joblessness. Here and there, though, a few states defied the national trend, at least for a while.

A category break can be described by a vertical line intersecting the horizontal axes at a specific unemployment rate. Whenever a state's trend line crosses this vertical line, the map symbol changes to reflect a transition in category. This change would be relatively meaningful if the state stayed in its new category for several years. But a state hovering around a class break might shift back and forth by very small amounts producing a trivial but distracting flicker on the map. When many states oscillate around a value, that rate, by and large, is not a good location for a category break.

To help map authors recognize good breaks lurking among bad ones, I have devised a scheme for (1) identifying trivial transitions as well as more substantial category shifts, and (2) graphing a index of visual stability and worthiness. High points on my graph represent potential breaks with comparatively few trivial shifts, whereas low points on the curve signify breaks likely to produce a needlessly busy display (Monmonier, 1994). Although the method requires map authors to experiment with various thresholds—how else to define 'trivial' and 'substantial'?—it can identify worthy breaks if there are any. However promising the approach, my

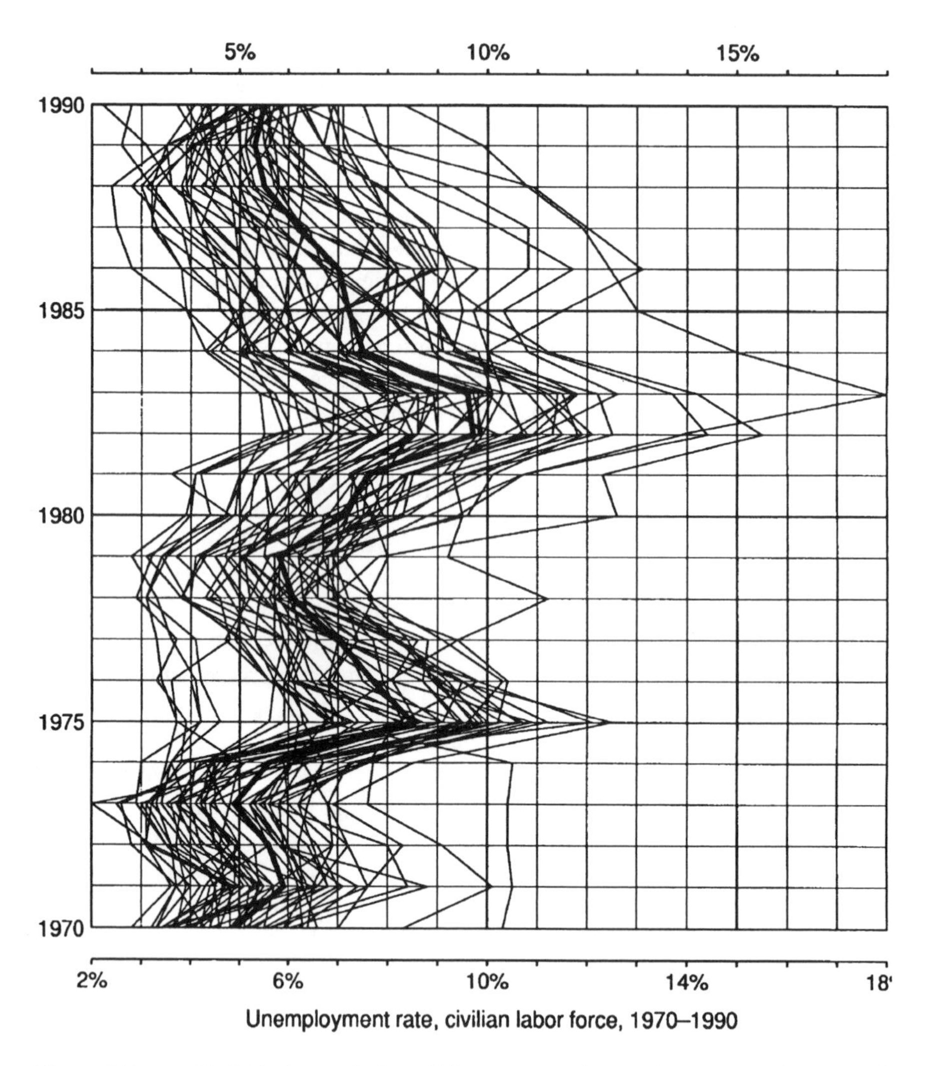

Figure 7.6 Statistical graph describing temporal trends in unemployment for the 50 states.

demonstration data (the unemployment rates in Figure 7.6) revealed no spectacular contrasts between a very good break immediately adjacent to a very poor one. But the results were not so disheartening as to warrant looking for more agreeable test data.

I could have approached the problem differently. A more heavy-handed solution is to identify and deliberately suppress all momentary jarring, trivial changes —a search-and-destroy mission yielding a less than rigorously exact dynamic map that, well, lies a little to avoid distracting the viewer from more salient changes in category. I don't like to admit it, but I find the idea cleverly appealing.

That's one strategy; here's another. Although developed originally to explain the need for minimum-change class breaks, the composite time-series diagram in Figure 7.6 suggests a dynamic linkage similar to scatterplot brushing. Figure 7.7 describes schematically a complementary pair of dynamic links between a temporal choropleth map and a composite time-series graph. Across the graph are two movable lines—I call them 'combs' because they act like very thin conditioning brushes. By moving the vertical comb back and forth along the graph's time axis, the viewer can vary the time displayed in the map. By moving the horizontal comb up and down along the value axis, the viewer can experiment with the single break defining a two-category choropleth map. Although the viewer might want to set the time comb on 'automatic' mode, and run the map dynamically through time like the "satellite loop" on the nightly TV weather report, interactive experimentation with the value comb can be useful in understanding a temporal data set as well as in identifying meaningful category breaks.

Another method seems equally promising. Instead of instantaneously describing individual years, the map window might describe pairs of successive years. With this modification, the cartographic window could informatively highlight states that conform to the national trend, states that anticipate the trend, and states that behave differently.

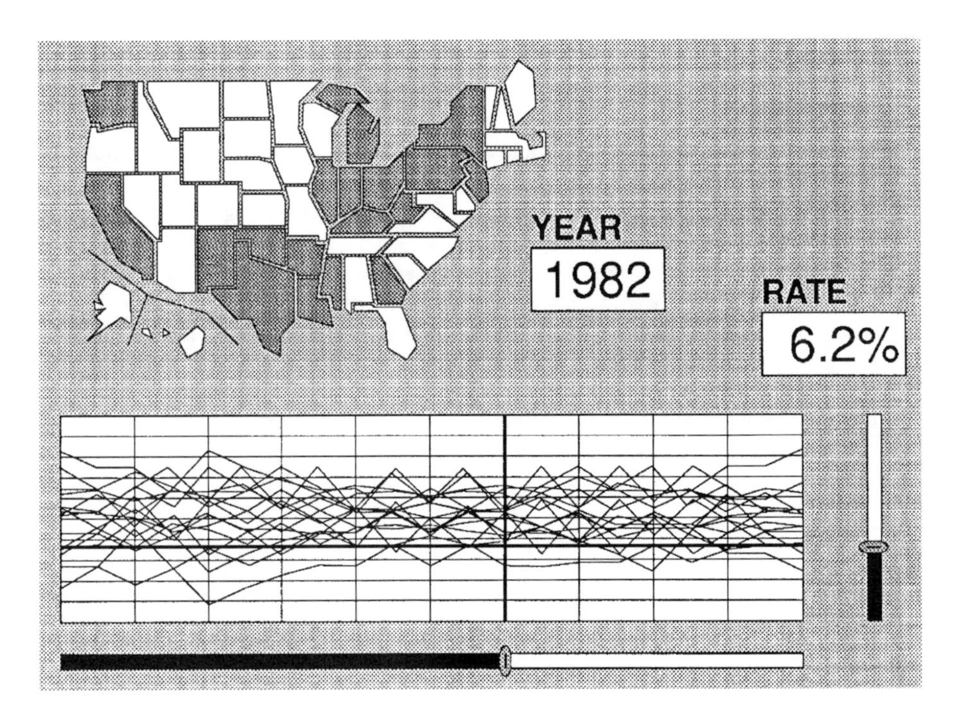

Figure 7.7 Proposed interface for the interactive display of complementary representations of a spatial-temporal variable.

CONCLUDING REMARKS

Complementary graphics may be either static or dynamic. For static graphics, complementarity usually requires careful attention to the juxtaposition or sequencing of the various views. Although sequencing and juxtaposition are also important for dynamic graphics, sequencing might include the piecemeal presentation of specially tailored categories as well as interactive displays that allow the viewer to explore information in geographic space, in attribute space, and for various instants or periods of time. The potential visual complexity of dynamic maps, whether interactive or merely animated, calls for careful attention to the issue of cartographic complementarity.

Electronic cartography has an important role in promoting cartographic complementarity. Although multiple complementary views do not depend on computers, adding additional maps or graphs to an electronic presentation does not trigger the print publisher's concerns about space and cost. Interactive electronic formats are less constrained by the cost-based parsimony common to paper maps and photomechanical reproduction. Even so, authored graphic narratives can introduce new users to the system and explain visual linkages and the use of combs and brushes.

REFERENCES

Jenks, G.F., and Caspall, F.C. (1971). Error on choroplethic maps: Definition, measurement, reduction. *Annals of the Association of American Geographers*, 61, 217-244.

Monmonier, M. (1982). Flat laxity, optimization, and rounding in the selection of class intervals. *Cartographica*, 19(1), 16-27.

Monmonier, M. (1991). Ethics and map design: Six strategies for confronting the traditional one-map solution. *Cartographic Perspectives*, 10, 3-8.

Monmonier, M. (1992). Authoring graphic scripts: Experiences and principles. *Cartography and Geographic Information Systems*, 19, 247-260, 272.

Monmonier, M. (1994). Minimum-change categories for dynamic temporal choropleth maps. *Journal of the Pennsylvania Academy of Science*, 69, 42-47.

Monmonier, M., and Gluck, M. (1994). Focus groups for design improvement in dynamic cartography. *Cartography and Geographic Information Systems*, 21, 37-47.

8

Tactile Mapping Design and the Visually Impaired User

Regina Vasconcellos

Department of Geography, University of Sao Paulo

INTRODUCTION

New technologies have introduced immense changes and improvements in the area of tactile map production. Cartographers will have to weigh the advantages and existing limitations of the different methods prior to producing tactile maps. This condition must be met before the cartographer can make decisions about components of map design and characteristics of the graphic language to be employed. Scale, degree of generalization, and choice of symbols greatly depend on the production method to be used, including the master construction and reproduction technique.

The findings presented in this chapter resulted from research undertaken at the University of Sao Paulo beginning in 1989. Following a first phase that was based on development of methodology and design, the research then concentrated attention on: 1) tactile graphics design, including analysis, construction and tests of various techniques and cartographic products; 2) tactile graphics evaluation that included definition of different levels of complexity according to prior spatial perception; and 3) tactile graphic use that involved preparation of training programs for teachers and visually impaired students.

The main purpose of this research has focused principally on the study and development of a tactual graphic language which will improve the teaching of geography and cartography to both blind and sighted children, while emphasizing an interdisciplinary approach directed at enhancing the role of maps in the school curriculum.

The research conducted in Brazil evaluated only two reproduction techniques: 1) the vacuum-forming with brailon plastic process, and 2) the silk-screen process using puff ink on paper, both of which were tested with visually impaired users. The results revealed ways to improve tactile map design in which the user's degree of sensory impairment, previous experience and map use skills were considered. The need for early training was stressed, and a program to introduce the tactile cartographic language was presented.

Technological and financial restraints had a relevant impact on tactile map production within the research program, and these restraints also affected map design in different ways. It is clear that the cartographer faces a challenging task when working with the myriad of variables involved in the design and production of efficient tactile maps for the visually impaired.

TACTILE GRAPHIC LANGUAGE, MAP DESIGN AND CARTOGRAPHY

The relevance of graphic representations today, in our everyday life and at school or work, has been discussed by the author in a previous paper (Vasconcellos, 1993). For the sighted user, visual perception is the most important channel for the conceptualization of space and the acquisition of spatial information and geographic knowledge. For visually impaired users, graphic representations are extremely important since they may need a map to find their way inside buildings. All types of maps should become available in the tactile format, including thematic, reference and mobility maps. Tactile maps and graphics are the only way for the blind person to obtain images of our world, and to become acquainted with its changing realities.

New political, social and economic facts, together with a variety of technological innovations, are bringing important changes to cartography at all levels. Taylor (1991) has presented an excellent analysis of this matter, calling attention to the need for new concepts in cartography, considering the social and cultural contexts without the predominance of the technological paradigm. Kanakubo (1993) has also pointed out the main theoretical issues facing cartography today. Figure 8.1 summarizes the author's view on the perspectives and dimensions related to cartography, which is outlined at the centre as a communication system. Within tactile cartography, each variable listed in this theoretical framework may have a different weight and role, depending on map purpose, map reproduction, and user-special needs. The variables highlighted in Figure 8.1, such as technology, communication and semiology, denote relevance to tactile map design and are, therefore, of primary importance. In the case of perspectives, for example, cultural and cognitive aspects might exert a great influence on choosing a better design. On the other hand, some of the dimensions of cartography, such as art and precision, have secondary roles when considered in the context of tactile map design. Also of secondary importance for tactile map design are the social and psychological perspectives, while economic and political perspectives might, in some manner, assist in defining tactile map production.

Culture may also be a factor influencing tactile map design. A good example is the comparison between the degree of reduction in size and generalization that is applied to map construction when directed to visually impaired users from different cultures. It is well known that visually impaired Americans, as distinct from their Japanese counterparts, have not developed the same levels of touch perception. This highly developed tactile sense among some oriental cultures permits them to read smaller symbols with details much closer together. The small braille

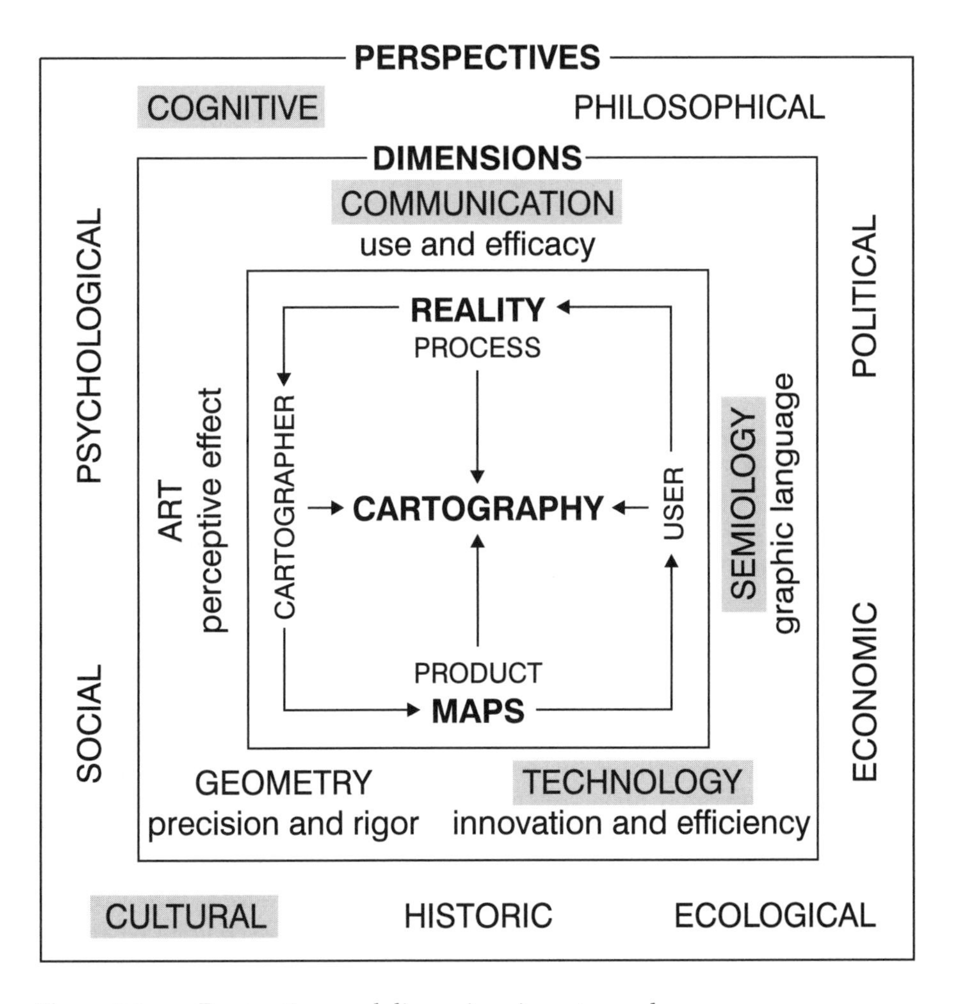

Figure 8.1 Perspectives and dimensions in cartography.

cell widely used in Japan, even with children, does not seem to work in other countries. While there has been a great deal of cross-cultural research on brain differences, natural abilities, multiple intelligence, and previous experience and opportunities, there is a noticeable lack of research related to tactile map design and use.

Another example of factors that can influence the existing rules of graphic representation is the nature and degree of visual impairment. One might think that dealing with a low vision user would be easier than dealing with a user who is totally blind. All past evaluations have been proven to the contrary, as tests have revealed several problems in the use of maps by partially sighted young people. They fail to rely on either vision or touch. Among the identified problems that contributed to difficulties in working with the young persons were lack of constancy in visual acuity, psychological and social problems, and cognitive restraints. While some can see a few colours and some can discriminate lines and points, others are unable. The extreme variation in visual abilities makes design decisions that promote effective communication very difficult.

Both the maker of tactile maps and the visually impaired user have specific needs. In order to communicate geographic information to the sight impaired, some of the design considerations normally avoided in cartography become qualities and conditions that promote good tactile maps. For example, the visually impaired user needs a much higher degree of feature generalization with the appropriate levels of omission, exaggeration, and distortion never imagined by the designer/producer of conventional cartographic products. Tactile cartography should be based on different concepts that follow other guidelines and distinct techniques, both in map design and production. One such guideline pertains to map use where the visually impaired has particular needs. In this case, a greater amount of prior training will be required for the sight impaired user unlike the sighted user who has had visual reinforcement from their environment since birth. Despite this requirement, blind users can develop extremely good abilities in decoding map keys. They are able to use both hands simultaneously, one on the map key, the other on the map. That is something that the sighted user is unable to do with their eyes.

Although there is very limited information on tactile mapping as compared to that for conventional cartography, the literature is expanding. Many authors have contributed significantly to the field of tactile cartography as editors of basic publications (e.g., Wiedel and Groves, 1972; Weidel, 1983; Schiff and Foulke, 1982; Nicolai, 1984; Tatham and Dodds, 1988; Ishido, 1989), and by carrying out relevant research or writing papers and books (e.g., James and Armstrong, 1976; Qingpu, 1991; Keming, 1991; Coulson et al., 1991). Several others have contributed greatly to the subject, although they are not cited in this chapter, including E. Berla, J. Gill, H. Vlaanderen, G. Jansson and G. Leonard. Tactile map design and use have been studied by several authors, among them: Franks and Nolan, 1970; Wiedel and Groves, 1972; Kidwell and Greer, 1973; James and Armstrong, 1976; Levi and Amick, 1982; Bentzen, 1982; Barth, 1987; Tatham, 1991, 1993; Edman, 1992; and Rener, 1993. Nor has there been an expressive literature for map design and use in visual form as a result of practical evaluations. Conventional cartographers have not been concerned enough about the requirements and perceptual limitations of the visually impaired. Despite this apparent lack of concern, some research has been carried out on the legibility and discrimination of line, point, and areal symbols for the visually impaired. There is no doubt that during the last two decades or so, tactile cartography has made considerable progress through programs of applied research on map design and use. Unfortunately, not all of the research considered all of the variables involved in designing and using a tactile map. The findings published to date help in providing design principles to direct further studies, and much remains to be done.

TACTILE MAP DESIGN AND USE: NEW APPROACHES TOWARDS CARTOGRAPHIC COMMUNICATION

Tactile maps are effective examples that stress the significance of the cartographic communication process studied widely by cartographers all over the world during the last 30 years. The questions what, how, and to whom summarize the essence of this process that begins with a decision to map a portion of geographic reality by a cartographer. Additional questions might also be added to those three basic ones, such as when, where, why and with what results. A diagram illustrating the process of tactual graphic communication was presented by the author in a previous work (Vasconcellos, 1991). Some stages have always been essential in order to achieve better communication for the blind or low vision user. Tactual graphic language that answers the question "how" is a relevant stage at which graphic signs have to be selected. The purpose of the step was to analyse the visual variables proposed by Bertin (1977) and applied in education by Gimeno (1980). The ideas of Bertin have not been closely studied and applied in English speaking countries due to a lack of translation for much of his work. Other authors, however, have presented papers on the topic of visual or graphic variables and include Green (1993), Pravda (1993), and Ucar (1993).

In the case of tactile maps, however, there are additional graphic variables that must be added to those already proposed by Bertin. One such variable this author has added is elevation (height), which can be combined with other variables in several ways and is related to size. One of Bertin's variables, colour, cannot be used in the tactual graphic language. It can be replaced, though, by use of different textures. In the same way colour is used for conventional maps, texture may be utilized to represent qualitative data, and it can also be used to express ordered information on quantitative maps. In order to select the correct variable, it is necessary to analyse carefully the nature of the data and to determine whether it is quantitative, ordered, qualitative, or differential.

Figure 8.2 shows Bertin's (1977) visual variables and the tactual graphic variables that incorporate the third dimension for use in tactile mapping (Vasconcellos, 1991, 1993). During the development of the tactual variables, graphic representations were constructed using the six tactile variables, that is, height, size, value, texture, shape, and orientation, that were then related to geographic points, lines, and areas. Testing of different materials and tools was then conducted to produce the tactile symbols to be used in the construction of maps, diagrams and drawings. Testing of the tactile symbols was then carried out with visually impaired persons. Results revealed a high efficacy associated with the tactile graphic language and its importance regarding the perception of space and the acquisition of geographic knowledge. Theoretical studies associated with practical experiences led to the derivation of a new model for tactile map production and use, illustrated in Figure 8.3. The basic idea of the model is to have a dynamic structure where all variables are interrelated and connected with one another.

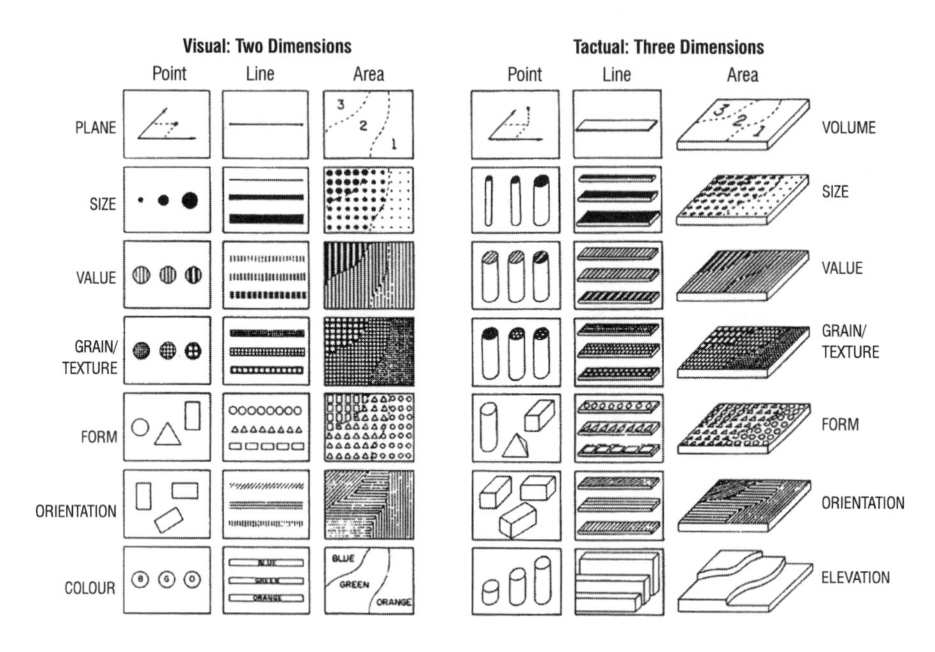

Figure 8.2 Visual and tactile graphic variables.

The proposed system stresses the role of the tactile map user and underscores the necessity for the user to provide feedback during all stages of the cartographic process. Map user involvement is important even during the initial stages of the tactile cartographic process. An example of this requirement is apparent in defining the purpose of the tactile map, where orientation and mobility have been added. Evaluation of tactile maps by the visually impaired user is a prerequisite for reviewing map design, construction, and reproduction.

Earlier models of cartographic communication have been based on two separate and distinct spheres, one of the map maker, the other of the map user. It must be stressed that in the case of the tactile cartographic process, both the map maker and the map user should have an active role, although the nature of that role will be different. These differences are defined as determinant factors and are listed in Figure 8.3. It should be noted that some of the factors are the same for both the map maker and the map user, such as creativity, motivation, and skills or natural abilities. Other factors are unique, such as the theoretical and technical knowledge required for the map maker, while psychological influences and sensory impairment impact on the map user. The full range of variables taken together in this diagram illustrate well the interrelatedness and complexity of the tactile cartographic process.

Just as the map maker was seen in a separate sphere opposed to the map user in early cartographic communication models, map production stages were also viewed as being separate and distinct. There was no provision for user feedback.

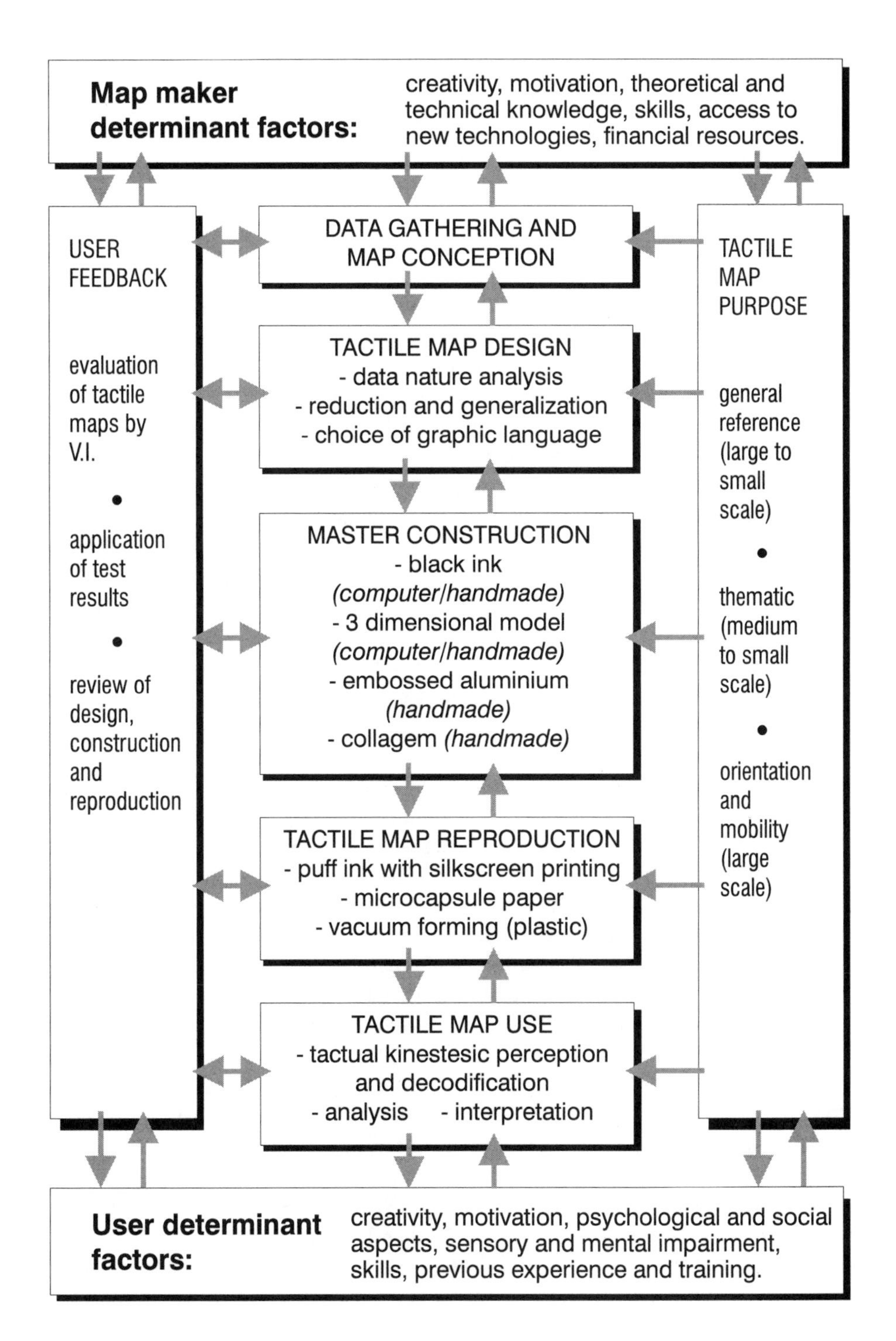

Figure 8.3 Tactile map production and use.

Map production stages, including map design, concluded in a finished product that only then was subjected to evaluation. This scheme can be observed in most of the models proposed previously. Perhaps the lack of user input during the process can be ascribed to the lack of applied research during the time when the cartographic communication paradigm was strongest in theoretical cartography. There have been critical reviews on this subject by Keates (1982) and by Woodward (1992) calling attention to the absence of substantive studies to back up theoretical assumptions.

For several years applied research has led this author to emphasize the role of user feedback while taking into account the complexity of all of the variables involved. Evaluation, a necessary addition to the cartographic process, has been added in order to produce better maps to suit the impaired user. For this reason, it is difficult to define a set of rules that could be applied in all situations. Some guidelines, of course, have been proposed and are being followed for improving both map production and map use. In each case, however, it is very important to be able to classify all the cartographic materials while evaluating the levels of complexity of the cartographic information, as well as the previous experience and training of the prospective user. To promote better tactile information communication, each map must be analysed regarding its design and the master construction technique and reproduction system that will be used. It often happens that one map can be suitable for one category of users, but not for another group, unless training and assistance during actual map use are provided.

The graphic materials constructed in Brazil were tested with over 200 visually impaired students. In addition, all the materials were evaluated by 80 special education school teachers. The results of these tests showed that it is difficult to reach agreement on the definition of guidelines, mainly due to individual variations and preferences and varying levels of skill related to map reading and understanding of graphic language. These difficulties are discussed by Hampson and Daly (1989), who arrived at similar conclusions. Cartographers are generally aware of the role of human factors and prior experience in the production and use of maps for the sighted user. But, when tactile maps are the issue and the users are visually impaired, the type and number of variables have a much greater impact on the cartographic process. Studies on developmental stages, mainly based on Piagetian theories (Gottesman, 1971; Hatwell, 1985), can assist in finding answers, but more research is necessary with the visually impaired. The aspects of cognition and language applied to the sensory impaired are the topics of an excellent book by Andrews et al. (1991). Other authors have made relevant contributions to understanding map use and communication, without mentioning visual impairment specifically. They include: Petchenik, 1979; Boardman, 1983; Anderson, 1987; Castner, 1990, 1991; Gersmehl, 1991; and Gerber, 1992. Newer digital technologies might bring the cartographic communication paradigm back into prominence, since new products on video monitors will have to be evaluated as a means to communicate spatial information. Cognitive and perceptual research will be needed in order to improve map design in all hardware and software forms.

FINAL REMARKS

Following practical experience with the use of tactile maps and graphics, it is possible to analyse the many problems and to suggest a few solutions toward improving the tactile cartographic process.

About communication:

- Tactile cartography helps to overcome the informational barriers for those who cannot see, thereby facilitating their access to our world dominated by images. It is very important to design, to reproduce, to distribute, and to teach how to use tactile maps.
- It is vital to detect failures during the entire process of cartographic communication. Such failures may occur at any stage such as design, master construction, or reproduction. User reaction should be considered at all levels as a means to review all decisions.

About map design:

- It is advisable to create and use conventions as much as possible to facilitate access to graphic information.
- The map key is very important for the visually impaired, who are expert at decoding a legend. Not only should keys be used on maps, but they should also be used even for drawings.
- Proper size reduction and level of feature generalization are vital in tactile maps. Size and distance between symbols are also important because tactual perception has a very different resolution and is not global or holistic since the blind user must assemble pieces of information in a serial fashion to form an image.
- Use of sufficient relief and contrast together with the use of redundancy helps to achieve better map design that incorporates the required amount of clarity. Accuracy and precision very often have a secondary role. It is sometimes necessary to produce a collection of maps rather than trying to add too much information to a single map.
- Decisions about the graphic tactile language depend on the data and its nature, on the type of master, on the reproduction method and on the special needs and background of the visually impaired user.

About map use:

- Cartographic materials must be classified according to levels of complexity, previous experience, age, and school grade.
- Activities and games can facilitate the process of learning cartography and geography.
- The visually impaired user needs personal assistance or good instructions to read maps.

About previous training:

- Basic geographical concepts, such as proportion, scale, point of view, location, and orientation, must be well understood before working with maps.
- The tactual graphic language has to be introduced to the user prior to map reading through exercises with the visual and tactual graphic variables.
- Relief models help children understand physical space. Such models are less abstract and should precede the introduction of maps.

About map reproduction:

- It is advisable to make maps in both visual and tactile formats, using colours when possible. In this case, information for low vision users should not be in relief to avoid noise for the blind user.
- Microcapsule paper and puff ink are techniques that should not be used with the novice user since the processes have low relief capabilities. They do, however, have strong points in their favour such as relatively low price per copy and the possibility of computer-generated masters.
- Thermo-formed plastics can have a much greater variety of textures and elevations, and are, therefore, a better choice to represent all kinds of information and data.

Applied research has shown that acquisition of visual and tactual graphic language depends on training if the knowledge is to be used properly to its full potential. Such a training program directed to the visually impaired user was outlined for the second phase of this research beginning at the end of 1992. Graphic materials, exercises and games were constructed to introduce the basic concepts selected, namely point of view, proportion, scale, distance, location and orientation. Introduction of all the tactile graphic variables (Figure 8.2) in the form of playing cards was also included in the program. The last stage of decoding and reading maps was incorporated into two exercises that facilitated practice in using a map key and a grid. The use of exercises (or games) such as the "full address" and the "building of a city" allowed the integration of all information and skills needed in map reading. The training program was tested with visually impaired students in 1993, and the same materials were then tested with sighted students and teachers during 1994.

A selected bibliography has been organized on the topics related to cartography, tactile maps, visual impairment and education. More than 700 citations were collected and grouped in several themes according to the above subjects. This bibliography will be constantly updated.

Future directions for the author's research will concentrate on new resources, such as multi-sensory cartographic materials. It will also focus on challenging tasks, such as discovering the limits of learning the graphic language through testing a few groups of sighted and visually impaired users over a period of several years.

REFERENCES

Anderson, J.M. (1987). The relation of instruction, verbal ability and sex to the acquisition of selected cartographic skills in kindergarten children. [Unpublished Ph.D. dissertation]. Madison, Wisconsin: University of Wisconsin-Madison.

Andrews, S.K., Otis-Wilborn, A., and Messenheimer-Young, T. (1991). *Pathways—Beyond seeing and hearing*. Indiana, Pennsylvania: The National Council for Geographic Education.

Barth, J.L. (1987). *Tactile graphics guidebook*. Louisville, Kentucky: American Printing House for the Blind.

Bentzen, B.L. (1982). Tangible graphic displays in the education of blind persons. In W. Schiff and E. Foulke (Eds.), *Tactual perception: A sourcebook*. New York: Cambridge University Press.

Bertin, J. (1977). *La graphique at le traitment graphique de l'information*. Paris: Flammarion.

Boardman, D. (1983). *Graphicacy and geography teaching*. London: Croom Helm Ltd.

Castner, H.W. (1983). Tactual maps and graphics—Some implications for our study of visual cartographic communication. *Cartographica*, 20(3), 1-16.

Castner, H.W. (1990). *Seeking new horizons: A perceptual approach to geographic education*. Montreal and Kingston: McGill-Queen's University Press.

Castner, H.W. (1991). Cartographic languages and cartographic education. *CISM Journal ACSGC*, 45(1), 81-88.

Coulson, M.R.C., Rieger, M., and Wheate, R. (1991). Progress in creating tactile maps from geographic information systems (G.I.S.) output. *Proceedings*, 15th International Cartographic Conference, Mapping the Nations. Bournemouth, U.K. 23 September-1 October, 1, 167-174.

Edman, P.K. (1992). *Tactile graphics*. New York: American Foundation for the Blind.

Franks, F.L., and Nolan, C.Y. (1970). Development of geographical concepts in blind children. *Education of the Visually Handicapped*, 2(1), 1-8.

Gerber, R. (1992). *Using maps and graphics in geography teaching*. Brisbane, Australia: International Geographical Union Commission on Geographical Education.

Gersmehl, P.J. (1991). *Pathways—The language of maps*. Indiana, Pennsylvania: The National Council for Geographic Education.

Gimeno, R. (1980). *Apprendre a l'ecole par la graphique*. Paris: Editions Retz.

Gottesman, M. (1971). A comparative study of Piaget's developmental scheme of sighted children with that of a group of blind children. *Child Development*, 12, 573-580.

Green, D.R. (1993). Wherefore art thou cartographer? Your GIS needs you! *Proceedings*, 16th International Cartographic Conference. Cologne, Germany. 3-9 May, 1, 1011-1025.

Hampson, P.J., and Daly, C.M. (1989). Individual variation in tactile map reading skills: Some guidelines for research. *Journal of Visual Impairment and Blindness*, 83(10), 505-509.

Hatwell, Y. (1985). *Piagetian reasoning and the blind*. New York: American Foundation for the Blind.

Ishido, Y. (Ed.) (1989). *Proceedings*, Third International Symposium on Maps and Graphics for Visually Handicapped People. Yokohama: Yokohama Convention Bureau.

James, G.A., and Armstrong, J.D. (1976). *Handbook on mobility maps*. Mobility Monograph 2. Nottingham, England: Blind Mobility Research Unit, Department of Psychology, University of Nottingham.

Kanakubo, T. (Coordinator) (1993). The selected main theoretical issues facing cartography. Japan: The ICA Working Group to Define the Main Theoretical Issues on Cartography for the 16th International Cartographic Conference. Cologne, Germany. 3-9 May.

Keates, J.S. (1982). *Understanding maps*. London: Longman.

Keming, H. (1991). Design and production of Tactual Atlas of China. *Proceedings*, 15th International Cartographic Conference, Mapping the Nations. Bournemouth, U.K. 23 September-1 October, 2, 707-716.

Kidwell, A.M., and Greer, P.S. (1973). *Sites, perception and the nonvisual experience: Designing and manufacturing mobility maps.* New York: American Foundation for the Blind.

Levi, J.M., and Amick, N.S. (1982). Tangible graphics: Producer's view. In W. Schiff and E. Foulke (Eds.), *Tactual perception: A sourcebook.* New York: Cambridge University Press.

Nicolai, T. (Ed.) (1984). *International conference on tactile representations for the blind—A documentation.* Berlin: Association of the Blind and Partially Sighted of the German Democratic Republic.

Petchenik, B.B. (1979). From place to space: The psychological achievement of mapping. *The American Cartographer*, 6(1), 5-12.

Pravda, J. (1993). Map expression, map semiotics, map language. *Proceedings*, 16th International Cartographic Conference. Cologne, Germany. 3-9 May, 1, 782-786.

Qingpu, Z. (1991). Present situation of tactile mapping in China. *Proceedings*, 15th International Cartographic Conference, Mapping the Nations. Bournemouth, U. K., 1, 435-439.

Rener, R. (1993). Tactile cartography: Another view of tactile cartographic symbols. *The Cartographic Journal*, 30(2), 195-198.

Schiff, W. (1982). A user's tangible graphics: The Louisville workshop. In W. Schiff and E. Foulke (Eds.), *Tactual perception: A sourcebook.* New York: Cambridge University Press.

Schiff, W., and Foulke, E. (Eds.) (1982). *Tactual perception: A sourcebook.* New York: Cambridge University Press.

Tatham, A.F. (1991). The design of tactile maps: Theoretical and practical considerations. *Proceedings*, 15th International Cartographic Conference, Mapping the Nations. Bournemouth, U.K. 23 September-1 October, 1, 157-166.

Tatham, A.F. (1993). How to make maps and tactile diagrams. Madrid: World Blind Union. September 1992-June 1993, pp. 30-34.

Tatham, A.F., and Dodds, A.G. (Eds.) (1988). *Proceedings*, Second International Symposium on Maps and Graphics for Visually Handicapped People. London: Kings College, University of London.

Taylor, D.R.F. (1991). A conceptual basis for cartography: New directions for the information era. *Cartographica*, 28(4), 1-8.

Ucar, D. (1993). A semiotical approach to typology of the map signs. *Proceedings*, 16th International Cartographic Conference. Cologne, Germany. 3-9 May, 2, 768-781.

Vasconcellos, R. (1991). Knowing the Amazon through tactual graphics. *Proceedings*, 15th International Cartographic Conference, Mapping the Nations. Bournemouth, U.K. 23 September-1 October, 1, 206-210.

Vasconcellos, R. (1992). Tactile graphics in teaching of geography. 27th International Geographical Conference *Technical Program Abstracts* (pp. 639-640). Washington, D.C.: International Geographical Union.

Vasconcellos, R. (1993). Representing the geographical space for visually handicapped students: A case study on map use. *Proceedings*, 16th International Cartographic Conference. Cologne, Germany. 3-9 May, 2, 993-1004.

Wiedel, J.W. (Ed.) (1983). *Proceedings*, First International Symposium on Maps and Graphics for the Visually Handicapped. Washington, D.C.: Association of American Geographers.

Wiedel, J.W., and Groves, P.A. (1972). *Tactual mapping: Design, reproduction, reading and interpretation.* University of Maryland Occasional Papers in Geography, No. 2. College Park: University of Maryland.

Woodward, D. (1992). Representations of the world. In R.F. Abler, M.G. Marcus, and J.M. Olson (Eds.), *Geography's inner worlds: Pervasive themes in contemporary American geography* (pp. 50-73). New Brunswick, N.J.: Rutgers University Press.

9

What Does That Little Black Rectangle Mean?: Designing Maps for the Young Elementary School Child

Jacqueline M. Anderson
Department of Geography, Concordia University

INTRODUCTION

In terms of the design of any map, cartographers have traditionally identified who is to use the map and for what tasks. Then, cartographers manipulate the map and map-related elements in a manner they believe will permit the user to employ the map for its intended purpose(s). But what do we really know about the cartographic needs, knowledge, and abilities of young children, nine years of age and under, as map users? This chapter is an attempt, using Québec as a case study, to identify the perceived map needs of the young elementary child, as well as the abilities of the young child to work with map symbols. The first part of the chapter examines the theoretical exposure of children (six to nine years of age) to maps in the early years of their formal school education. This is accomplished by looking at the development and nature of social studies in Québec's school curriculum, and presenting a sample of map activities taken from current instructional material. In the second portion of the chapter, some empirical findings about the understanding of symbolization for children five to six years of age are presented.

ELEMENTARY EDUCATION IN QUÉBEC

General Background

In Canada, education is a provincial domain. Within Québec, the Ministère de l'Éducation (M.E.Q.) is responsible for "determining the goals of education, general educational objectives and the course of study for each subject and level" (Québec, Ministère de l'Éducation (M.E.Q), 1979: 88). The responsibility of implementing the course of study lies with the various school boards. In Québec, young students' exposure to maps is primarily associated with the Social Studies program—"A course of study . . . that includes sociology, geography, history and politics" (Webster's, 1990: 1575). Prior to any detailed look at students' exposure to "maps", some general points pertinent to Québec's education and social studies are presented. Although many five year old children attend Kindergarten, generally for half-day classes, Québec's compulsory education starts at six years of age (Grade 1) and continues until the pupils have completed Grade 6 (11 to 12 years of age). Québec's system of education is unusual in that it is one of the few places in North America which has a confessional public school system: Protestant, Catholic and Other (Alliance Québec, 1982). Within the confessional school boards both French and English sectors may be present. Increasingly, however, within the English sector the elementary schools are following French immersion programs. In many of these programs, rather than restricting French to a specific time slot where the language itself is taught as a subject, French is used as the medium of instruction for a variety of subjects. Social Studies is one of these subjects.

Social Studies in the Elementary School—Historical Background

In Québec, "social studies" has changed dramatically over the last 35 years. Prior to 1971, history and geography were seen as separate disciplines, one or the other of which was included in the Elementary Curriculum for a given academic year. In 1971, however, the Ministère de l'Éducation published "New Orientation Proposed for the Teaching of the Social Sciences in the Elementary School," a brochure designed as a "temporary measure to bridge the gap between the existing and the planned social studies programmes" (M.E.Q., 1983: 3). In the brochure, it was recommended that no distinction be made between history and geography, the two subjects being "considered as elements of a more comprehensive area of study: the Social Sciences, applied to the observation of *Reality* from a physical, historical, economic social or human point of view, the observation of what is generally referred to as the Environment or 'Milieu' " (M.E.Q., 1971: 4). The brochure led school boards to re-examine their teaching practises with the result that, "In many schools, programmes and textbooks were abandoned; in some cases, history and geography courses were simply eliminated, while in others programmes were re-orientated towards a study of the milieu" (M.E.Q., 1983: 3). The 1971 recommendation was accompanied by a list of possible topics and guiding principles for each academic year but, due to a lack of appropriate instructional guidelines and materials, the recommendation and document were implemented in immensely different ways. As a result, in the 1970s it was recognized that a more precise definition of the concepts and skills involved was required, along with a more detailed description of the program's goals and content. Therefore, the new social studies program of the 1980s was laid out in terms of teaching objectives and learning content.

Today, the domain of social studies is seen as an opportunity and means of developing the skills and concepts students require for a basic understanding of the world in which the students live. At the elementary level, the Ministère describes social studies as, "the study of man's interaction with his environment, with a view to helping the child become aware of the concepts of time, space, and society, and to introduce him to the history and geography of Québec and Canada" (M.E.Q., 1983: 8).

An examination of Québec's elementary social studies program reveals that it is composed of two cycles. The First Cycle (Grades 1-3) concentrates on the gradual exploration of the local environment: the Immediate Environment (Grade 1); the Local School Environment (Grade 2); and Local Environment—the Community (Grade 3). As a result, based upon *their own experiences* students develop an awareness of self and the geographical realities of the community. Cycle Two (Grades 4-6), introduces students to the geographical, economic, and social features together with their interrelations of: a) their Region (Grade 4); b) Québec (Grade 5); c) and Canada (Grade 6). For each of the grades it is recommended that two hours per week be devoted to the program. There is also the recognition that, in many areas, the social studies program overlaps other elementary programs, for example the physical education program, in its mandate to develop a child's sense of space. The elementary program paves the way for the history, geography and economics programs in the secondary school.

First Cycle (Grades 1-3)—Awareness of the Concept of Space

Because of the map's inherent property of spatial representation, in Grades 1 to 3 "maps" are employed to help to develop an awareness of space. Developing an awareness of space is seen as enabling the student "to situate himself in his physical environment and to form a mental picture of physical space" (M.E.Q., 1985a: 6). According to Picard (1979), as cited by the Ministère, an awareness of the concept of space is expressed in six different ways: spatial representation, location, distance, spatial association, spatial distribution, and spatial relationship. The first of these dimensions, spatial representation, is seen as involving "a number of operations by which the individual processes information obtained from his spatial environment into an organized series of representations . . . [which] may be stored either internally (mental notes) or externally (drawings, maps, models, written reports)" (M.E.Q., 1985a: 6). Based on Picard's six dimensions, it is not surprising to find that the Ministère has identified that, "developing an awareness of the idea of space requires the concepts and skills involved in determining location, distance, relationships to the environment, and spatial distribution patterns, in acquiring and representing data, and interrelating local and broader environments" (M.E.Q., 1983: 8). The Ministère also acknowledges that the learning of social studies requires the use of a number of special abilities, one of which is "map reading." Categorized (for Grades 1 to 3) as a "technical skill," map reading is defined as, "the ability to decode the representation of a particular space on a map, to locate different features on a map and to establish relationships between these features" (M.E.Q., 1985a: 8). Orientation, classified as another technical skill, is seen as "the ability to situate oneself in space using the landmarks and the cardinal directions and to locate one place in relation to another on a map using the cardinal points" (M.E.Q., 1985a: 8).

Nearly one third (32 percent) of the current content of the First Cycle is devoted to meeting the general objective of developing student's awareness of the concept of space. To meet this general objective, a series of terminal objectives, each of which are composed of several intermediate objectives, are identified by the Ministère. Many of the terminal and intermediate objectives, in order to establish the conditions under which a proposed learning activity should be accomplished, include in their wording the expression "using" (M.E.Q., 1983: 17). The terminal and intermediate objectives which involve "using maps," identified in relation to the general objective "awareness of the concept of space," appear in Table 9.1.

Cycle Two (Grade 4)—The Region

In Grade 4 social studies, the area of study is the student's "Region" (M.E.Q., 1984). Frequent reference to the use of maps is made in connection with the general objectives which relate to developing: a) a basic knowledge of the geography of their region and; b) an awareness of the economic activities of the region. The technical abilities which the Ministère links to Grade 4 map work are (M.E.Q., 1990c):

a. To read a map legend.
b. To identify the Poles, the Equator, and the Northern and Southern Hemispheres on a globe and world map.
c. To recognize the shape and location of a specified area on a globe and outline map.
d. To read and use the scale on a map.
e. To indicate the location of one place with respect to another on a regional map, using the cardinal and intermediate points.
f. To give the directions required to go from one place to another using a regional map as well as the cardinal and intermediate points.

In Cycle Two maps are also seen to play a role in the development of cognitive skills, for example, "TO DETERMINE how some of the physical features of a given area are related, using photographs, and maps of different landscapes in his/her region" (M.E.Q., 1990c: 16).

Curriculum Materials for Grades 1-4

The Ministère de l'Éducation produces numerous documents related to the social studies. In addition to documentation which defines the domain (M.E.Q., 1990a; M.E.Q., 1990c), learning objectives (M.E.Q., 1990b), and curriculum guides (M.E.Q., 1983; M.E.Q., 1985a), the Ministère has produced guidelines for the preparation of materials for use in the general education sector (M.E.Q., 1985b). The criteria and examples contained in the document are very general. For example, in the remarks accompanying the statement that the format of the material must be suitable for the learning process, are the points: a) "colour should be avoided and

Table 9.1 Awareness of the concept of space—
Terminal and intermediate objectives "using maps"

Terminal Objective

1.3 To describe through fieldwork the route taken to go from one place to another in the local school environment, **using a plan or map**, identifying landmarks, and giving directions followed.

Intermediate Objective

1.3.2 **To mark on a plan or map the principal landmarks** of the local school environment **using symbols.**

1.3.3 **To identify on a plan or map landmarks** in the local school environment which are **represented by conventional symbols.**

1.3.5 To compare the distance of alternative routes between two places in the local school environment **using a plan or a map.**

1.3.7 **To identify** the same **physical and man-made features** of a given area as shown in a vertical aerial photograph, and as **represented on a map or a plan.**

1.3.8 **To decode conventional symbols** including pictographs and colours.

1.3.9 To identify the **four cardinal points on a map or a plan.**

1.3.11 To **compare different distances on a plan or map**, using the naked eye, string or strip of paper.

Terminal Objective

1.4 To situate the local environment in relation to the physical and man-made features of surrounding areas **using a plan or map.**

Intermediate Objective

1.4.1 To **name the physical and man-made features** which characterize the local environment **using a plan or a map.**

1.4.2 To **give the location of the physical and man-made features** which characterize the local environment **using a plan or a map.**

1.4.3 To identify the four **cardinal points on a map or a plan.**

1.4.4 **To identify** the same **physical and man-made features** of a given area as shown in a vertical aerial photograph, and as represented **on a map or a plan.**

1.4.5 **To decode conventional symbols** including pictographs and colours.

Terminal Objective

1.5 To identify some similarities and differences between his local environment and other environments, using photographs or other aids.

Intermediate Objective

1.5.3 To introduce the child to representations of the earth **using a globe.**

Source: M.E.Q., 1983, p. 22-24.

The text in **bold** is the addition of the author.

other processes used (dots, hatched lines) in diagrams or graphs so that students who are colour blind, or who have similar visual handicaps, will not be at a disadvantage"; and b) "The choice of graphic process (character size and width, screen, outlines, colours, symbols, etc.) should be functional rather than aesthetic and should facilitate identification, grouping, analysis and synthesis of the learning elements" (M.E.Q., 1985b: 29). Few guidelines are given for the preparation of graphics, and the word map is never mentioned.

The educational materials available are evaluated by the Ministère who release lists of approved items (M.E.Q., 1993; M.E.Q., 1994). These materials are classified into two groups: instructional packages and reference works. Instructional packages are defined as those that treat most of the compulsory objectives of a course of studies for a given year. They consist primarily of a student's textbook that is usually accompanied by a teacher's guide. Reference works are identified as those materials used in a classroom on a regular basis for the duration of a course or program, for example, an atlas. Approved instructional material in French is available for each of the Grades 1 through 4: a student's textbook and teacher's guide for each grade and an atlas (Carrière, 1985) for Grade 4. Approved instructional material in English is available only for Grades 1 and 4. For Grades 2 and 3, the school boards use materials which may have be produced by the board itself (Protestant School Board of Greater Montreal (PSBGM), 1986a; PSBGM, 1986b; PSBGM, 1986c; and PSBGM, 1986d).

Map Activities for Grades 1-4

The Ministère regards the documents it produces as guides, the ideas and recommendations of which the school boards are urged to expand and modify, "in line with the actual conditions in each school as experienced by each individual teacher" (M.E.Q., 1985a: 1). As a result, within each school board the actual implementation of a program's general, terminal, and intermediate objectives depends upon the knowledge and interest of the teacher and the materials available. The following descriptions of map activities, gleaned from instructional material used in the Anglophone sector, are included to provide some details of students' exposure to, and expected use of, maps during the early grades. In Grade 1, the kind of representation of their immediate environment that a student encounters is similar to the sample created as Figure 9.1. This simple plan view of a classroom includes selected items of furniture (e.g., desks, chairs, floor mat, bookcases, wastepaper basket and blackboard), the walls, door, and windows. Such plans, which are in colour, lack a legend. One argument to explain this omission could be linked to the use of annotations. However, whereas the annotations for the basic abstract cardinal directions are prominent on such plans, the concrete features employing shape and a colour (often unrelated to their colour in reality), are left unexplained. As a mapping activity, the students may be encouraged to collectively prepare an actual plan of their own classroom, using cardboard to represent the various classroom elements. Then, on a teacher prepared reduction, to carry

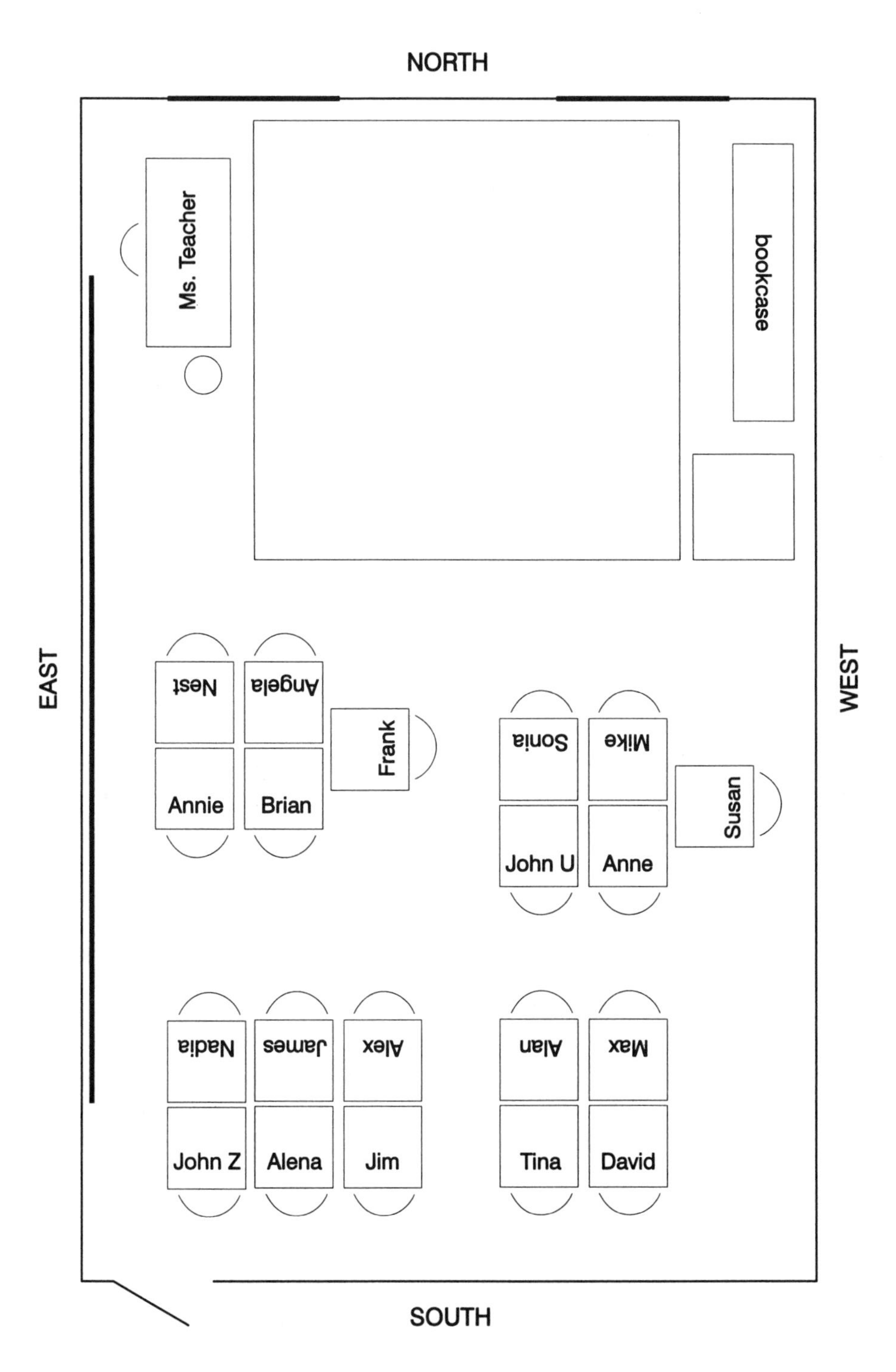

Figure 9.1 Sample of a school room plan.

out tasks such as locating and identifying where they and their friends sit in the class (Lambert and Picard, 1989).

Grade 2 students' map activities revolve around the neighbourhood and the globe. Through actual observation in their environment, the students are encouraged to become familiar with the street names and the nature and location of stores in the immediate school neighbourhood. Then, working in the classroom with a teacher-generated neighbourhood street plan (containing an indication of north), the students can become involved in activities such as: naming the streets; pasting their illustrations of the stores in the appropriate locations (diagrammatic representation); and giving directions. The student's work with globes is designed to reinforce the idea that, while a globe is a model of the earth, it can be represented on a flat surface on which one can distinguish between land and water areas and locate the continents (PSBGM, 1986a).

Students in Grade 3 are concerned with their community. Using commercially available materials such as vertical air photography, plans and maps, students are encouraged to undertake activities which include: decoding conventional map symbols; symbolizing local landmarks on a plan or map; identifying physical and cultural features on aerial photography; relating the features identified on the photography to their representation on a plan or map; comparing, on a map or plan, the relative length of alternative routes between two places; identifying the four cardinal directions on a map; and using the cardinal points to give the location of landmarks relative to the school (M.E.Q., 1983). Figure 9.2 is the map portion of an assignment given to evaluate student understanding of symbols in a legend (PSBGM 1986c: 23-24). The students were asked to use designated colours to colour the map symbols (see the map legend) and then to use the map to answer a series of questions. The actual questions appear at the end of the paper.[1]

In Grade 4 the area of study is the student's home region (M.E.Q., 1984). Many of the social studies activities included in the instructional text (Picard, 1991) refer to maps, requiring the text to be used in conjunction with commercially available maps (e.g., Québec road map, map of the home region, and map showing the Administrative Regions of Québec), and outline maps. These materials permit students to perform tasks such as giving cardinal directions for the location of their region within Québec, and locating and plotting features mentioned during the study of their region (e.g., main cities, major roads, the boundaries of administrative regions). On their regional map, students are also expected to examine and understand physical features (e.g., relief, hydrography, and vegetation) as well as locate and identify cultural and economic information (e.g., transportation networks and service centres). The kind of map activity students are encouraged to become involved in is illustrated by the following exercise which is designed to assist the student in learning about the transportation routes in their area: "Using a road map of Québec, compare the communication routes in your region with those in other regions of Québec. Are there many divided highways in your region? Why do you think that is so?" (Picard, 1991: 32).

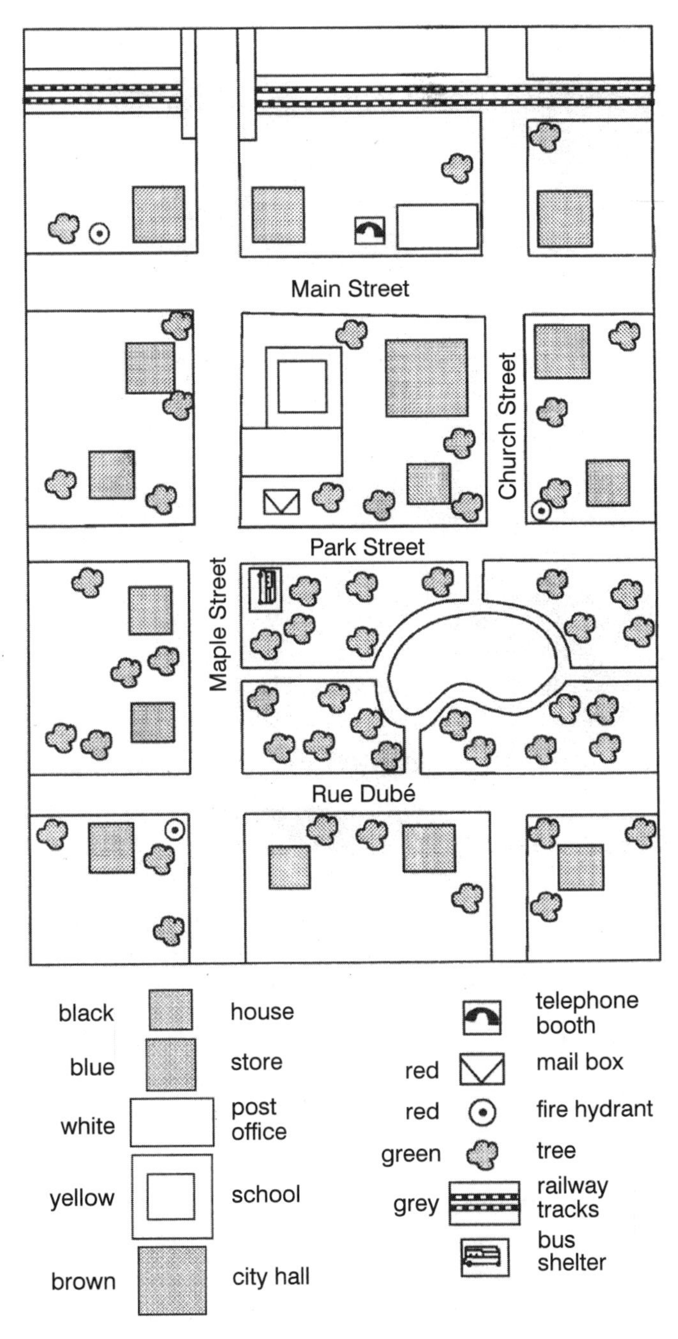

Figure 9.2 Assignment used in Grade 3 to evaluate student's understanding of symbols in a legend (60 percent of original scale; legend was originally located in a single column to the left of the diagram) (Source: PSBGM, 1986c, p. 23).

After reviewing the documents of the Ministère and educational materials, I am left with the impression that Grade 4 students should be map literate, provided that they executed the suggested activities involving the large scale and thematic maps. Why, then, do so many students in the higher grades appear to have difficulty with maps?

Summary—Social Studies, Mapping, and the Young Elementary Child

Given the provincial responsibility for education within Canada, the history, development and current nature of social studies in Québec may not reflect a situation which exists elsewhere in Canada. I believe, however, that Québec's case study does provide some insight into educators' attitudes to, and use of maps in the early elementary grades. I believe it clearly illustrates the fact that educators do **not** see the map as an important form of graphic representation in its own right. They perceive a map primarily as a medium, an intermediate tool, to understand the concept of space. This mirrors the perception of many psychologists who, until recently, used maps to explore their subject's understanding of the concept of space, without any consideration of their subject's knowledge and comprehension of a map as a medium for expression. The ability to work with maps is viewed as a given, particularly in Grade 4. The attitude taken towards pictures is not reflected in maps. Pictures as instructional aids are seen as valuable only if "they are used properly . . . to learn through pictures it is also necessary to know how to interpret them" (M.E.Q., 1985a: 18). Although the Ministère recognizes that learning to read maps is a basic "technical" skill, little attention appears to be paid to either developing students' understanding of the concepts associated with maps, or the comprehension of **how** a map represents space. From Grade 1, references to different points of view are encountered. But, in a map context, little mention is made of the other concepts required for an understanding of this complex graphic, that is, scale, classification, and abstract symbolization. Verification of whether these concepts are treated in one of the core areas, such as language, arts or mathematics, awaits a future investigation.

On paper, the Ministère's theoretical objectives which incorporate "maps" (M.E.Q. 1990b; 1990c), curriculum guide (M.E.Q., 1985a), and suggested activities in the instructional material which "use" maps, are impressive. What happens in reality? The author is unaware of any study which, for each grade level, has attempted to provide answers to questions such as: What mapping materials are used by the students? What types of mapping activities are the students exposed to? And, How much time is devoted to map users in social studies? From personal communication with teachers, I suspect, however, that many children (particularly those in Grades 1 and 2) receive little exposure to either maps or mapping as the mapping activities in these grades present the problem of obtaining or creating suitable teaching materials.

What would be the response of children in Grades 1 to 4 to the question, What Does That Little Black Rectangle Mean? To successfully answer this question, the student must appreciate that the rectangle is an abstract symbol, to decode the meaning of which requires both knowledge of and an ability to use a map key (Anderson, 1992). In Québec's social studies, from Grade 2, there is the expectation that students learn to work with map symbols (although the technical skill of *reading a map legend* is first mentioned formally in the intermediate objectives associated with Grade 4!). Nevertheless, within a grade level, I suspect that student's responses to the question would be nearly as diverse as the answers between grades. The greatest understanding of the abstract symbolic notation system would be exhibited by those students who had been: introduced to the basic map concepts in a meaningful manner; exposed to a variety of map products; and successful in accomplishing meaningful map activities.

YOUNG CHILDREN'S UNDERSTANDING AND COMPREHENSION OF THE GRAPHIC LANGUAGE—Introduction

In this part of the chapter, the emphasis is on exploring the young child's familiarity with the use of the graphic language in a map context. The introduction, which includes some general research findings related to children's familiarity with maps and their mapping abilities, precedes a report of the findings of an empirical study with six year old children. This study was designed to provide some insight on Kindergarten children's understanding and use of components of the graphic language. The components examined include the nature and role of shape, colour, and size in symbol identification.

In a recent study with 150 Grade 1 children, prior to any formal mapping instruction, I asked questions about what they thought a map was, what maps show, their use, and what maps they had seen. I was unprepared for the range of their responses (Anderson, 1992).[2] Many said they had seen maps in their daily environment, locations mentioned included in cars, train and bus stations, in hospitals, and in books such as dictionaries and the bible. Map users identified by the children, in addition to parents and siblings, included firemen, police, airline pilots, robbers and archaeologists! Yet, although they saw the function of maps primarily as means of navigation, very few perceived themselves as an active map user.

Within the cartographic community, there appears to be a growing concern and interest in the development of graphicacy in children together with the identification of what constitutes an appropriate cartographic design for products for this user group (Canright, 1987; Gerber, 1993). Maps use a symbolic notation system. But what notation system should be used for the young child? A simple query that in turn generates many other questions for which we have no answers. Most of the maps which children encounter in their daily lives are planimetric representations. Yet commercially produced materials for the young (and not so young!) elementary child employ pictorial symbols and often exclude a key or legend. In designing maps that contain primarily pictorial symbols, are we doing young students a disservice? How easy is it for a child to "unlearn" using picture symbolization and replace it with an ability to work with an abstract symbolic notation system? Since young children are aware of the vertical perspective long before it appears in their drawings, would it not be easier to assist the young children from an early age to learn to deal with an abstract notation system?

The body of literature on the mapping abilities of children less than nine years of age is small, but growing. Since the early 1980s there has been an increasing number of studies suggesting that young children are more aware of maps than previously thought (Blades and Spencer, 1987; Liben and Downs, 1986; Panckhurst, 1989); and young children are capable of learning to understand and work with a varied range of map materials, provided they are exposed to meaningful materials, and appropriate instruction and tasks (Anderson, 1987; Atkins, 1981). In a recent study by Liben and Downs (1993), there is the suggestion that it may be the geometric (spatial) rather than the representational (symbolic) space-map correspondences which poses problems to elementary children.

Although there has been an attempt to classify the types of errors young student make in map symbol identification (Liben and Downs, 1986), little research has been conducted into what, for a child, characterizes a particular map symbol. Is it the symbol's colour, shape, size, the feature's function, or some combination of these? For a young child, is the parameter of size secondary to shape or colour when more than one variable is used to symbolize a feature? Answers to such questions are important, particularly if students from the age of seven may be required to use commercially produced materials in mapping activities associated with their local neighbourhood, community, region, and so on.

MAP SYMBOLS AND KINDERGARTEN CHILDREN

The Study

One component of the study which involved 108 kindergarten children, two thirds of whom received six 20 minute lessons of basic mapping instruction, was designed to test map reading skills (Anderson, 1987).[3] The map reading tasks required the students to identify and locate point, line, and area symbols on two, commercially printed, colour test maps. Figure 9.3 (pictorial) and Figure 9.4 (abstract) are black and white renditions, at a reduced scale, of the test maps.

During both of the individual interviews with each child (each map was tested separately), two approaches were used to obtain information—spontaneous symbol identification and directed symbol identification. In spontaneous symbol identification the student was asked to, "Point to things which you can name on the map, and name them for me." In the directed symbol identification approach, the child was asked to point to a particular symbol (e.g., "Point to a gate on the map"), or respond to the question "What is this?" 'This' was a located map symbol which was identified by an arrow or box drawn on clear acetate placed on top of the map. Following a response the child was then asked, How do you know this is a ______?[4]

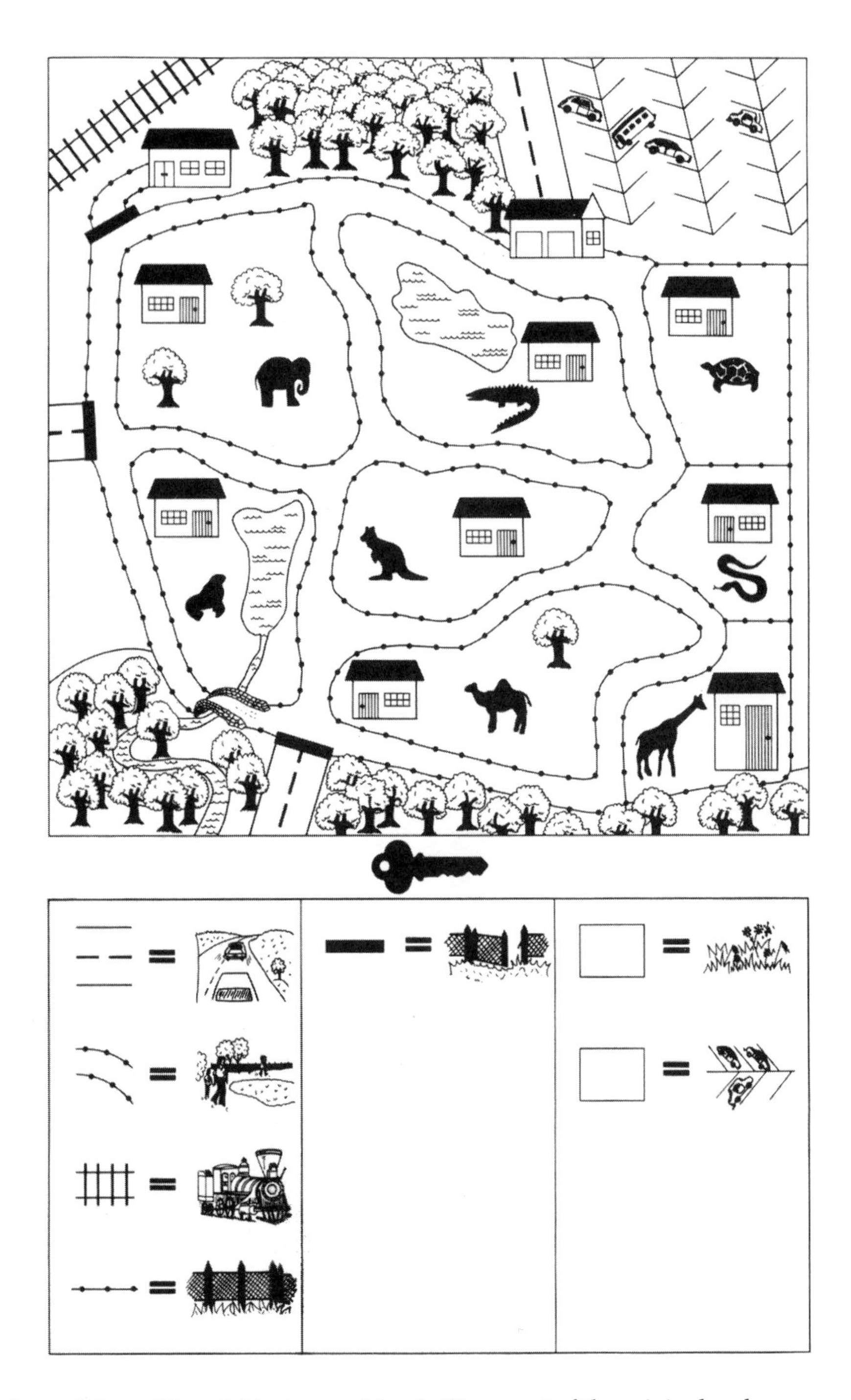

Figure 9.3 Pictorial test map. Map is 70 percent of the original scale (Source: Anderson, 1987, p. 287).

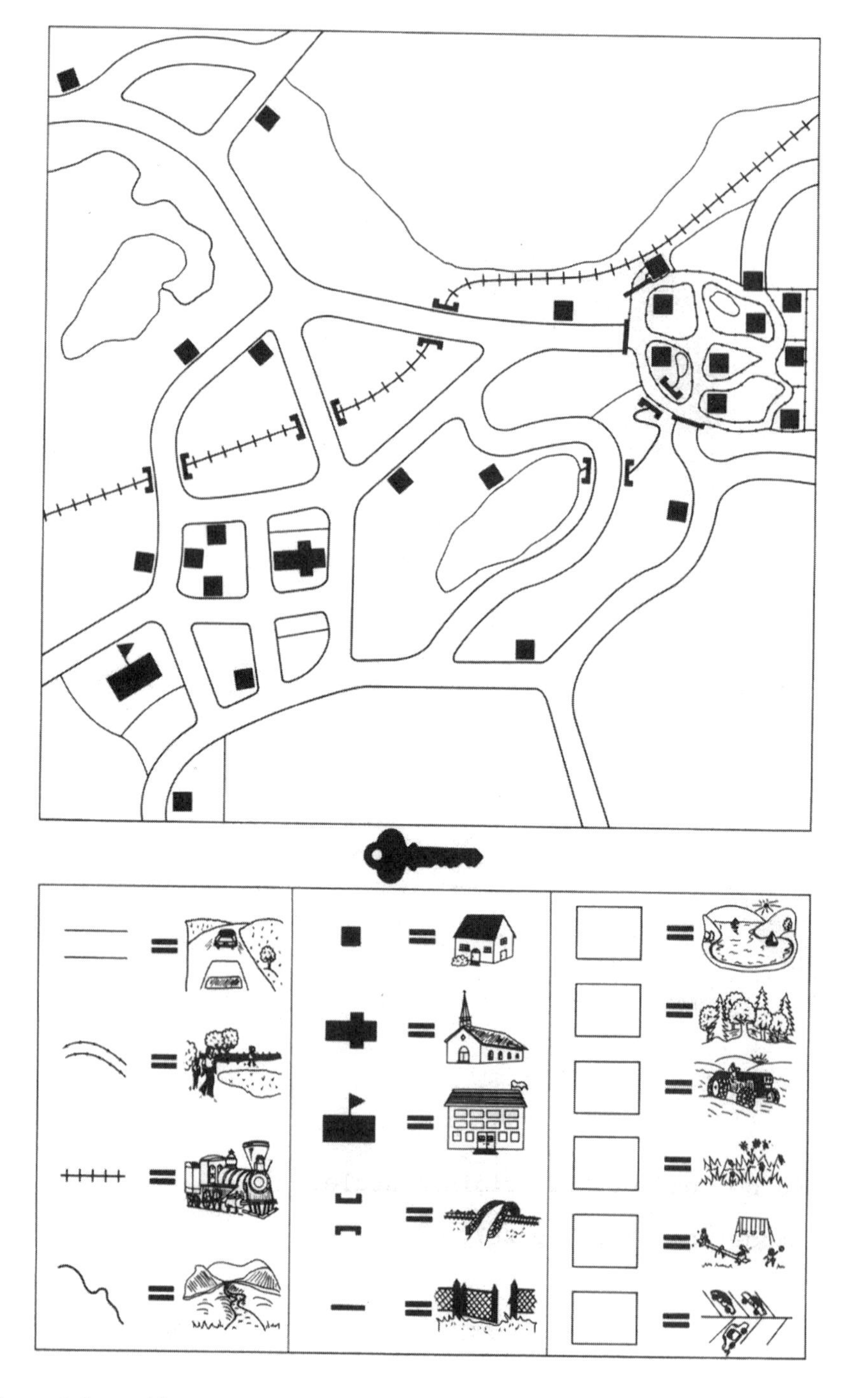

Figure 9.4 Abstract test map. Map is 70 percent of the original scale (Source: Anderson, 1987, p. 300).

Symbol Identification

When the kindergarten students were asked how they were able to identify a particular symbol on the pictorial map, all children frequently replied, "because it looks like that." Beyond this response, the children provided a variety of answers for both the pictorial and abstract maps. Generally these could be classified into the following categories: descriptive properties, shape, colour, size, function, personal experience, use of the map key, and "do not know." Different categories appeared to be important for different symbols. Some symbols were identified on the basis of a single category while for others more than one category was used. Excluding the categories "do not know" and "use of the map key," examples of responses reflecting each of the categories mentioned above follow. The children's comments are verbatim.

The *descriptive property* reflects how the symbol appeared on the map, for example, the tree on the pictorial map was identified because of the "leaves and trunk." Many children identified the "waves" (also called "those up and down lines," "these things here," "zig zags," "curls," and "bumps") as a major factor in identifying the lake on the pictorial map.

From the children's responses it appears that *shape* may mean different things to different children. A house on the pictorial map was identified because it was a "... square and a triangle." The house on the abstract map was often identified as a "square." One student identified the church on the abstract map as "T's". On the pictorial map, the animals were the first elements identified. Most of these were correctly named. Only the seal created problems—it was identified as a dog, frog, seagull, crab, monkey, beaver, and a dolphin. Although the children seemed aware of the term seal, they were unable to match the black shape provided to that word.

The visual dimension of *colour* was very important for the identification of the "tree," "the place where there are many trees" (abstract map), and the lake. Many children gave "because it is green" as the primary reason for correctly identifying the tree symbol. Similarly, "because it's blue" was the reason given for correctly identifying the lake symbol on the pictorial map.

However, colour identification was not without its ambiguities. When children who had correctly identified the tree symbol were asked the colour of the tree trunk on the pictorial map, few identified it correctly as black. Most thought the trunk was brown. Did the children need only the dominant colour cue, which in this case is the green of the foliage, or is colour combined with another category, for example, shape? This raises the intriguing question of whether there is a hierarchy of categories which children employ in the identification of cartographic symbols.

Size (which really meant length) was used by the students to identify "a river" on the abstract map. However, many students erroneously identified the larger of the two lakes as the river "... because it's big" or "because it's longer" (than the smaller lake). Sometimes the children's responses were a combination of more than one of the defining categories. A frequent grouping was description and colour, for which examples included "green leaves," "green grass," and "blue water."

In addition to the "visual" categories, the *function* of the element symbolized was frequently given as the reason for identifying a particular symbol: "because cars go on it" (road); "because trains go on it" (railway track); "because they're blocking off" (gate); and "because lines keep the cars out so the animals will not get them" (fence on the pictorial map). Personal experiences were often recounted as explanations for what a symbol represented, for example, "roads [are] brown in Jamaica," "because I have an apple tree in my back yard," "I saw a railroad track and we went over the railroad." One child identified the church symbol as a cemetery as it "is like the one my mother goes to."

Symbols—Interrelationship of Visual Cues

In designing map symbols, cartographers use the visual variables of shape, size, and colour to introduce contrast between the symbolized features. Although some symbols may use one variable, e.g., an unfilled square, most map symbols involve the two or all three variables. Take for example the gate symbol shown on the pictorial map. This symbol uses the three visual cues: shape (rectangle), colour (black), and size (small to a child, yet large in relation to other symbols on the map, e.g., the railway tracks). Is it possible to distinguish, for the young child, which of a symbol's dimensions is the most important in the task of identifying a particular map symbol?.

In the directed identification tasks the same request produced a range of responses from the children. To the request for identification of the lake, answers included "blue," "water," "lake." A similar range of responses was identified for the house on the abstract map: "square," "box," "black," "building," "house." The childrens' responses suggested that for pictorial point map symbols with a horizontal perspective, shape is a more important characteristic than colour, and this might explain why many of the animals were correctly identified. The seal, the only black animal, was the symbol that a large number of the students (20 percent) failed to identify correctly. Shape is a powerful cue, but confusing when a clear prototype is lacking.

Although the shape of many pictorial point map symbols could be recognized, abstract map symbols presented problems. These problems may relate to the complexity of the abstract symbol. When asked to identify the gate on the pictorial map, one child identified the roof of the house adjacent to the zoo. The symbol dimension of colour was correctly identified, but the dimensions of size and shape were ignored. The lines down the centre of the road were also identified as the gate. This response illustrates attention to the shape and colour, but not to size.

The children's responses also suggested that shape is the important variable for line symbols on both the pictorial and abstract maps. Both the road and railway were identified as being lines, although the colour "grey" (which was sometimes identified as white) also appears to have been important to some children in the labelling of the road. One child identified the road on the abstract map by a process of elimination as it "... did not have those lines [railroad cross ties] on it." Because of the nature of the feature they represent, line symbols have a similar shape on the pictorial and abstract map. Therefore, consistency of shape may be one reason why Kindergarten children found the majority of line symbols easier to identify than point or area symbols on the map test. For area symbols on the maps there is no dimension of shape, since the feature represents a finite space, and colour appears to be the most important decision criterion. However, the colour is one of association, for example, the pale green (representing grass) on the abstract map was identified as a place where there are many trees "because it's green."

Taken together, these findings suggest that although shape may be an important variable in symbol identification, its significance varies with the nature of the map symbol (point, line, area) and representation (pictorial, abstract).

Symbol Identification—Sources of Error

Problems in Symbol Identification

As part of their study investigating children's understanding of maps, Liben and Downs (1986) attributed error in the identification of map symbols to the following: problems of scale; reification; the identification of elements out of context; inconsistency of classification; and problems of perspective. In my study I found the Kindergarten students made many of the same errors mentioned in Liben's and Downs's study. Therefore, I have adopted the same terminology to structure the examples of errors included here.[5]

PROBLEM OF SCALE. Although kindergarten children are able to deal perceptually with differences in area (Anderson, 1987), they seemed unable to demonstrate a similar facility when verbally labelling point symbols. For example, the alligator was also identified as a lizard, crocodile, dinosaur, and dragon. Similarly, the lake on the large-scale pictorial map was labelled as a sink, puddle, swimming pool, lake, stream, river, sea, and ocean. These examples demonstrate that, for the kindergarten child there can be semantic confusion in relating a word to its graphic referent.

PROBLEM OF REIFICATION. The term "reification" refers to the literal interpretation of the symbol, for instance, a road coloured red on a map is interpreted as being red in reality (Liben and Downs, 1986). Reification is considered a sign of pre-operational thought. Evidence of reification occurred in individual symbol identification. However this thinking was most obvious in response to a question that arose at the start of testing. After correctly identifying the house symbol on the pictorial map, one child added "but the kangaroo can't live in that house because it will break up all the furniture." Of the 76 children who were asked "Can the kangaroo live in that house?", 33 percent responded "yes" and 67 percent said "no." While a small number of the "no" responses related to the ideal habitat for a kangaroo, most were related to the fact that the symbolic kangaroo was "too big" to fit into the house symbol. This finding has implications for map design. How many designers pay attention to the relative sizes of pictorial symbols on maps designed for pre-school children?

PROBLEM OF IDENTIFICATION OF ELEMENTS OUT OF CONTEXT. Students sometimes identified a map element correctly, but subsequently would provide an answer out of context. As an example, one child correctly identified the road on the abstract map, but when asked to identify another outlined symbol (a lake) the child responded "a balloon . . . [because] . . . it's round and [there is a] string for you can hold it" (the string being the adjacent river).

PROBLEM OF INCONSISTENCY OF CLASSIFICATION. The same symbol would initially be identified as one thing and later as something else. For instance, the railway line on the pictorial map would be initially correctly identified as a railway line. But later, the railway lines would serve as the answer to the request "point to a fence on the map."

PROBLEM OF PERSPECTIVE AND SCALE. The symbols on the pictorial map were portrayed from a number of perspectives including horizontal, vertical, and oblique (drawings in the map key). Some children had difficulty differentiating between these perspectives, for example, the railway tracks were identified as a fence. The gate on the pictorial map was correctly identified, but for the wrong reason—as the horizontal black map symbol was perceived as "a fence post."

All symbols on the abstract map except the school had a vertical perspective. Nevertheless, confusion between perspectives persisted. For instance, several children correctly identified the bridge, not because the conventional abstract symbol on the map was recognized, but because half of the abstract symbol could be construed as a bridge in horizontal perspective because it "goes up and down." The same type of false logic resulted in the curved portion of the road in the top, north west portion of the abstract map being identified as a bridge.

COLOUR. In addition to the sources of error identified by Liben and Downs (1986), I found that some children had a problem with the colours on the maps. There are several cases of insufficient contrast in value on the maps; for example, on the abstract map, the pale grey of the roads was for some indistinguishable from the white surround of the map and several students found it indistinguishable from the dark grey of the car park (located below the railway, north of the zoo). The green also presented a problem to one subject who was not able to differentiate between the light green representing the grass and the dark green representing the trees.

Symbol Identification—Individual Differences

Another finding of this research supported an observation of Liben and Downs (1986) that there are great individual differences among students. Since one of the initial criteria for selecting children for the study was either their high or low verbal ability, it was to be expected that there would be differences in their vocabularies. Examples of differences in vocabulary were seen in the children's labelling of the animals. Most children responded in general terms, for example, a "snake" or "camel." However, some identified these animals respectively as "cobra" and "dromedary."

Individual differences were observed in other areas, such as the strategies the children used for responding to various symbol identification questions, for example, in the amount of attention paid to detail by the Kindergarten child. In the map key picture of the park (adjacent to the abstract fence symbol), one child identified what were really tufts of grass as "the ducks on the pond." Several children identified the "grass" on the pictorial map and said "but it has no flowers," (a reference to the fact that the grass symbol in the key contained flowers). In his explanation of why the pictorial house symbol was a house, one child's description included reference to the doorknob.

Another interesting example of individual differences was provided by the distinct ways individual children responded to the question on the pictorial map "how do you know that it is a place where there are many trees?" In addition to the verbal responses, such as "because I see all the green," children employed a variety of estimation and counting strategies. One child, pointing in succession to two different areas of the map, stated "because there is just a little bit and here there is a lot." Another child stated "I counted them." Several students painstaking and methodically tried to count the individual trees (some were correct, and others not). Some students made a wild guess at a number, while one responded "because there is not one, there is 100 . . . actually 20 to 15."

Map Symbols and Kindergarten Children—A Summary

Symbols are the language of maps. For the successful communication of map information it is essential that the meaning attached to a symbol by the map designer is the same as that interpreted by the map user. If cartographers become involved in designing maps for a young audience, they will need to acquire an understanding of the young child's familiarity with the nature and use of the graphic language. Presently, relatively little is known about how young children attach meaning to the visual cues used for point, line, and area symbols in a map context. It is a complex and challenging topic which requires much more detailed investigation. The observations and the various statements made by the children in connection with symbol identification illustrate that both the factors of experience (knowledge base) and the level of development are responsible for the answers.

CONCLUSIONS

The long-term involvement of cartographers in map design, particularly in the design of thematic maps, is being questioned as current trends in computer technology have led to an increasing number of predictions that future maps will be generated, not by cartographers, but by individuals with little or no cartographic training. If this prediction should come to pass, can the profession survive? I believe the answer is yes, but only if we are willing to become involved in the role of assisting the next generation of map producers and users to become cartographically literate. This will be not be an easy quest. The Québec case study revealed that, while educational directives encourage children's use of maps from an early age, little attention is paid to developing student's comprehension of a map's components and how a map functions. Part of this oversight can be attributed to the fact that so little is known about the needs, knowledge and abilities of young map users. The recognition and acknowledgement of this situation are important first steps. The next stage, carrying out appropriate research, will enable the creation of a knowledge base that can provide guidance to educators (concerned with understanding the process of becoming map literate), and those engaged in producing maps for the young child, in both conventional and multimedia formats. While some of you may still be puzzling over the title of this chapter, others may be wondering how successful the Kindergarten students were in identifying the "gate." One third of the students using the abstract map (40 percent on the pictorial map), most using the map key, were able to identify that a "gate" was shown by "that little (to the child) black rectangle." There is hope for graphic literacy.

REFERENCES

Alliance Quebec. (1982). Working Paper ll: Education. *In Papers on English Language Institutions In Québec*. Montréal: Alliance Québec.

Anderson, J. (1987). The relation of instruction, verbal ability, and sex to the acquisition of selected cartographic skills in Kindergarten children. Unpublished Ph.D. Thesis, University of Madison, Wisconsin.

Anderson, J. (1992). The role, nature, design and use of the map key by young children. Unpublished paper presented at the 27th International Geographical Congress. Washington D.C. August 9-14.

Atkins, C. (1981). Introducing basic map and globe concepts in young children. *Journal of Geography*, 80, 228-33.

Blades, M., and Spencer, C. (1987). The use of maps by 4 - 6 year-old children in a large-scale maze. *British Journal of Developmental Psychology*, 50, 19-24

Canright, A. (1987). Elementary schoolbook cartographic: Creation, use, and status of social studies textbook maps. Unpublished Ph.D. Thesis, University of California, Los Angeles.

Carrière, J. (1985). *Atlas des jeunes Québécois*. Montréal: Centre Éducatif et Culturel.

Gerber, R. (1993). Map design for children. *Cartographic Journal*, 30, 154-158.

Lambert, G., and Picard, L. (1989). *On my street*. Montréal: Centre Éducatif et Culturel.

Liben, L., and Downs, R. (1986). Children's comprehension of maps: Increasing graphic literacy. Pennsylvania State University. Final Report to the National Institute of Education. Grant NIE-G-83-0025.

Liben, L., and Downs, R. (1993). Understanding person-space-map relations: Cartographic and developmental perspectives. *Developmental Psychology*, 29, 739-752.

Panckhurst, F. (1989). The Acquisition of Cartography in Preschool Children. Unpublished Ph.D. Thesis, University of Wellington, Wellington.

Picard, J. (1979). Sciences humaines au primaire (2e cycle). Unpublished document, Research Paper 78-37, report of phase 1. Sherbrooke.

Picard, J. (1991). *Explore your region*. Montréal: Centre Éducatif et Culturel.

Protestant School Board of Greater Montreal (PSBGM) (1986a). *Teachers resource book, level 2*. Montréal: PSBGM. Code PSBGM 058355.

Protestant School Board of Greater Montreal (1986b). *Masters for duplication, level 2*. Montréal: PSBGM. Code PSBGM 058352.

Protestant School Board of Greater Montreal (1986c). *Teachers resource book, level 3*. Montréal: PSBGM. Code PSBGM 058353

Protestant School Board of Greater Montreal (1986d). *Masters for duplication, level 3*. Montréal: PSBGM. Code PSBGM 058348.

Québec. Ministère de l'Éducation (M.E.Q.) (1971). *New orientation proposed for the teaching of the social sciences in the elementary school*. Québec, Ministère de l'Éducation. Code 680A.

Québec. Ministère de l'Éducation (M.E.Q.) (1979). *The schools of Québec, policy statement and plan of action*. Québec: Ministère de l'Éducation. Code 49-1070A.

Québec. Ministère de l'Éducation (M.E.Q.) (1983). *Elementary school curriculum, social studies, history, geography, economic activity, and culture*. Québec: Ministère de l'Éducation. Code 16-2600A.

Québec. Ministère de l'Éducation (M.E.Q.) (1984). *Framework for the development of a regional curriculum guide. Social studies, history, geography, economic activity and culture, elementary*. Québec, Ministère de l'Éducation. Code 16-2601A.

Québec. Ministère de l'Éducation (M.E.Q.) (1985a). *Social studies, history, geography, economic activity, and culture, first and second cycle*. Québec: Ministère de l'Éducation. Code 16-2600-01A.

Québec. Ministère de l'Éducation (M.E.Q.) (1985b). *General specifications manual for the preparation of basic instructional materials general education. Guide for publishers. producers, and authors*. Québec: Ministère de l'Éducation. Code 16-4311-23A.

Québec. Ministère de l'Éducation (M.E.Q.) (1990a). *Definition of the domaine, social studies, first cycle, elementary*. Québec: Ministère de l'Éducation. Code 16-2685A.

Québec. Ministère de l'Éducation (M.E.Q.) (1990b). *Distribution of the learning objectives for the first cycle of the elementary social studies by grade*. Québec: Ministère de l'Éducation. Code 16-2688A

Québec. Ministère de l'Éducation (M.E.Q.) (1990c). *Definition of the domaine, social studies, second cycle, elementary*. Québec: Ministère de l'Éducation. Code 16-2689A.

Québec. Ministère de l'Éducation (M.E.Q.) (1993). *List of instructional materials approved by the Minister of Education, preschool and elementary levels 1993-1994*. Québec: Ministère de l'Éducation. Code 16-405236A.

Québec. Ministère de l'Éducation (M.E.Q.) (1994). *La matériel didatique approvê pour l'éducation préscholaire et l'énseignement primaire 1994-1995*. Québec: Ministère de l'Éducation. Code 12-8004.

Webster (1990). *Webser's illustrated encyclopedic dictionary*. Montréal: Tormont Publications.

ENDNOTES

1 Questions accompanying Figure 9.2.

Which do you find difficult?

1. Draw the symbol that represents a fire hydrant.
2. On which streets do you find a bus shelter?
3. How many stores are there on this map?
4. What colour is the school?
5. Is the telephone booth nearer the Town Hall or the Post Office?
6. Which road has the most trees?
7. Name the roads which are crossed by the railroad.
8. How many houses have their front door on Park Street?
9. For those people who live at the corner of Park Street and Church Street, is it a shorter distance to post a parcel at the post office or to put it in the mailbox?
10. Is Main Street a residential or a commercial street (service area)?

2 The empirical data referred to in this paper comes from research conducted with children attending Anglophone schools on the Island of Montréal.

3 The lessons introduced the cartographic transformations of plan view, scale, and symbolization; the location and navigational use of a map; and the use of an alpha-numeric grid. Students were shown that a spatial arrangement of objects could be represented as a model, a small scale picture presentation, and a plan view. The concept of a map key was introduced. During the lessons the children made and used a variety of symbol picture keys as the concept of maps was extended to include the classroom and school environments, and a community.

4 The following features were used in the map reading tasks:

a) *Pictorial map*

Point: animal house+; tree*; gate*
Line: road+; fence*; railroad*
Area: pond/lake+; many trees*; parking lot*

b) *Abstract map*

Point: house+; bridge*; gate*
Line: road*; railroad+; river*
Area: many trees*; lake+; parking lot*

* Child asked to label an isolated map symbol.

\+ Child requested to point to the symbol on the map.

5 The subjects in this study were students from Kindergarten, Grade 1, Grade 2, and a small number of adults. It is unclear, however, from Liben's and Downs's study which of their comments related to which of the age groups.

10

Gender Differences in Map Reading Abilities: What Do We Know? What Can We Do?

Mark P. Kumler
Barbara P. Buttenfield
Department of Geography, University of Colorado

INTRODUCTION

Cartographers are faced with a great number of design decisions when producing a map. While there is a well-established body of literature on the basics of good map design, little of this work has focused on designing maps for successful use by both genders. This void exists despite a wealth of research in psychology and related fields that has shown that males and females have different learning styles, different abilities in interpreting and relaying spatial information, and different approaches to representing such information. This chapter summarizes some of the major findings in these other fields, reviews what little is known specifically about gender differences in map reading, reports some anecdotal observations of 124 student sketch maps, and suggests possible directions for future map design research.

PREVIOUS WORK

For a review of previous work relevant to these questions, we shall rely primarily on two thorough reviews in psychology and geography (Maccoby and Jacklin, 1974; Self et al., 1992), and the path-breaking work in cartography by Gilmartin (Gilmartin and Patton, 1984; Gilmartin, 1986). The interested reader is encouraged to use these works to identify others of particular interest.

The psychology literature contains thousands of articles that address the differences between the genders. One of the most exhaustive reviews of the early literature in this field—through the early 1970s—is found in Maccoby and Jacklin's *The Psychology of Sex Differences* (1974). These authors compiled summaries of over 1400 experiments involving gender differences, and they interpreted the wealth of literature to suggest four basic psychological differences between the sexes: women have superior verbal abilities, men are more aggressive, and men have superior skills in both mathematics and visual-spatial tasks. While their sweeping generalizations continue to be the subject of debates today, their compilation of so many studies remains valuable. Some of their assertions remain unproven, despite studies focused directly upon them. For example, they claimed that "The male advantage emerges in early adolescence and is maintained in adulthood for both kinds of tasks [visual-spatial tasks that involve disembedding, and those that do not]" (Maccoby and Jacklin, 1974: 94; also cited in Gilmartin and Patton, 1984). More recent studies, however, have suggested that such differences either increase with age (McGee, 1979) or decrease with age (Beatty, 1988; Stumpf and Klieme, 1989).

Maccoby and Jacklin's work is significant for another reason: their frank discussion of authors' biases. They describe themselves as feminists (p. 12), and add that they doubt whether complete objectivity is possible for anyone engaged in such work, whether male or female (p. 13). Such useful warnings led one of the authors of this work to seek a co-author of the opposite gender, and we have done our best to cancel out each others' biases, where they could not be completely suppressed.

In the general geographic literature, there has been precious little. With the exception of a brief work in the *Professional Geographer* (Brown and Broadway, 1981), in which the authors reported that females were less accurate in their assessments of inter-town distances from their own mental images, nothing appeared until 1984, when Gilmartin and Patton (1984) published a paper in the *Annals of the AAG* that continues to serve as the most thorough review of gender differences specific to maps. Gilmartin and Patton surveyed the leading English-language cartographic journals from 1964 to 1982 and found only two perceptual studies that reported whether any gender differences were observed (Morrison, 1974; Phillips and Noyes, 1982); in both cases the differences were statistically insignificant.

Gilmartin and Patton's work included a report on the results of five map use experiments designed to determine whether differences existed for both thematic and general purpose (road) maps, and for both young and adult populations. In examining 4th graders' abilities to recall thematic spatial information about a fictitious country from a brief text, they found no significant differences between boys and girls when no maps were involved, and a slight, but insignificant, improvement in the boys' scores when supplementary maps were included. A similar test of college-aged subjects resulted in men performing significantly better on tests both with and without maps, but women benefiting more than men by the maps' presence. In the other set of experiments involving route planning and symbol identification, 1st and 3rd grade boys achieved better scores than girls in both grades and on both tasks, but the difference was only significant with 1st grade boys identifying symbols. Similar tests of college-aged students yielded no significant differences. Gilmartin and Patton concluded that since no gender differences were found in most of the map-use skills examined, we should not extend automatically the "males have superior spatial abilities" premise to include map-reading. Their observation of age-related differences (young girls having less success with maps, but the differences disappearing with age) ran contrary to the prevailing developmental theories put forth by psychologists, and the authors note that geographers should be cautious about blindly accepting the results of psychological research.

A follow-up study by Gilmartin (1986) examined the effects of subjects' use of mental image maps. Whereas no gender differences were observed in one group that used no maps, or in a second group that was told to "visualize mentally the information contained in the reading" (p. 338), males outperformed females in the third group which used prepared maps to supplement the text.

The sole remaining example in the mainstream cartographic literature appeared in *The American Cartographer* in 1987 (Chang and Antes, 1987). In this study, the authors tested 105 subjects in an introductory undergraduate psychology course. The test consisted of basic questions about reference (thematic), topographic, and street maps. The authors found that males performed significantly better on the reference and topographic maps, but they found no differences on street maps.

Most recently, a review of nearly 200 studies on the general topic of gender-related differences in spatial abilities appeared in *Progress in Human Geography* (Self et al., 1992). The authors divide the literature into three classes based on the different theories being tested: "deficiency [theory] is primarily biologically based and is often tied to hypotheses such as right and left hemispheric specialization in the brain [...], or differences in the oestrogen or androgen content of the human body"; "difference theory argues that sociocultural factors such as early childhood training and expectations, parental and institutional expectations and experience, differential course taking, and cognitive strategies used, can account for most of the differences used"; and "inefficiency theory argues that spatial abilities are approximately the same in both sexes, but that performance and behaviour indicators often favour males" (Self et al., 1992, p. 316). They argue that spatial ability encompasses at least three dimensions—spatial orientation, spatial visualization, and spatial relations—and then praise Gilmartin's work as one of the first to consider all three dimensions. Self et al. review numerous related works in cognitive psychology and sociology, and conclude with several concerns about previous studies and recommendations for future ones. They highlight five areas of concern: the validity of the tests, the representativeness of the subject populations, the need for examining intrapopulation variability in the test responses, relevance to geography, and the implications for recruitment or retraining of the work force.

The gist of the existing literature is that there are still many questions that need to be answered about the specific causes, and types, of gender differences in map reading, and great care must be taken when designing an experiment to investigate such differences. In this light, we present some of our own findings and suggest some very specific design issues that need more study.

STUDENTS' SKETCH MAPS

One of the authors has acquired a collection of 124 sketch maps prepared by undergraduate students on the first days of three introductory courses involving maps. These sketch maps were not collected as part of any scientific study, and thus the sampling and collection procedures have many shortcomings; they may, however, provide some interesting ideas for future study in more rigorously designed experiments.

The sketch maps were produced by students at the end of an opening day's lecture in three different undergraduate courses at the University of Colorado-Boulder. Two were sections of a course titled Maps and Mapping, and were composed largely of second and third year students, the majority of who had declared a major in either Geography or Environmental Conservation (EC). The other course, titled Cartography I, consisted mostly of fourth-year Geography or EC majors. Of the total 124 students, 52 (42 percent) were female.

The collection procedures were similar in all three situations. After approximately 45 minutes of introductory, opening-day material—an overview of what would be covered in the course, a review of the syllabus, the mechanics of grading, and so on—the students were asked to fill out an index card with their name, phone number, college level and major (if decided), and their reasons for taking the course. Then they were told to turn the cards over and draw a map. The students were told to imagine that they had a friend visiting them—a friend unfamiliar with the local area—who needed a map to get from the student's local home to this building, such that the two could meet after class for lunch. The students were instructed to draw a simple sketch map that their friends could follow to get to the building. They were assured that their sketch maps would not affect their grade in the course, and that they would never be linked with their names in any way.

The students had received no formal instruction in map design in the preceding minutes in the classroom, and few had any prior formal course work in cartography. The subjects did, however, probably have an above-average interest in geography or cartography, as they had elected to take this course; and, given that the topic of the course was maps, they may have taken the task more (or less) seriously than normal. Examples of the sketch maps appear in Figures 10.1 and 10.2.

In each course, selected sketch maps were used in the subsequent lecture to illustrate some basics of map reading or map design. Copies of the maps were projected onto a screen to illustrate some of the concepts needed to interpret even such simple maps—orientation, scale, symbolization of landmarks and linear routes, and so on. The maps were then largely forgotten until this study surfaced, at which point the authors re-examined the maps more carefully.

The students' genders were easily determined from their names on the fronts of the cards (and the instructor's familiarity with the subjects, in the case of gender-unspecific names). This arrangement also aided the data collection process, by reducing the likelihood of bias in the analysis with the genders remaining unknown as the information was compiled.

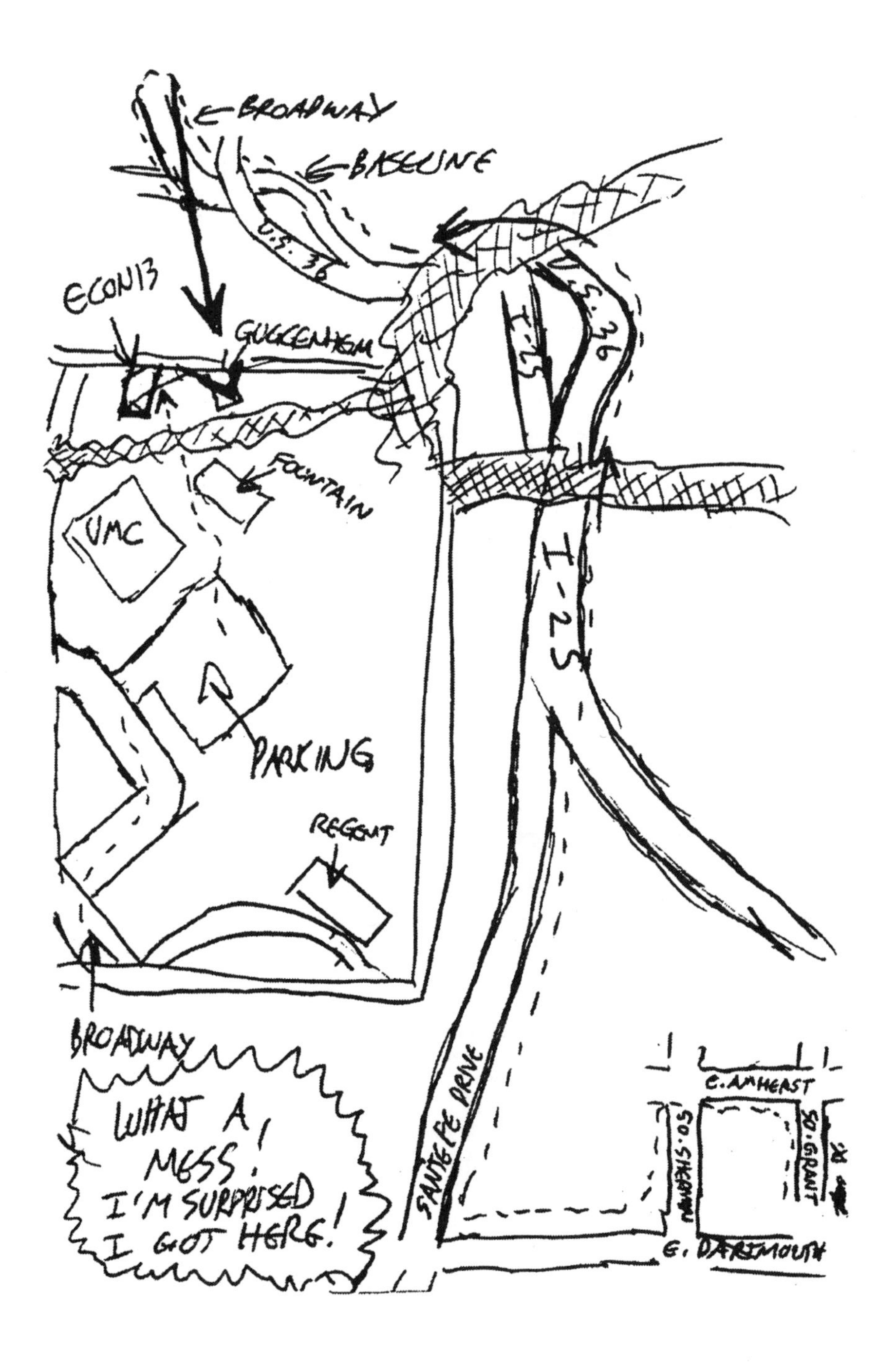

Figure 10.1 Example sketch map illustrating use of an inset map and multiple line styles.

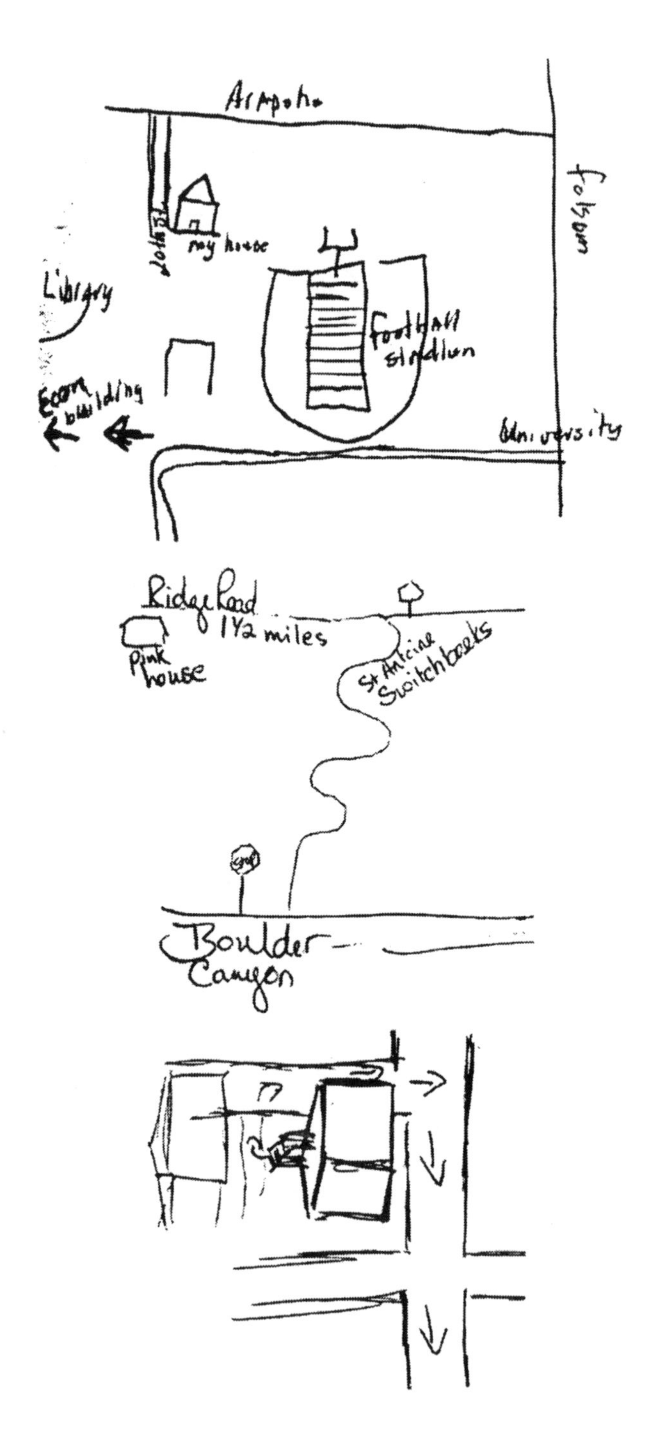

Figure 10.2 Pieces of three sketch maps including some symbols in perspective view.

RESULTS

On the basis of previous studies and a preliminary examination of the sketch maps, several variables were identified for tabulation. These included three on orientation: whether north was at the top edge of the card (indicated by the orientation of the text), whether the orientation was indicated in any way, and, if so, whether north orientation was indicated correctly; whether any explicit reference was made to the map scale; whether insets or locators were used; how many types of line symbols were utilized; whether any symbols were drawn in perspective or profile, instead of plan view; and several other variables, all summarized in Table 10.1.

Table 10.1 Summary of observed differences in subjects' sketch maps.

	males	*females*	*both*	*male rate minus female rate*	*male rate as ratio of female rate*
total maps	72	52	124		
north at top	32 (44%)	14 (27%)	46 (37%)	18%	1.20
north indicated	27 (38%)	17 (33%)	44 (35%)	5%	1.06
north indicated incorrectly	3 (4%)	5 (10%)	8 (6%)	-5%	0.65
incorrect norths, as % of indicated	(11%)	(29%)	(18%)	-18%	0.38
any reference to scale	3 (4%)	5 (10%)	8 (6%)	-5%	0.65
use of inset or locator	3 (4%)	0 (0%)	3 (2%)	4%	1.72
fill patterns used for some areas	5 (7%)	7 (13%)	12 (10%)	-7%	0.72
symbolization of lines and routes					
all identical	30 (42%)	17 (33%)	47 (38%)	9%	1.10
2 levels	38 (53%)	27 (52%)	65 (52%)	1%	1.01
3 or more levels	4 (6%)	8 (15%)	12 (10%)	-10%	0.57
symbolization of buildings (other than endpoints)					
none	35 (49%)	24 (46%)	59 (48%)	2%	1.02
generic	21 (29%)	16 (31%)	37 (30%)	-2%	0.98
approximate sizes/shapes	14 (19%)	10 (19%)	24 (19%)	0%	1.00
detailed footprints	2 (3%)	1 (2%)	3 (2%)	1%	1.15
perspective views	2 (3%)	2 (4%)	4 (3%)	-1%	0.86
anything in perspective	8 (11%)	13 (25%)	21 (17%)	-14%	0.66
text in ALL CAPS	17 (24%)	10 (19%)	27 (22%)	4%	1.08
stoplights indicated	3 (4%)	3 (6%)	6 (5%)	-2%	0.86

Differences in performance were assessed by comparing both the absolute and relative differences in the rates of occurrence. The percentage of males including a specific feature was compared to the percentage of females who used the feature; the ratios of these percentages were also examined to reveal differences in rate that were small in absolute terms but large in relative terms. This simple analysis was deemed more appropriate than an exhaustive assessment with tests of statistical significance for such anecdotal data.

Males were much more likely (44 percent vs 27 percent) to orient their maps with north at the top. This could reflect a number of things, including the possibility that males are more accustomed to working with traditional maps, or that females may be more likely to put the destination at the top of the map. While males had only a slightly greater rate for indicating the orientation (with North arrows or compass roses), they were much more likely to do so correctly: three males indicated north incorrectly, whereas five females indicated north incorrectly.

Females were more likely to state something about the scale of their sketch maps (10 percent vs 4 percent). However, in all eight cases where the subjects said anything about scale, it was of the "Not to Scale" variety. Females may be more attune to the need for a map scale.

Males were more likely to use inset maps (4 percent vs 0 percent), although the small number of occurrences (three) makes any interpretation of this data suspect.

Females were slightly more likely to use fill patterns for areal features (13 percent vs 7 percent).

Females were also more likely to use a variety of line symbols. While roughly equal proportions used exactly two types of line symbols—typically simple solid lines for streets and dashes or arrowheads for the recommended route—females were less likely to use only a single line style (33 percent vs 42 percent) and were much more likely to use at least three different line styles (15 percent vs 6 percent). Females may be more likely to put different types of linear features on their maps (not checked), or they may be more attune to the need to distinguish different types of routes.

All subjects symbolized the starting and ending buildings in some manner, typically with simple rectangles, stars, or views of the buildings. About half of the subjects included additional buildings as landmarks along the route, and these symbols were examined. The most common approach was to use "generic" symbols that varied little in size, shape, or orientation. About one quarter of each group used more advanced symbols that varied in size and shape, and 6 percent of each group included detailed building footprints and/or perspective views of the landmark buildings. For the landmark buildings there were no significant differences between the genders.

When examining the maps for the symbolization of the landmark buildings, we noticed that in many cases the subjects had used perspective views of the endpoint buildings and/or other non-building features, such as mountains, bicycles (to symbolize a bike path), stairs, and stop signs. This led to the collection of a separate field of information—whether any symbols on the maps were drawn in perspective. Here was where the most dramatic difference between the genders appeared. Females were more than twice as likely (25 percent vs 11 percent) to illustrate some features in this way.

SUGGESTIONS FOR FURTHER RESEARCH

Given the existing literature and our observations of these sketch maps, we suggest two specific map design issues that need more thorough study: the use of perspective views for symbolizing landmarks, and the orientation of route maps with the destination at the top.

Perspective Views

In the sketch maps considered in this study, females were much more likely to illustrate landmarks with perspective views. They may prefer to see maps with such symbols, and, in turn, might make better, more accurate use of such maps. We are planning a controlled experiment to test this hypothesis more thoroughly.

We are designing a test booklet that will contain six maps of simple street patterns and a selection of landmarks. The subjects will be asked to complete several basic map-reading exercises with each map, including typical tasks such as "Describe a direct route from the high school to the bank", "Which gas station is closest to the intersection of Broadway and Pearl?" The six maps will have similar street patterns and distributions of landmarks, but they will differ in how they symbolize the landmarks. Two will use generic "pictographic" icons, two will use building footprints, and two will use perspective views. The maps will be ordered in the test booklets such that no two adjacent maps use the same symbology. A preliminary edition of the perspective symbol map of Boulder, Colorado is reproduced here as Figure 10.3.

To determine whether one symbolization is more effective than the others, and to make the tasks as realistic as possible, the six maps will be of only two different real-world areas. Three maps based on the major streets and landmarks of Boulder, Colorado will be interspersed with three similar maps of Buffalo, New York. Only one map of each triplet will be correctly titled and traditionally oriented (north at the top); the others will be mirrored or rotated and will be given fictitious titles, street names, and landmarks. All maps will have "north" indicated as being at the top of the page. The interleaving and fictionalization are intended to minimize any learning that might occur during the course of the testing. Half of the test booklets will be prepared with the maps in reverse order, to allow verification that the order of presentation has negligible effect on subjects' performance.

The subjects will be selected from undergraduate non-geography classes at the University of Colorado at Boulder and the State University of New York at Buffalo.

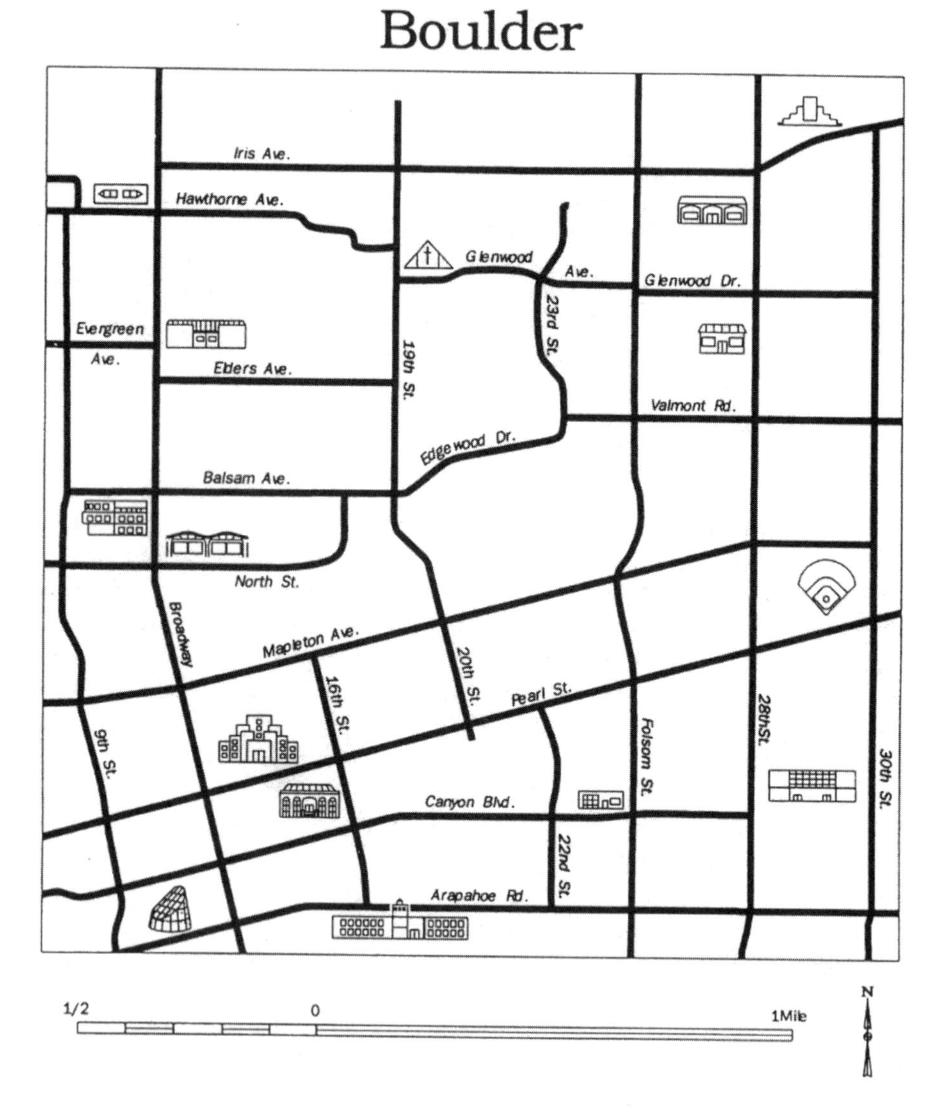

Figure 10.3 Preliminary map of a simple street pattern with perspective symbol landmarks.

The intention is that each subject will be unfamiliar with five of the maps, but familiar with the one map with the local label and the familiar perspective views of the landmarks.

While the titles of the maps and names of the landmarks will vary between the three editions of each map, the routes the subjects are asked to describe will be topologically identical, thus permitting a comparison of the effectiveness of the symbols alone.

Orientation of Route Maps

This is a more difficult design issue, at least with static printed maps, given the fixed orientation of the text. The advent of dynamic, interactive computer-drawn maps-on-demand has opened up the possibility of maps that are customized to suit particular needs. Map readers may be given the option of having their route maps oriented with north at the top, or the destination at the top edge, and all text could still be properly oriented. Indeed, many of the in-vehicle navigation systems currently reaching the market give the user the option of either orientation; while it might be of interest to examine which users prefer the different orientations, it also reveals the fact that the issue may be moot as such dynamic maps become increasingly common.

SUMMARY

The existing literature on gender differences in map reading abilities leaves many questions unanswered. It is widely agreed that there are relevant biological and environmental differences between the gender groups, but the extent to which these are responsible for real differences in map reading is still not clear. As cartographers, we may wish to design maps for successful use by both groups; further study is needed to determine how exactly we might achieve this. Two promising areas are the use of perspective views for representing landmarks and the use of alternate map orientations.

ACKNOWLEDGEMENTS

The authors acknowledge Christi Seemann and Jonathon Kopp for their assistance in gathering background literature and the preparation of test maps. At the time this paper was written, Barbara Buttenfield was a faculty member of the State University of New York-Buffalo.

REFERENCES

Antes, J.R., McBride, R.B., and Collins, J.D. (1988). The effect of a new city traffic route on the cognitive maps of its residents. *Environment and Behavior*, 20, 75-91.

Beatty, W.W. (1988). The Fargo Map Test: A standardized method for assessing remote memory for visuospatial information. *Journal of Clinical Psychology*, 44, 61-67.

Beatty, W.W., and TrÜster, A.I. (1987). Gender differences in geographical knowledge. *Sex Roles*, 16, 565-590.

Brown, M.A., and Broadway, M.J. (1981). The cognitive maps of adolescents: Confusion about inter-town distances. *Professional Geographer*, 33, 315-325.

Chang, K-T., and Antes, J.R. (1987). Sex and cultural differences in map reading. *The American Cartographer*, 14, 29-42.

Gilmartin, P.P. (1986). Maps, mental imagery, and gender in the recall of geographical information. *The American Cartographer*, 13, 335-344.

Gilmartin, P.P., and Patton, J.C. (1984). Comparing the sexes on spatial abilities: Map-use skills. *Annals of the Association of American Geographers*, 74, 605-619.

Maccoby, E.M., and Jacklin, C.N. (1974). *The psychology of sex differences*. Stanford: Stanford University Press.

McGee, M.G. (1979). *Human spatial sbilities: Sources of sex differences*. New York: Praeger Press.

Morrison, A. (1974). Testing the effectiveness of road speed maps and conventional maps. *The Cartographic Journal*, 11, 102-116.

Phillips, R.J., and Noyes, L. (1982). An investigation of visual clutter in the topographic base of a geological map. *The Cartographic Journal*, 19, 122-130.

Self, C.M., Gopal, S., Golledge, R.G., and Fenstermaker, S. (1992). Gender-related differences in spatial abilities. *Progress in Human Geography*, 16, 315-342.

Stumpf, H., and Klieme, E. (1989). Sex-related differences in spatial ability: More evidence for convergence. *Perceptual and Motor Skills*, 69(3), Part 1, 915-921.

11

Design Issues to be Considered When Mapping Time

Irina Vasiliev

Department of Geography,
State University of New York College at Geneseo

INTRODUCTION

Cartographers make maps to show where things are and what happens in different places. Maps are used to locate objects and places, and to show the relationships of activities and events to geographic spaces. Maps are graphic representations of spatial relations of physical environments and of activities in space. Often, the information portrayed on a map is not simply spatial—where things are —but also incorporates a temporal dimension—when these things are. Human, animal, and environmental activities and events in geographic space are usually concerned with a change of some kind: movement from place to place, or differences of phenomena from time to time in a place. Even when there is no change being represented, and a map shows simple location of objects in space, there is always a time at which this location of objects is true. This implicit temporal dimension is taken for granted and not emphasized. And, when the location of these objects changes, the map is no longer useful for its original purpose.

If there were no time, there could be no change, so in attempting to understand changes in various activities, the temporal dimension must be included in any discussion or representation of these activities. Therefore, there is a representation of time on any map that is concerned with change. From this point of view, spatial information must be termed spatio-temporal information. However, just as there are many different types of spatial information, there are also different types of temporal information. And, just as there are different methods of graphically representing spatial information, the same is true for the temporal aspect of this information. The representation of different types of spatio-temporal information is dependent on various types of graphic symbols.

TIME, SPACE, AND MAPS

Space is easily understood. It is the two or three dimensions that define where things are. The relationships between elements in and to space are what define geography. Time, however, is a more difficult concept because it has no physical characteristics. We cannot grasp it; we can only know that it exists, and even this knowledge is based on metaphors or analogies, not on a concrete objective substance. The metaphors that we use for the understanding of time depend on change or movement. The most basic of these is the movement of the sun and other celestial bodies across the sky. These movements fostered the development of the measurement of time, and therefore, clocks and their system of time measurement have become the definition of time for us. (For more on the history of time measurement, see Aveni, 1989; Fraser, 1987; and Landes, 1983.)

However, there is no single type of time. We find time useful in many ways, and it is these different uses of time that are of interest here. Different types of time present different information about our world and what we want to know about it. To understand how geographers have used time, this research examined many maps to see what kinds of temporal information were represented. This examination resulted in five categories of temporal information:

1) *Moments*: the dating of an event in space;
2) *Duration*: the continuance of an occurrence in space;
3) *Structured Time*: the organization or standardization of space by time;
4) *Time as Distance*: the use of time as a measure of distance; and
5) *Space as Clock*: spatial relations as a measure of time.

Each category contains certain graphic symbols that are used to represent the temporal information. These symbols are based on the point-line-area symbols with which cartographers show map information, but are extended into the spatio-temporal dimension by various graphic means. These symbols are summarized in the figure at the conclusion of this chapter.

1) Moments

This is the simplest of time forms to be found on maps. A moment in time is generally expressed using a system of time measurement. The most accepted system of time measurement is our familiar one that consists of 60 seconds per minute, 60 minutes per hour, 24 hours per day, 7 days per week, about four weeks or 30 days per month, 12 months per year, and the start of the system, Year One, the year of Christ's birth. Other systems of time measurement exist, but these are not in general use, and they all still depend on the day as the basic unit of measure (Zerubavel, 1989). The differences lie in the divisions of the day and the number of days used to express larger units of time.

An event is dated by noting the time (as defined by the system of time measurement) at which it occurred. This date is then recorded on a map using a time label in the location of the occurrence. A description of the event must be represented on the map as well. The type of event dictates the cartographic symbology used to describe it. A point symbol would indicate the location of the event, thus describing its placement in space. Other symbols or text would be used to describe the event itself; for example, a symbol of crossed swords is used to indicate a battle with the time label next to it, indicating the time of the battle's occurrence. Text is added giving the name of the battle or the armies involved.

In this category of time on maps, time labels are static. They represent a moment of time; they stop time along its continuum to show an event in one place, at one time. However, time labels as graphic symbols can be used to show more complex temporal information. They may be attached to line symbols, as in roads or rivers, and to area symbols, as in floods or urban sprawl.

2) Duration

Events occur in a time and a space. The continuation of this occurrence is its duration, and it is associated with change. On maps, duration may be viewed in several ways: a) how much time it takes for an event to occur; b) what happens within a certain period of time; and c) what changes from one time to another, in specific time intervals. As in the previous category, a known system of time measurement is employed in understanding each of these manifestations of duration.

a) Maps that show how long it takes for an event to occur in a space have a starting point and lead the map reader to the end with a series of graphic symbols that emphasize the changes associated with the event. For example, the flight path of an airplane in trouble would be mapped using a base map of the area over which the airplane flew, lines or arrows indicating the direction of flight, and time labels at points indicating the start of trouble, times at which the flight pattern changed, and the time of the crash-landing. The spatial extent of the event is evident by the area covered by the depiction of the flight path. The temporal duration is indicated by the difference between the starting and ending times of the time labels. The mapping reveals the process and duration of the event through the use of various graphic and temporal symbols, such as extended time labels at arrows and points.

b) On occasion, the time period of an event and the space in which it occurs are known. In this case, the map shows the changes that occurred within a certain period of time. Maps of this sort are used often in textbooks and atlases. A significant period of time is chosen and the events that make this time period significant are mapped. Take, for example, a map of a war. The time period of a war is known. The information portrayed on the map shows what happened during the course of this time period: the battles that were fought, the movements of troops, the positions of generals at different times during the war, and so on. Again, the graphic symbols used for the representation of this information include arrows, lines, points, and time labels to indicate direction and extent of movement, and temporal and spatial locations. In this case, the duration of the event is given, and the map illustrates the processes within this time period.

c) The representation of the continuation of a process can be shown at regular time intervals. Maps of this type of duration depend on a sequence of maps. It is difficult to represent many spatial and temporal changes of areal information on one map. A sequence of maps that uses the same base map, but shows the areal changes that occur in regular temporal intervals is useful for events that occur over a large area and are themselves areal in nature, such as forest fires and floods. To show the changes in cloud cover over an area, for example, the group of maps, one per day, would show the location of the cloud cover on each of the days in the series. Not all duration is areal in nature. The movement of population or diffusion of ideas can be represented using points with time labels to show how long something was in one place and lines and arrows with time labels to show how long and in what direction movement occurred. A good example of this is the centrogram that shows the movement of the centre of population from one census year to another for a century or so. Time labels at points that are separated by a regular time interval, in this case the 10 year census interval, indicate the assumed 10 year duration at each point of the centre of population.

Mapping duration results in the simplification of the continuous nature of change into discrete moments, points frozen in time, in order to better illustrate in a static manner the continuance of the event being mapped. Trying to show any change or movement in a static manner reduces to the stopping of motion in time and then representing the whole of the process by its parts. With increasingly available technology for cartographic animation, maps of duration can better represent the changes or continuance of events. The graphic symbols involved in this representation, however, stay the same. Arrows, lines, points, areas, and dated labels are still the more useful cartographic conventions.

3) Structured Time

Organizing space seems to be a basic human activity, one done in a number of ways: property ownership, agricultural or hunting potential, common land administered by bureaucracies made for the purpose, and the scheduling of events in appropriate places. All of these depend on some temporal dimension to be successful. Twenty-year mortgages, planting seasons, highway construction budgets, and train schedules are just a few examples that depend on time to organize their spaces. And for time to be useful in these kinds of endeavours, time itself needs to be organized.

The standardization of time came about because of the need for large groups of people to be synchronized in their time measurement system. Not only did they have to use the same system, but they had to be coincident with it. The hands of clocks in different locations had to point to the same numbers at the same time. Standardizing time was a geographical problem as well as a temporal one. Global space had to be divided in such a way that synchronized, coincident time would actually make sense to the people required to use it.

The idea of dividing the world into equal time zones has existed for hundreds of years before the economically mandated standard time zones of the late 1880s came into being. Anyone with an understanding of the world as a rotating globe could see that the sun could not be in the same place in the sky everywhere at once. However, it was only with the increase in the speed of transportation that it became necessary to organize time. In effect, this was meant to regulate the sun, to make sure that the sun was in one place for a certain length of time before moving over to the next place, where it was to stop for the same length of time. (For more on Standard Time, see: Bartky, 1989; Holbrook, 1947; and Stephens, 1989.)

Soon, the sun was no longer necessary for clock time, and factors such as economic dependence and business efficiency dictated the boundaries of standard time zones. Time zone boundaries are very much like many internal political boundaries; they exist, but rarely in a true physical form. Fences and guard stations delineate some countries from others. But states, provinces, counties, and other administrative areas do not usually need the gross physical structures to manage the boundary crossings. Signs along highways are enough to let one know that a county or state line was crossed. The same is true for time zones, although less often. A traveller is not always informed of the time change that had occurred as the vehicle crossed from one zone to another.

On maps, however, these same political boundaries that have no physical manifestations on the ground may be drawn in bright colours or bold symbols. Time zone maps do the same. At the turn of the century, when time was being structured to organize people's activities, maps were accompanied with instructions that let the user in on the understanding of standard time and its benefits. The time zones were accentuated by colour or heavy lines and labels, often in the form of a clock face that indicated the time in each zone, one hour apart from its neighbour. Instructions on the map or in the margins told the user how to read the map and why standard time was important to society.

Other forms of structured time on maps use time to organize space in ways other than to synchronize clocks. Routine chores that are performed over space are scheduled and organized with the help of maps. For example, the United States Post Office in the 1880s produced maps that showed the frequency of mail delivery on different routes. A map of routes in a state would show the post offices, the roads linking them marked with distances, and, in colour codes, how many times a week the mail was delivered on each route. Colour could be replaced with different line forms to indicate the same information. Dotted lines, dashed lines, and dot-and-dash combinations would serve the same purpose. Again, this kind of temporal information uses time labels to indicate time, based on a structured temporal system. The labels indicate the frequency with which an activity is performed. The standard measures are days and weeks, and the time intervals between deliveries are based on these units.

Organized space is essential to efficient human interaction. And space cannot be well organized if its time is not. Maps that combine these two elements are very useful in the execution of spatial organization.

4) Time as Distance

Using time as a measure of distance is a natural process. Long before there were clocks to subdivide the day, the rising and setting of the sun served as a rhythmic marker to human activity. Native peoples of North America, for example, used the number of "sleeps," night rest periods, to tell how long journeys were from one place to another. In modern times, we often indicate distance by how much time it takes to get somewhere: 15 minutes to the mall, four hours to New York City, and so on. This use of time for measuring space has been translated to cartographic form. An 1837 Iowa Indian map (Lewis, 1987) of tribal migrations between the Mississippi and Missouri Rivers shows the routes with dots—each dot representing an overnight encampment. Counting the dots gives the distance in number of days of travel that each migration route takes. A modern version of this idea is the ubiquitous road atlas map that shows the distances between major cities and along Interstate Highway segments in both miles and time—hours and minutes. Time and distance are synonymous along the linear dimension.

Areas may also be represented with time as the distance measure. The use of isochrones—lines of equal temporal distance from a central point—is helpful in seeing how far someone or something gets in space in equal amounts of time. The diffusion of insects, diseases, and ideas over an area can be done using isochrones. Each line demarcates the extent to which the phenomenon had spread in some time period. The lines may be drawn at equal intervals, like contour lines on a topographic map, or they may indicate significant distances, in irregular intervals. In this latter case, the line would be labelled with a time that tells when the phenomenon reached the area. These isochrones are drawn on a base map that closely resembles reality. Geographic space stays true while the isochrones meander around barriers to represent the temporal distance that was traversed.

Another use of isochrones distorts geographic space to fit temporal distance. In this case, the isochrones resemble a bull's eye, with concentric circles, at equal temporal intervals, emanating from a central point. The geographic space is pushed and pulled to fit the concentric circles.

Chronological representations of this sort, however, are very specific to the starting point of the time on the map, the centre of the concentric circles. Distance in time can be measured only from the centre point to all other points. Nothing can be inferred about distances in relation to other starting points. To calculate distance on multi-use maps of this type, one needs not only information on travel times from all points to all other points, but one must also be able to recalculate the map projections to conform to each starting point. Maps of this type can be useful, however, and, with the use of computers, may well be worth producing.

5) Space as Clock

In this last category of time, space is used as a measure of time, either as a clock or a calendar. A map of the spatial relations of temporal objects incorporates certain cultural understandings of both space and time that we do not confront on a daily basis. A map that is a calendar is a useful item in some societies and a frivolous, decorative work in others.

For pre-conquest Meso-American peoples, time was an important shaper of space. Here, spatial relations were used as a measure of time. Time organized the space which, in turn, was used to express its temporal influence. The Maya were preoccupied with time cycles and their correct arrangements. It was important to the Maya to be able to predict the sun's movements precisely. Their fundamental unit of time was the day, the word for which meant also "sun" and "time." The day was not a unit measure of time, but was the "manifestation of the cycle of the sun . . . time is the sun's cycle itself" (Aveni, 1989: 193). The cardinal directions of east and west, then, became not only where the sun came from and where it went, but were the moments in time when the sun changed direction in order to be able to repeat its performance. Time and space were one and the same—where the sun went at the end of the day was also the time at which it did that. These ideas were carved into the stone glyphs used by the Maya in their written language. A four-pointed floral design symbolized day, time, and sun, with each of the points designating the extreme points of the sun on the horizon. This representation of time on the glyphs extended to Mayan architecture and the organization of their space according to the sun's movements, creating a calendar from their structures. The sun/time structured their space at the same time as it served to represent time. (For more details, see Leon-Portilla, 1988.)

The Aztecs and Incas in later pre-conquest Meso-America continued to use the Mayan system of time/space measurement. An excellent example of this inseparability of time and space is shown on the Feyérváry screenfold, a document from central Mexico in the post-classical period, in which a map of the world is drawn according to the ritual calendar and its temporal ordering of space. Rather than being geographically correct, as we would understand it, the map is temporally correct. Each petal of the design shows a variety of information, such as year, colour, and deity. Surrounding the petals are symbols for each day of the 260-day Meso-American year (Harley, 1990: 7-9). It was more important to represent space as a calendar with time symbology than it was to get from one place to another. The map of the world was the calendar; the calendar was the map.

A three-dimensional example of a literal "space as clock" should be mentioned in this section. This is an object called a "Globe Clock." Clock makers have spent much time incorporating various playful and representational elements into their designs of clocks. Traditional Black Forest cuckoo clocks use birds and people and animals to tell us the hours, while monumental clocks include entire parades of creatures, prophets, devils, and angels to let us know that time is passing. A variation of this is a world globe around the equator of which is built a clock to tell us the time at different parts of the world, in real time.

A modern-day version of a globe clock can be found in electronic atlases, a good example of which is *The Small Blue Planet: The Electronic Satellite Atlas*, from Now What Software in San Francisco. A portion of the atlas is a chronosphere, an interactive map of the world on which is depicted the 24-hour day, using dark and light areas. (This map is a very nice example of animation.) The map reader may tell the time and day "from where you are to where you want to go. It registers accurate time zone information for the world map as well as automatically adjusting for North American Daylight Savings." By using a view of the Earth from outside of it, and by using sunlight as the indicator of time passing (as in real life), the map reader can gain a good understanding of the relationship between time, the Sun, and the Earth; this is something that is fascinating to those viewing it, especially in the years since the standardization of time and the industrialization of our lives. It is a view of the world that brings home to the viewer just how far from nature we have strayed. It is also a view of the world that very much shows how space can be used as a true clock.

CONCLUSION

As we have seen, cartographers have been mapping temporal information along with the spatial dimensions for many years. Understanding the kind of time that needs to be represented on a map helps the cartographer in designing and using graphic symbols to portray this temporal information. Figure 11.1 summarizes most of the symbolization used to represent spatio-temporal information.

Time labels at points serve to place the "when" of an event in a place within a standard system of time measurement. The points themselves indicate where an event occurred, both in space and time.

Lines and arrows are useful in indicating both spatial and temporal directions of events. In conjunction with time labels, these arrows and lines may be used to show the duration of an event. Duration can also be represented by the use of sequential maps on which the changing phenomenon is shown at different times, one time to a map.

Special symbols can be devised to represent standard time. Clock faces are useful for time zones. Different colours or textures applied to points or lines serve to represent varying frequencies of scheduled events.

Isochrones are both temporal and spatial in extent. They are useful in measuring distance in terms of time. Their intervals are indicated with time labels, when the intervals are irregular, or they themselves indicate a regular temporal interval, previously established as the temporal measure with which distance between points is calculated.

And, finally, maps and globes can be calendars and clocks, showing spatial and temporal relations based on certain cultural understandings of these things, as in the Meso-American case, where the map is also a calendar. In the form of both a globe clock and a chronosphere, the map user is brought back to the basic understanding of the relationship between the Earth, the Sun, and time.

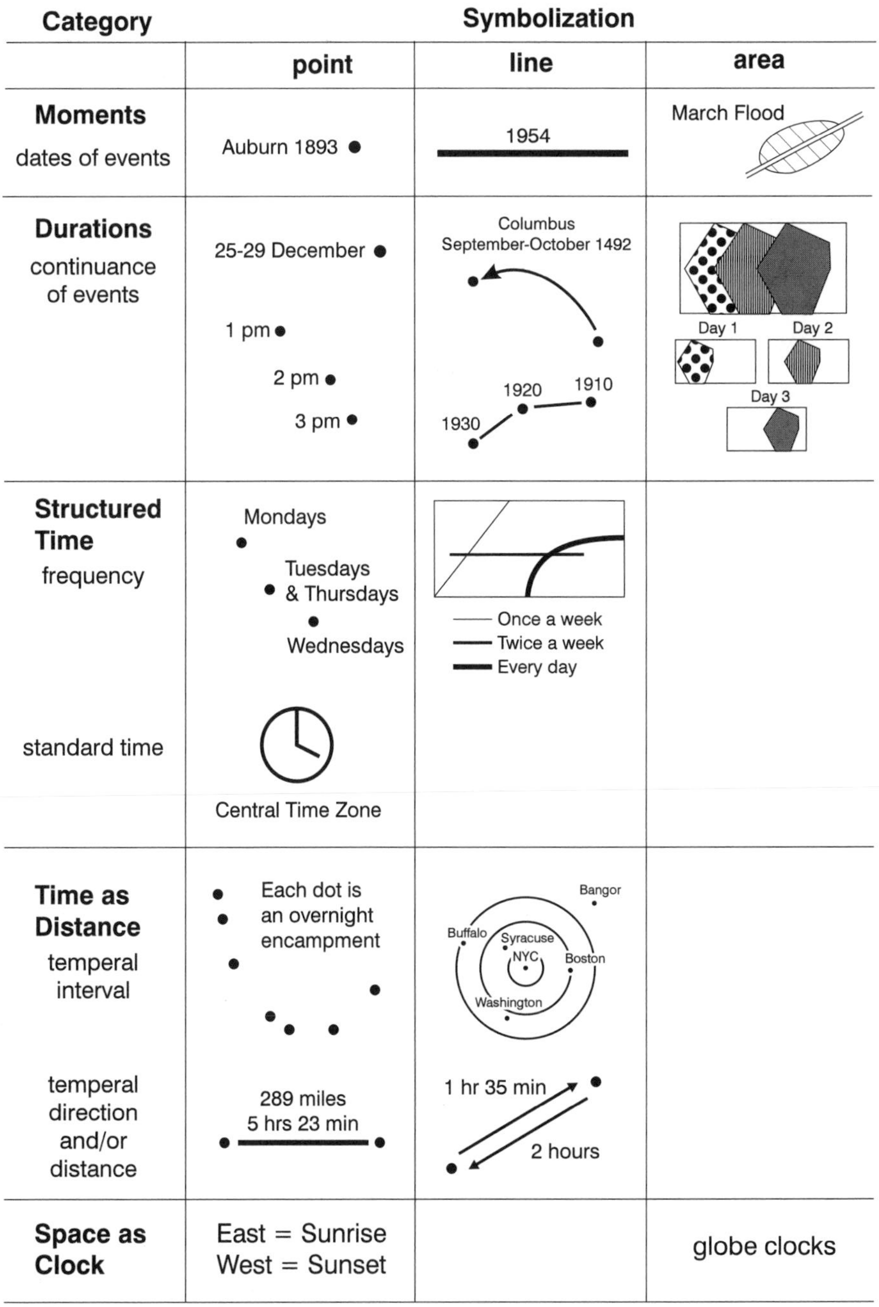

Category	Symbolization		
	point	**line**	**area**
Moments dates of events	Auburn 1893 •	1954	March Flood
Durations continuance of events	25-29 December • 1 pm • 2 pm • 3 pm •	Columbus September-October 1492 1920 1910 1930	Day 1 Day 2 Day 3
Structured Time frequency	Mondays Tuesdays & Thursdays Wednesdays	Once a week Twice a week Every day	
standard time	Central Time Zone		
Time as Distance temperal interval	Each dot is an overnight encampment	Bangor Buffalo Syracuse NYC Boston Washington	
temperal direction and/or distance	289 miles 5 hrs 23 min	1 hr 35 min 2 hours	
Space as Clock	East = Sunrise West = Sunset		globe clocks

Figure 11.1 Symbolization used to represent spatio-temporal information.

REFERENCES

Aveni, A. (1989). *Empires of time. Calendars, clocks, and time.* New York: Basic Books, Inc.

Bartky, I.R. (1989). The adoption of standard time. *Technology and Culture,* 30(1), 25-56.

Fraser, J.T. (1987). *Time, the familiar stranger.* Amherst, MA: University of Massachusetts Press.

Harley, J.B. (1990). *Maps and the Columbian encounter.* Milwaukee: University of Wisconsin.

Holbrook, S. (1947). *The story of American railroads.* New York: Crown Publishers.

Landes, D. (1983). *Revolution in time.* Cambridge, MA: Harvard University Press.

Leon-Portilla, M. (1988). *Time and reality in the thought of the Maya.* Norman: University of Oklahoma Press.

Lewis, G.M. (1987). Indian maps: Their place in the history of plains cartography. In F.C. Luebke, F.W. Kaye, and G.E. Moulton (Eds.), *Mapping the North American plains* (pp. 63-80). Norman: University of Oklahoma Press.

Stephens, C. (1989). 'The most reliable time': William Bond, the New England railroads, and time awareness in 19th century America. *Technology and Culture,* 30(1), 1-24.

Zerubavel, E. (1989). *The seven day circle: The history and meaning of the week.* Chicago: University of Chicago Press.

Re-examining the Cartographic Depiction of Topography

Roger Wheate

Faculty of Natural Resources and Environmental Studies, University of Northern British Columbia

INTRODUCTION

The visual depiction of the land surface has long been one of the cartographer's most challenging tasks, since topography poses several complications not present in the symbolization of other map elements. Most fundamentally, it contains a third dimension that varies continuously, as opposed to the general symbolization of points, lines and areas, which do not. Furthermore, topography has several mappable elements, such as slope, shape and height, which yield multiple methods of depiction.

The continuous nature of topography promotes a selection of depiction techniques which results in visual domination of the cartographic image by the depiction method chosen. This has varied over the centuries according to the technologies available and the map purpose. It is possible to recognize three such eras in the history of topographic mapping:

1. *Renaissance era (1450-1800)*

Topographic rendering was limited to stylized drawings based upon observation rather than measurement. These included 'sugar-loaf' mountains, hairy-caterpillar hachures and bird's eye view sketches. This era ended at the beginning of the nineteenth century as surveying instruments enabled the more accurate depiction, first of slopes through hachures, and then elevation through contours.

2. *Analogue (Traditional) era (1800-1980)*

By the middle of the 19th century, contouring, that is isarithmic lines of equal elevation, had firmly replaced hachures as the standard technique, though its supremacy as a means of encoding quantitative data about the terrain remains partly offset by its shortcomings in 'imageability', or the ability of map users to visualize the surface. This is variable depending on the complexity of the terrain, contour interval, scale and users' experience with topographic maps.

The visual drawbacks of contour renditions led to other methods being added, such as hypsometric tints and shaded relief. The former is generally limited to smaller scales and creates design dilemmas in tint gradation and contrast: tints can conceal other cartographic information and provide no extra information about the landscape. Shaded relief is the most visual and least abstract method of terrain depiction, since it simulates an illuminated three-dimensional topographic model. However, in traditional cartography, it is expensive to render and requires a high degree of artistic ability and time investment.

3. *Digital era (1980-present/future)*

The rapid development of digital mapping technology and the encoding of cartographic data, where available, have created a store of topographic data and modified the constraints and relative advantages between the different depiction techniques. While contouring can be reproduced at different scales without re-scribing due to generalization algorithms, many of the traditional problems associated with hypsometric tints and shaded relief have been lessened or removed. Most geographic information systems can manipulate digital elevation data to produce shaded relief and layer tints, as well as create further options such as slope and aspect mapping (see Figures 12.1a-d).

DESIGN CONSIDERATIONS

In the 1970s cartographic research compared relief depiction methods, aiming mostly at how they affected (negatively) the perception of other cartographic information (e.g. Phillips et al., 1975). The general conclusion was that although shaded relief (and layer tints) added a more easily visualized component, this was offset by an increased amount of background 'noise' that could conceal other symbols, particularly in areas of darker shading or tints. In contrast, Castner and Wheate (1979) hypothesized that shaded relief, nevertheless, contained further significant advantages based on its continuous nature, versus the discrete presentation of contour surfaces. Their hypotheses were:

a. shaded relief enables a faster interpretation and visual processing of the relief surface;

b. shaded relief can act as a meaningful structure in searching for terrain-related and terrain elements;

c. shaded relief can act as an organizational background or reference for all map elements, even those without a direct association with the terrain.

Test maps depicted identical design except that relief was portrayed either with contours or shaded relief. A third map type with no relief representation other than selected spot heights was used as a 'control' for the effect of relief renditions as visual noise. Subjects were asked to perform a variety of map use tasks, involving map reading, analysis, or interpretation. These were further subdivided according to whether the targets represented elements not easily related to the

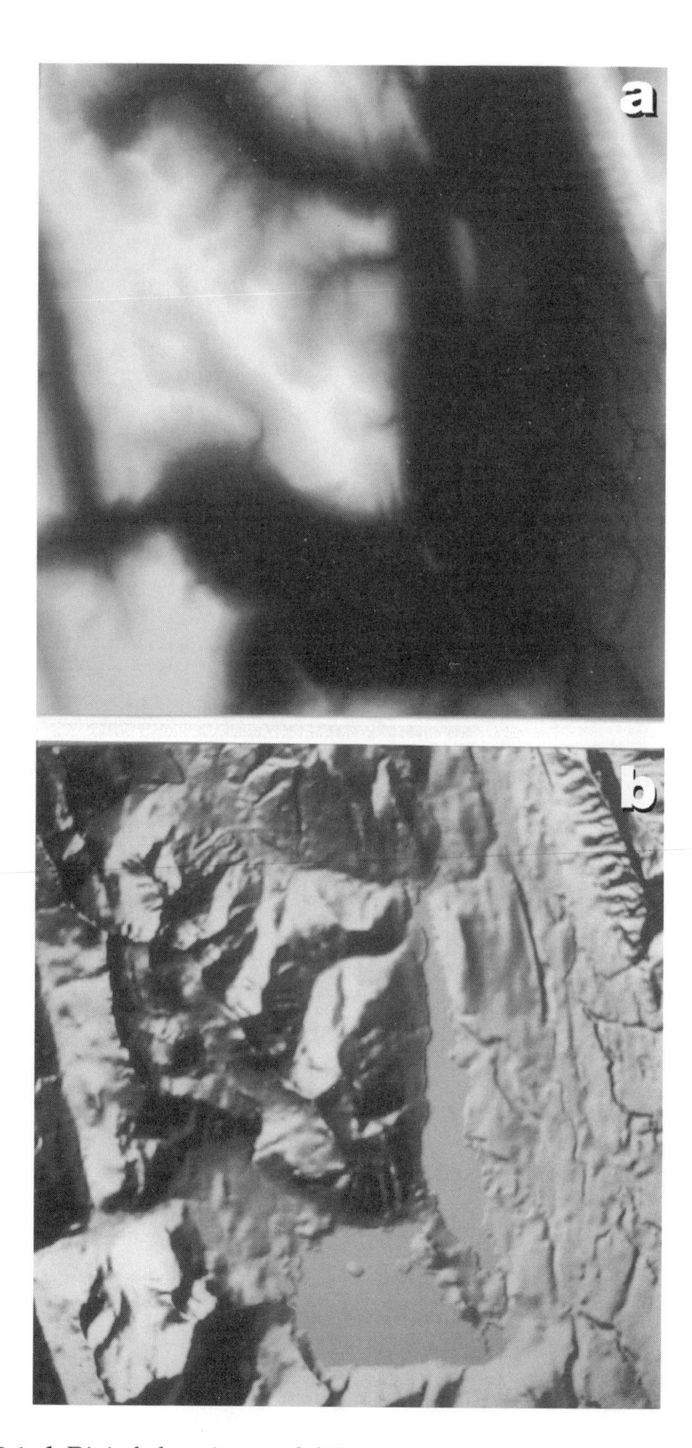

Figure 12.1a-b Digital elevation model, Kananaskis Valley, Alberta (elevation and shaded relief).

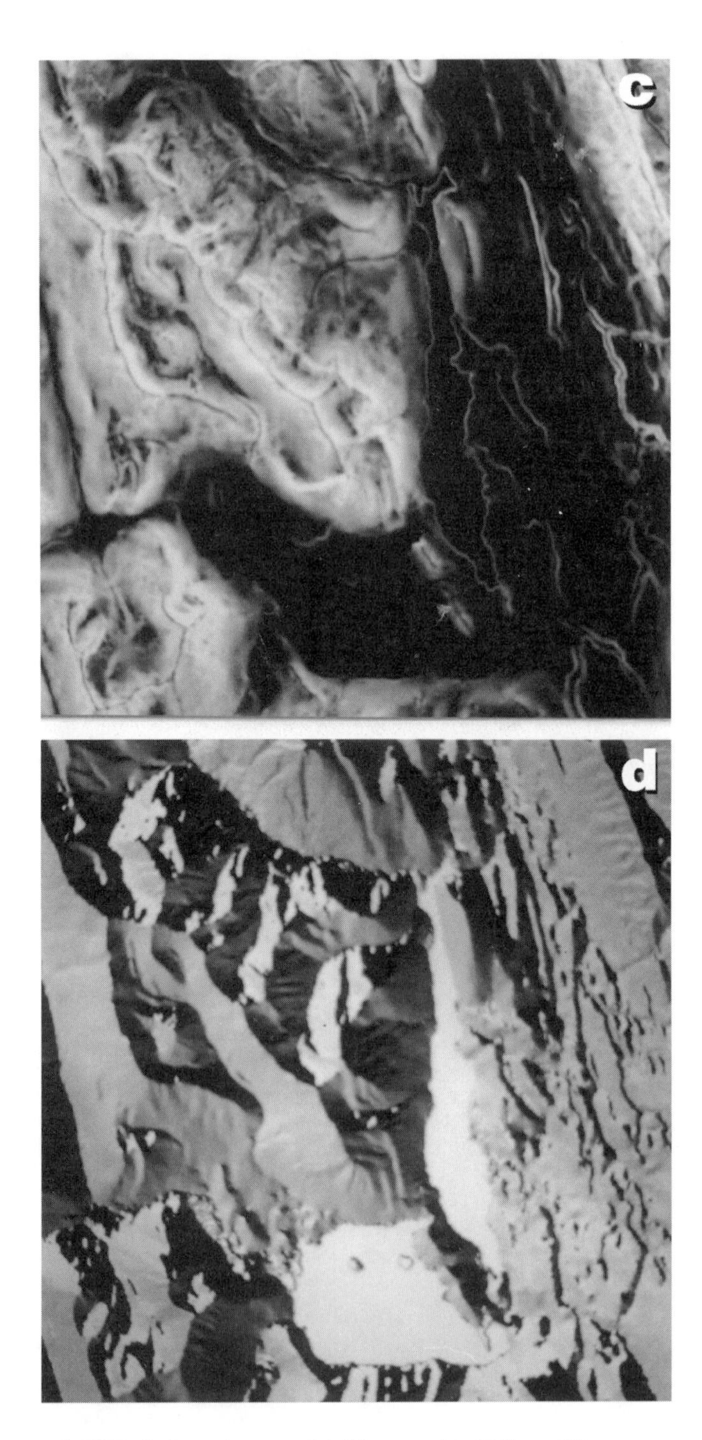

Figure 12.1c-d Digital elevation model, Kananaskis Valley, Alberta (slope and aspect).

terrain (e.g. dams, campgrounds), elements associated with the terrain (e.g. fire lookouts), and the terrain itself (heights and slopes).

In the case of the latter two, search times were significantly faster with the shaded relief map versions than with contours, although tasks requiring absolute height estimation were not included. When subjects were asked to relocate all items, terrain-related or otherwise, they did so faster with shaded relief than with contours.

As a result, the three hypotheses were accepted. Shaded relief, in addition to providing a more imageable terrain surface, also enabled non-terrain tasks to be completed more efficiently despite the implicit higher level of noise presented by a continuously varying background tone. The authors explained this with reference to eye movement studies and the principles of visual perception: shaded relief could be processed with peripheral vision, while contour line work required focal attention. In figure-ground terms, contours compete with other point and line symbols for focal attention, while shading acts as a background on which other map elements may lay. As a result, the shaded relief rendition could lead viewers' eyes ahead, while still mentally processing previously visualized areas.

This did not work immediately with symbols that lay in darker (SE facing slopes) areas, as they initially could not be perceived peripherally. However, once located, the fabric of the relief rendition acted as a 'memorable'structure around which the map user's 'mental map' could be organized. We concluded that while large scale topographic maps may be used by the scientific community at large for specific purposes that require numeric height information, other types of map users and maps, particularly recreation-oriented, would benefit from the inclusion of shaded relief wherever possible. These conclusions have become even more pertinent in the 1990s, with the expansion of recreational activities and associated map products.

PRODUCTION CONSIDERATIONS

The problems in finding a practioner of shaded relief at reasonable cost limited the application of shaded relief in traditional cartographic map production. As a result, some maps displayed a modified type of shading, where areas are either shaded or not, with only two grey tones. This kind of rendition was partially successful in giving the map user an idea of where steeper and higher lands lay, but lost the organizational aspect of the continuously shaded version.

In the digital era, some production problems of shaded relief have been removed; however, some vector based mapping systems may have difficulty in importing raster based shading images. The availability of digital elevation data is uneven in scales and geographic distribution. Production agencies might not be able to afford either the digital files, storage space, or the software required to produce automated shading. Where they are accessible, shading programs can be extremely versatile in their ability to select lighting source zenith and azimuth, and can also be combined with other images, or data layers, such as slope and aspect (Moellering and Kimerling, 1990).

Digital elevation data is gradually becoming more available for a variety of mapping scales. In (western) Canada, provincial data are produced at 1:20,000 and 1:250,000 scales, with larger and smaller scales supplied by municipal and federal sources respectively. Smaller scale mapping can draw on sources such as the Digital Chart of the World at 1:1 million and specialized products such as Mountain High Maps, a collection of artistically shaded backgrounds of mountain areas for use with desktop mapping packages such as Aldus Freehand and produced by Digital Wisdom Inc.

PHILOSOPHICAL CONSIDERATIONS

Traditionally, the widespread use of contours could be further justified because the elevation information encoded in them formed the source for all other rendition techniques, including shaded relief, layer tints, illuminated 'Tanaka' contours, perspective blocks, and orthographic relief. Additional details might be interpolated into a shaded relief rendition between the contour interval, but only by inference from the contour information, or re-examination of aerial photographs.

The digital database, in contrast, consists of either a regular grid of height values in raster systems, or a series of values and vectors in triangulated irregular networks in vector systems. A digital elevation model (DEM) can be produced through a number of methods, which include manually digitizing contours from existing maps, stereo-photogrammetry or applying photogrammetric principles to stereo digital images from satellite or airborne sensors. The DEM now represents the 'parent' or source for all derivations including elevation contours and shaded relief. Unless the DEM was produced by digitizing contours, shaded relief may actually contain additional information compared to contours. It is also possible to induce elevational information from shading by applying 'shape from shading' algorithms (Horn and Brooks, 1989).

PERSPECTIVE VIEWS

Three dimensional perspective views have been used to depict landscapes since the Renaissance era through bird's eye views. Manually, they are extremely time-consuming to produce, with a high degree of artistic license, interpretation, and little flexibility in the view selected. Digitally, these renditions are limited more by the availability and cost of software and data, and can incorporate a range of data layers 'draped' over topography. The image designer may select the viewer's azimuth and zenith angles, horizontal and vertical points of observation. These provide a powerful tool for managers in resource management and for recreational users in visualizing a targeted scene, but have not so far become a standard feature in map production. This may be due to selected mapping software having limited GIS functions, the uneven availability of DEM's, or lack of awareness of this potential by map producers and users.

Two examples are illustrated here, reproduced in monochrome. Figure 12.2a shows Thematic Mapper (TM) data draped over a 100 metre resolution DEM for Yoho National Park, B.C., and can be compared with Figure 12.2b, the same park area in a manually drafted perspective view. In both, lighter areas portray bare rocks, while forested areas are darker. Parks planners and the public alike find these views helpful in visualizing the landscape.

Figure 12.3a shows the Ann and Sandy Cross Conservation Area, near Calgary, Alberta. This rolling area has been ranched for a century, but was recently donated to the Nature Conservancy for primary use in educating young groups about natural environments. Topography was not incorporated in Figure 12.3a since only 50 foot contours were available from 1:50,000 maps. Monochromatic reproduction in the brochure could have caused confusion between either contours, or shading and non-topographic symbols. After visitors complained that they didn't realize the area was so hilly, we added a DEM draped with an aerial photograph (Figure 12.3b). Despite the low resolution of the DEM, visitor response was very favourable to this visual element, and a colour version is planned for display and incorporation into future brochures.

CONCLUSIONS

GIS technology has enabled new possibilities in topographic depiction on general purpose maps, whose potential has been realized only recently. High quality shaded relief renditions should be incorporated into general map design given the perceptual advantages described in the 1970s and the obvious visual appeal, ideally in combination with the quantitative information embedded in contours. Map sheets can be augmented with one, or a set of perspectives to further enhance the map as a tool for graphic communication to use the language of the 1970s, or 'visualization' as the term of the 1990s.

REFERENCES

Castner, H.W., and Wheate, R.D. (1979). Re-assessing the role played by shaded relief in topographic scale maps. *The Cartographic Journal*, 16(2), 77-85

Horn, B.K.P., and Brooks, M.J. (Eds.) (1989). *Shape from shading*. Cambridge, MA: MIT Press.

Moellering, H., and Kimerling, A.J. (1990). A new digital slope-aspect display process. *Cartography and Geographic Information Systems*, 17(2), 151-159.

Phillips, R.J., De Lucia, A., and Skelton, N. (1975). Some objective tests of the legibility of relief maps. *The Cartographic Journal*, 12(1), 39-46.

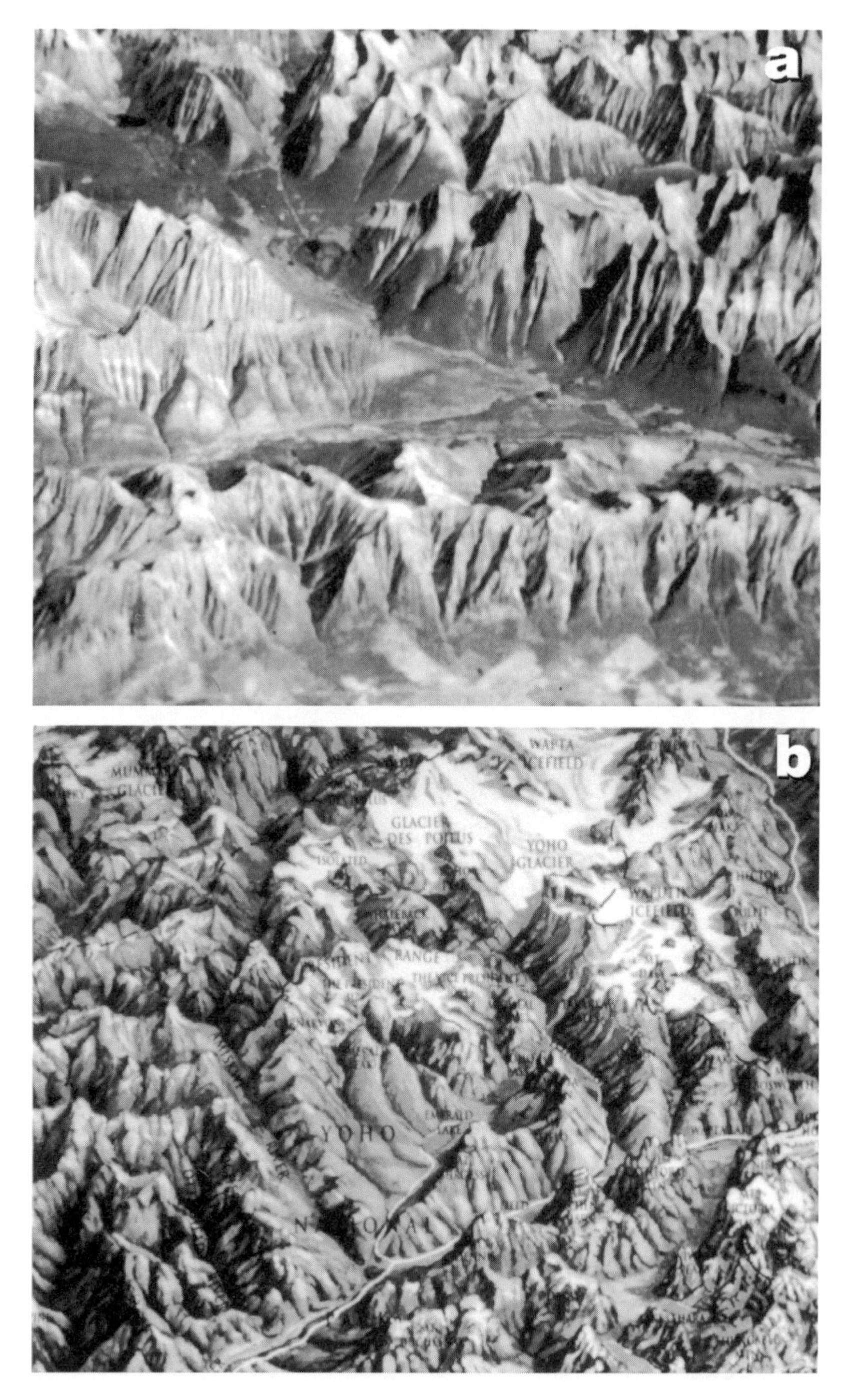

Figure 12.2 Perspective views of Yoho National Park, British Columbia (reprinted with permission from "Panorama Map of the Canadian Rockies", Altitude Publishing Canada Ltd.).

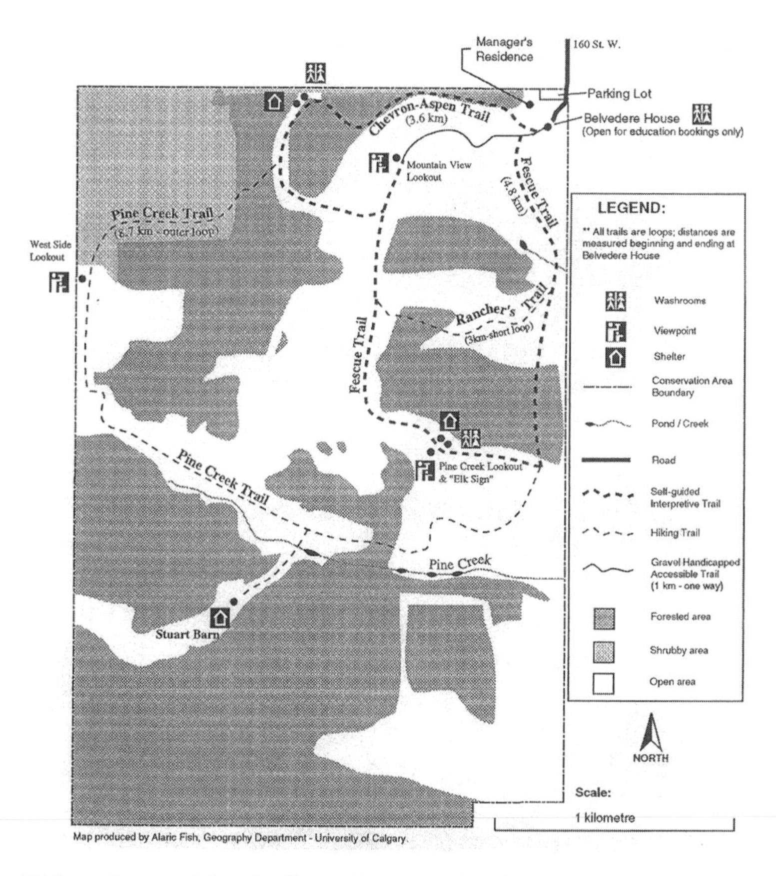

Figure 12.3a Ann and Sandy Cross Conservation Area.

Figure 12.3b Ann and Sandy Cross Conservation Area (from south-east).

13

Cartographic Symbolization Requirements for Microcomputer-Based Geographic Information Systems

Janet E. Mersey
Department of Geography, University of Guelph

INTRODUCTION

Many of the symbolization techniques used by cartographers today to portray spatial variations in geographic phenomena were first developed over a century ago. Since that time thematic cartography has become the focus of a substantial body of scholarly research, and has accumulated a rich collection of theoretical and empirical literature. While cartography is a discipline accustomed to adapting to technological innovations, the tremendous popularity of microcomputer-based mapping evidenced during the last decade has had some dislocating effects on the traditional relationships among cartographers, map users, and manufacturers of cartographic equipment. The result of these changes has been a potential loss of map quality and effectiveness, as both software developers and users may be unfamiliar with fundamental cartographic principles for data encoding and map design.

The recent emphasis on scientific visualization as a powerful tool for exploratory spatial data analysis has drawn attention to the graphic quality of GIS output. Although updates of GIS software usually include improvements in map display features, programs still vary considerably in their capabilities for representing data using standard thematic mapping symbolization techniques. The restricted choice of thematic mapping options characteristic of many GIS programs may be a conscious decision on the part of the software developer, or it may simply reflect an unfamiliarity with the needs of the user community. It is the intent of this paper to clarify those needs. The most commonly used cartographic techniques for displaying both qualitative and quantitative spatial data are identified, based on a content analysis of conventionally-produced thematic atlases. Approaches vendors may adopt to incorporate these required symbolization and graphic design features into their products are also addressed.

CARTOGRAPHIC DISPLAY AND GIS

In the early phases of GIS development, shortcomings in output quality were often overlooked as attention focused on developing new and complex analytical functions. With the maturation of the industry, and with parallel developments in graphic illustration software, user expectations have increased. While cartographers may have become accustomed to referring to choropleth maps as hatch maps, area fill maps, reclassified maps, or patch maps, the lack of meaningful ways for dividing data into classes, or alternative methods for symbolizing data, are viewed as serious deficiencies. In an article on map design and GIS, Donnelly writes: "Mapmakers should not lament the loss of aesthetic quality found in maps made the old way. They should work to build systems that can automate and replicate their best endeavors" (Donnelly, 1991: 34). An important step toward achieving this goal is to provide software developers with the information they need to accommodate the requirements of thematic map production.

As GIS software becomes more widely adopted by industry, government, and educational institutions for data analysis, more and more published maps will be produced by these programs. In fact, many GIS programs are now marketed primarily as digital mapping systems (Green, 1993: 92). As Ahner notes "organizations find themselves fitting their mapping requirements to the available GIS" (Ahner, 1993: 50). Despite arguments from GIS proponents that GIS maps are intended only as quick intermediate views of the data rather than final polished maps, the value of any GIS as a visualization tool will be greatly compromised if these quick views fail, for design reasons, to communicate the true nature of the spatial information they encode.

The importance of incorporating established principles of cartographic symbolization and design into GIS software is now well recognized in the literature (Donnelly,1991; Green, 1993; Ahner, 1993), and Buttenfield notes that this is a current focus of many GIS vendors as well (Buttenfield, 1992: 33). The thematic atlas content analysis described below may assist vendors in designing more robust display modules for their GIS programs that include techniques widely used in thematic cartography.

THEMATIC ATLAS CONTENT ANALYSIS

The eight thematic atlases selected for this study were published in North America between 1979 and 1991. Table 13.1 provides a complete reference for each atlas. With the exception of the *Historical Atlas of Canada*, they are all regional atlases focusing on particular provinces or states. Two criteria were particularly important in choosing atlases for this report: first, the atlases had to be produced by traditional manual techniques; and secondly, each had to contain a large number of thematic maps utilizing a rich selection of symbolization methods. The atlases thus reflect the variety of mapping techniques cartographers would choose to employ for such a project without any software or hardware imposed constraints.

Table 13.1 List of atlases examined. The reference codes are used to refer to the atlases in subsequent tables.

Reference Code	*Atlas*
A	***Historical Atlas of Canada, Vol. III.*** Donald Kerr and Deryck W. Holdsworth (editors). Toronto: University of Toronto Press, 1990.
B	***Atlas of Newfoundland and Labrador.*** Gary E. McManus and Clifford H. Wood. St. John's, Newfoundland: Breakwater Press, 1991.
C	***L'inter Atlas—Les ressources du Quebec et du Canada.*** Pierre Paradis, Jean Raveneau and Yves Tessier. Montreal: Centre Educatif et Culturel Inc., 1986.
D	***Atlas of British Columbia: People, Environment and Resource Use.*** A.L. Farley. Vancouver: The University of British Columbia Press, 1979.
E	***The California Water Atlas.*** William L. Kahrl. Sacramento: California Department of Water Resources, 1979.
F	***The Maritime Provinces Atlas.*** Robert J. McCalla. Halifax, N.S.: Maritext Ltd., 1991.
G	***Manitoba Atlas.*** Thomas R. Weir (editor). Winnipeg: Department of Natural Resources, Province of Manitoba, 1983.
H	***Atlas of California.*** Michael W. Donley, Stuart Allan, Patricia Caro, and Clyde P. Patton. Culver City, California: Pacific Book Center, 1979.

Although the range and number of themes varied from atlas to atlas, all covered a broad selection of information relating to the region's human, physical, and economic characteristics. Typically several pages (most commonly a two-page spread) were devoted to each theme or subject listed in the table of contents. These related pages or plates contained several maps as well as numerous non-map elements such as graphs, tables, diagrams, photographs and text blocks. A distinguishing characteristic of a map plate is that it is designed as a unit with the multiple display elements presenting an integrated interpretation of the subject.

Each atlas was carefully examined page by page and counts were kept of the methods by which information was displayed. The number of reference and thematic maps appearing in each atlas was tabulated along with the number of graphs, illustrations, tables, text blocks, and map insets. The number of times a particular qualitative or quantitative symbolization technique occurred was also recorded. These counts significantly exceed the number of thematic maps per atlas since a single map might employ several symbolization methods (i.e., a map which combined proportional circles with a choropleth method would be counted twice—once for each method). Table 13.2 presents some overall statistics: the eight atlases contained a total 1954 maps, 96.42 percent of which were thematic as opposed to general reference maps. Quantitative data was portrayed more frequently than qualitative data in most atlases, by a ratio of about three to two overall. In a recent study of school atlases, Young found that only 18 percent of the atlas maps utilized some sort of quantitative symbolization, but this number increased with the grade level of the intended audience (Young, 1994: 14). In the present study, the atlases are intended for adult audiences although at least one (*L'interAtlas*) has been used in schools.

Table 13.2 Number and type of maps examined.

Atlas	*A*	*B*	*C*	*D*	*E*	*F*	*G*	*H*	*Total*	*Percent*
Map plates or themes	66	22	39	60	43	40	114	127	511	—
Reference maps	11	1	8	4	0	10	5	31	70	3.58
Thematic maps	331	93	171	121	46	147	322	653	1884	96.42
qualitative methods	227	73	144	65	41	90	106	338	1084	—
quantitative methods	321	54	98	134	49	112	375	437	1580	—
TOTAL MAPS	342	94	179	125	46	157	327	684	1954	100.00

CARTOGRAPHIC SYMBOLIZATION METHODS USED IN THE THEMATIC ATLASES

Qualitative Techniques

As Young notes, qualitative information is generally much more straightforward to display cartographically than is quantitative data, as little data processing or complex encoding decisions are required (Young, 1994: 14). Table 13.3 lists six methods that were used to symbolize qualitative data in the eight atlases examined, in order of frequency of occurrence. Of the 1,084 instances of qualitative symbolization, nearly one third employed a simple thematic area shading technique (referred to as bounded area mapping by Young). Here homogeneous regions, such as salmon spawning grounds or areas of spruce budworm defoliation, are given a unique patterned or coloured area symbol. Two of the other methods listed in Table 13.3 also depict areal data. Regional outlines portray the same type of information as shading by thematic areas, except that no colour or pattern is applied to the area; instead the bounding line itself serves to symbolize the uniform region. When the regions depicted conform to administrative divisions such as census districts (a qualitative "choropleth" map) the technique is referred to as shading by enumeration areas. In total these three methods for representing areal data comprise nearly 60 percent of all the instances of qualitative symbolization used in the atlases.

Table 13.3 Frequency of qualitative symbolization techniques.

Qualitative Methods	*A*	*B*	*C*	*D*	*E*	*F*	*G*	*H*	*Total*	*Percent*
Shading by thematic areas	21	8	63	31	11	23	31	146	334	30.81
Nominal point symbols	93	25	31	19	11	24	34	73	310	28.60
Shading by enumeration areas	65	5	28	0	5	23	28	67	221	20.39
Flow maps	40	12	17	3	11	15	5	41	144	13.28
Regional outlines	8	22	5	12	3	4	8	6	68	6.27
Weather symbols	0	1	0	0	0	1	0	5	7	0.65
TOTAL	227	73	144	65	41	90	106	338	1084	100.00

Nominal point symbol mapping is an appropriate and popular choice for showing the location of point features such as downhill skiing facilities, active and abandoned mines, or offshore oil and gas wells. The site of each occurrence is assigned a small mimetic or abstract symbol whose meaning is identified in an accompanying legend. This was the second most common mapping technique for symbolizing qualitative data, used in 28.60 percent of the cases.

All the thematic atlases contained examples of qualitative flow maps to symbolize linear information such as explorer's routes or nature trails. Linear data tend to be less common than either areal or point data, and this method was used in only 13.28 percent of the qualitative displays.

Quantitative Techniques

The majority of the maps in these thematic atlases utilize some type of symbolization to communicate spatial variations in the magnitude of quantitative data. Because the nature of statistical information can be so diverse (it can relate to points, lines, or areas and be continuous, discrete, smooth, or complex), representing it effectively can be a challenge even for experienced cartographers. Compared with qualitative mapping, a wider range of standard symbolization methods has been developed to reveal the "statistical surface" inherent in quantitative georeferenced data. Table 13.4 lists the ten basic techniques employed in the atlases, but because there are many design options within each method, a quick glance through the atlases gives the impression that many more techniques have been used.

Table 13.4 Frequency of quantitative symbolization techniques.

Quantitative Methods	*A*	*B*	*C*	*D*	*E*	*F*	*G*	*H*	*Total*	*Percent*
Proportional pt. symbols	173	19	52	52	13	37	68	155	569	36.01
Choroplethic	23	3	3	35	3	14	127	176	384	24.30
Isarithmic	13	12	19	9	8	16	77	41	195	12.34
Data values	9	11	16	6	11	11	21	27	112	7.09
Non-proportional pt. symbols	20	0	0	15	6	13	47	4	105	6.65
Flow maps	25	5	7	11	5	20	6	10	89	5.63
Dot mapping	32	3	0	2	0	1	26	16	80	5.06
Dasymetric	25	0	0	4	0	0	2	8	39	2.47
Perspective surfaces	0	0	0	0	3	0	1	0	4	0.25
Cartograms	1	1	1	0	0	0	0	0	3	0.19
TOTAL	321	54	98	134	49	112	375	437	1580	100.00

As Table 13.4 indicates the most common type of quantitative technique employed was proportional point symbolization, comprising 36.01 percent of all the cases of quantitative data representation used in the eight atlases. The appearance of proportional point symbol maps can vary considerably depending upon the design choices the cartographer makes for the method's parameters, and whether single or multivariate data are being displayed. Because of the great range of maps of this type, the method is examined in more detail in the next section of this paper, where it is further broken down into sub-types.

Choropleth mapping ranks second in popularity (24.30 percent) followed by isarithmic mapping (12.34 percent). As discussed below choropleth mapping is featured in most GIS programs, although proportional symbol and isarithmic mapping are often neglected. Together these three methods constitute nearly three quarters of all the quantitative methods used.

One could argue whether the next technique, displaying data values, is really a thematic mapping method at all because it consists of merely writing the actual data values right on the map in a kind of "spatial table". Non-proportional point symbolization is a rather uncommon technique where the hue or value of a point symbol is changed to connote magnitude rather than its size. This method appeared frequently in the *Atlas of Manitoba* where it was combined with choropleth mapping.

Flow maps, dot maps, and dasymetric maps are all well established symbolization techniques in conventional cartography and are used where appropriate in these atlases. The minimal use of dot maps in five of the eight atlases is surprising, since it is usually promoted in cartographic textbooks as an effective method to communicate the spatial arrangement of discrete phenomena. Dent notes a decline in the number of dot maps drawn in the past several decades, which he attributes to the time, expense, and considerable research that the manual construction of dot maps requires (Dent, 1993: 158).

Perspective maps and cartograms account for less than 1 percent of the quantitative mapping techniques used. These types of maps are very difficult to produce by manual methods but can be effective at communicating magnitude variations. GIS offers great potential here for expanding the display options available to cartographers. Many GIS programs can quickly generate three dimensional views of an area from any vantage point in space.

Proportional Point Symbol Techniques

As mentioned above, proportional symbol maps can take on a wide range of guises depending upon the design choices made by the cartographer when applying the method to a particular set of data. Table 13.5 shows nine variations of this type of mapping that were exhibited on the atlas maps.

Proportional point symbols can be used to represent attribute data at specific sites or aggregated over areas. The method involves scaling the length, area, or volume of the point symbols according to the magnitude of the data values being represented. The shape of the symbol is another important design variable. Multivariate applications can be accomplished either by subdividing the symbol representing some total value, or by placing several symbols side by side on the map.

Proportional area scaling was far more common than length or volume scaling, with circles (40.95 percent) and squares (11.95 percent) the dominant shapes used. In only 5.80 percent of the cases was another shape chosen for proportional area scaling. An example would be changing the size of a mimetic symbol, such as a telephone, to represent the number of long distance calls made per region.

Table 13.5 Frequency of proportional point symbol techniques.

Proportional Symbol Methods	*A*	*B*	*C*	*D*	*E*	*F*	*G*	*H*	*Total*	*Percent*
Circles	66	5	40	29	0	13	31	49	233	40.95
Sectored Circles	40	2	0	6	0	9	22	22	101	17.75
Squares	7	0	0	7	1	4	2	47	68	11.95
Divided Squares	6	3	0	0	1	1	0	2	13	2.28
Bars	19	5	2	1	2	3	11	3	46	8.08
Divided Bars	20	0	0	0	2	0	2	10	34	5.98
Graphs as pt. symbols	6	1	1	3	2	3	0	3	19	3.34
Other (e.g. pictoral)	1	3	9	5	0	4	0	11	33	5.80
Volumetric pt. symbols	8	0	0	1	5	0	0	8	22	3.87
TOTAL	173	19	52	52	13	37	68	155	569	100.00

An additional 17.75 percent of the cases utilized proportional circles which were subdivided into sectors or slices. The size of each slice represents the amount made up by a particular subtype of the phenomena. For example, the size of the circle might represent total lumber production per area, and the size of the slices the amount of hardwood versus softwood lumber. The method can be extended to accommodate any number of slices. Squares were less commonly subdivided in this fashion.

Scaling the length or height of a rectangular symbol according to data values results in a bar or histogram map. Bar maps (both univariate and multivariate) made up 14.06 percent of all proportional point symbol cases. This technique is particularly useful when another method, such as choroplethic, is used on the same map, as narrow bars tend to obscure less of the background than circles or squares. This method can be extended to multivariate data by placing an entire line or bar graph at each map location. This technique was used in all but one atlas and often displayed climatic data.

Volumetric point symbols (usually cubes or spheres) were not very common. Not only can such symbols be tricky to prepare by manual methods, but research indicates that most map readers have difficulty estimating the relative sizes of such symbols correctly (Dent, 1993: 172).

Multivariate Mapping

A significant proportion of the atlas maps used several qualitative or quantitative mapping techniques in combination to display more than one thematic variable on the same map (see Table 13.6). Multivariate mapping can be accomplished either by using the same symbolization method more than once (i.e., bivariate choropleth mapping) or by combining different methods (i.e., proportional symbol mapping and dot mapping). In a study of atlas users Hocking and Keller found that a large number of adults thought that atlas maps contained too much information making them cluttered and difficult to read (Hocking and Keller, 1992: 115). Graphic designers such as Tufte, however, promote the use of information-rich graphics that allow for detailed examination and comparison of data (Tufte, 1983: 168).

Table 13.6 Frequency of single and multivariate maps.

Variables Per Map	*A*	*B*	*C*	*D*	*E*	*F*	*G*	*H*	*Total*	*Percent*
Single variable	109	61	106	47	18	89	162	499	1091	57.91
Two-variable	112	24	49	41	8	41	118	115	508	26.96
Three-variable	64	6	14	18	9	13	33	23	180	9.55
More than three variables	46	2	2	15	11	4	9	16	105	5.57
TOTAL	331	93	171	121	46	147	322	653	1884	100.00

Although there was a great variation among the atlas maps as to their level of complexity, some were extremely detailed, encoding over three variables simultaneously. The ease with which multivariate maps can be interpreted by users is integrally linked to the graphic design of the map. Careful selection of elements such as colour, line width, type size, and patterns can greatly assist in making even a complicated graphic clear and legible. Successful production of multivariate maps by GIS will require that the program not only has the capacity to utilize different mapping techniques on different data layers simultaneously, but also that it allows for the flexible manipulation of graphic design elements such as colour and type.

NON-MAP ATLAS CONTENT

Atlases employ other visualization tools in addition to maps to assist users in understanding geographic themes. These include graphs, diagrams, tables, sketches, air photos, satellite images, and photographs. When quantitative data are involved, graphs in particular can be an important complement to map analysis. By abstracting information from its spatial context, graphs can serve to emphasize anomalies and relationships among the attributes symbolized on a map (MacEachren et al., 1992: 125).

Graphs were very widely used in the eight thematic atlases—in fact, three of the atlases contained more graphs than maps (see Table 13.7). Encoding information effectively in graph form entails many of the same challenges as thematic mapping—an appropriate graphing method must be selected for the data, and graphic design elements such as line width and type must be manipulated so as to enhance the legibility of the graph.

Table 13.7 Frequency of maps and non-map elements in the map plates.

Plate Elements	*A*	*B*	*C*	*D*	*E*	*F*	*G*	*H*	*Total*	*Percent*
Maps	342	94	179	125	46	157	327	684	1954	3.82
Graphs/ diagrams	573	58	162	31	122	241	150	106	1443	2.82
Text blocks	166	48	99	1	91	7	94	119	625	1.22
Illustrations	37	38	147	70	115	29	1	17	454	0.89
Tables	19	0	33	16	33	18	6	37	162	0.32
Map insets	43	2	5	14	7	1	9	43	124	0.24
TOTAL	1180	240	625	257	414	453	587	1006	4762	9.31
No. of map plates	66	22	39	60	43	40	114	127	511	—

Text was the second most common type of supplemental tool used. The content analysis only considered text which was integrated with the maps and appeared within the plate frame. Many researchers, including Rittschof et al., have shown that users learn significantly more theme-related facts from a map when it is used in conjunction with pertinent text (Rittschof et al., 1993: 92). Keeping the text relevant, succinct, and in close proximity to the map also helps, as Hocking and Keller note that users tend to ignore lengthy text on separate atlas pages (Hocking and Keller, 1992: 114). Although word-processing and desktop publishing programs have sophisticated capabilities for handling text layout and design, mapping programs may also benefit from these features if large amounts of text are to accompany maps. Utilities such as justification, margination and spell-checking become important if text is to play a larger role than titling and labelling map locations.

Table 13.7 indicates that on average a single map plate contains 9.31 separate elements. The effective organization of all this material with the plate frame is an important part of atlas design. The position, size, orientation, and hierarchical order of all the map and non-map elements must be carefully tailored to create a balanced, legible, and integrated whole. Automating this task may be further complicated if different graphic image file formats are used for the various plate elements.

DISPLAY CAPABILITIES OF MICROCOMPUTER-BASED GIS PROGRAMS

Thematic Mapping with GIS

The content analysis of these eight thematic atlases clearly reveals the rich and varied display options cartographers have developed with manual technology for visually communicating spatial information. A switch to automated methods will, of course, entail much more than simply replicating these methods; GIS brings with it the potential for new approaches to data analysis involving such tools as animation, interactive querying, and three dimensional representations. Nevertheless, the need for well designed static displays will continue, and hardcopy maps are the main output from most GIS projects.

The logical question upon completing the content analysis is this: are microcomputer based GIS programs capable of creating the types of thematic maps cartographers most frequently use to convey complex spatial data? With over 100 PC-based GIS programs currently on the market (GIS World, 1992: 13-16), the answer is not an unambiguous yes or no. However, some generalizations can be made based on the literature, GIS textbooks, comparative software reviews, and practical experience.

Perhaps the best indication of the advanced state of computer mapping is the observation that *all* of the qualitative and quantitative techniques used in these atlases have in fact been automated. The limitations of using GIS for thematic mapping relate largely to the narrow range of symbolization options available within any specific GIS program, and with the lack of flexibility for controlling certain design parameters.

Qualitative Mapping with GIS

Most GIS programs can represent qualitative data with little difficulty. Shading map areas with patterns or colours, placing point symbols at specified locations, and differentiating line types for qualitative flow mapping are techniques well within the range of most PC-based GIS programs. If the selection of available patterns, point symbols or line types is limited, some design compromises may have to be made. Buxton suggests that for many marketing functions, GIS software should allow for customized icons (such as company logos) to be imported and displayed at site locations (Buxton, 1992: 240).

Quantitative Mapping with GIS

Quantitative mapping differs from qualitative mapping in that information portrayed on the map is often derived from the original or raw data as it was collected by some type of statistical manipulation. This data processing step may be simple (i.e., converting absolute values to percentages) or involved (i.e., computing potential erosion for map areas based on a mathematical model). GIS is ideal for these types of statistical processes which can be extremely time-consuming and error-prone when done manually. The ability of a GIS to perform complex computations on large databases in seconds is a tremendous attraction for map makers.

As illustrated above, a greater variety of symbolization techniques have been developed for mapping quantitative data in comparison to qualitative data. Unfortunately few, if any, GIS programs offer all of the quantitative techniques summarized in Table 13.4. The capabilities of 13 microcomputer-based GIS programs for creating five common types of quantitative thematic maps are listed in Table 13.8. The information for this table is based largely on a review article by Lang (Lang, 1993: 188-189) and the *1993 International GIS Sourcebook* (GIS World, 1992: 13-16). It is not meant to be all inclusive, and because of frequent program updates, the information is subject to change.

Table 13.8 supports the notion that simple choropleth maps are the mainstay of most GIS programs. Both raster and vector based programs have the capability for reclassifying cells or polygons according to some quantitative attribute; in fact, this operation is fundamental to many GIS query and overlay analyses. Choropleth mapping was one of the first thematic mapping techniques to be automated and Slocum and Egbert note that "univariate choropleth maps have become the most popular form of thematic display in digital cartography" (Slocum and Egbert, 1991: 172). Keller and Waters agree stating that "the bulk of all thematic microcomputer mapping packages concentrate on choropleth mapping" (Keller and Waters, 1991: 104).

In the GIS literature, relatively few texts or articles even comment on thematic mapping symbolization methods, focusing instead on the analytical components and applications of GIS. Two exceptions are textbooks by Star and Estes (1990) and Aronoff (1989). Both these texts indicate that choropleth and isarithmic maps are the data presentation techniques most commonly required in GIS analyses. Neither book mentions proportional symbol mapping, although Keller and Waters note that this method is "rarely supported" (Keller and Waters, 1991: 104). Recall that proportional symbol mapping was the most often used quantitative mapping method in the eight thematic atlases. Only 9 of the 13 programs in Table 13.8 provide this technique. Fewer still allow for sectored circle mapping, a popular proportional symbol mapping type used in the thematic atlases. Dot mapping is available in seven of the programs, but the placement of dots is usually random within regions, and the resulting maps are less than optimal. Although GIS analysts claim to have a frequent need for isarithmic displays, only six programs offer this technique. Because of the intensive computations demanded by some interpolation routines, isarithmic mapping is more common on workstation or mainframe platforms.

Table 13.8 Thematic mapping capabilities of selected microcomputer-based GIS programs.

Program (Manufacturer)	*Choropleth*	*Dot Mapping*	*Proportional Symbols*	*Sectored Circles*	*Isarithmic*
Atlas * GIS Strategic Mapping)	•	•	•		
GISPlus (Caliper)	•		•	•	•
Hunter GIS (Hunter and Assoc.)	•		•		
Idrisi (Clark Univ.)	•				•
MapBox (Decision Images)	•				•
MapInfo (MapInfo)	•	•	•	•	
MapViewer (Golden Software)	•	•	•	•	
PC Arc/Info (ESRI)	•	•	•	•	•
Quikmap (Axys Software)	•		•		
Spans Map (Intera Tydac)	•	•	•	•	•
StatMap III (GeoVision)	•	•			
Tactician (Tactics Int'l.)	•	•	•	•	
TerraSoft (Digital Resource Sys.)	•				•
TOTAL	13	7	9	6	6

Sources: Lang (1993), pp. 188-189. *GIS World* (1992), pp. 13-16. Information regarding Hunter GIS, MapBox and TerraSoft was also gleaned from conversations with vendors and/or promotional materials.

Only 2 of the 13 programs reviewed (PC Arc/Info and SpansMap) have the capability for producing all five types of thematic maps. Both of these programs are considered "high-end" tools with hefty price tags and significant user training requirements, although SpansMap can be purchased and operated separately from its parent program, Spans. If isarithmic mapping is not necessary, MapInfo, MapViewer, and Tactician are less expensive, easier to learn options. These programs may, however, be weaker in analytical GIS capabilities.

Manipulating Mapping Parameters

Once a particular symbolization method has been selected, critical decisions must still be made as to exactly how the method should be applied to the data. In choropleth mapping, for example, the cartographer must decide how many data classes to use, how to divide the data into discrete ranges, and how to symbolize each class of information in a meaningful and logical fashion. Here again programs vary widely. Cartography textbooks frequently cite Jenks' optimum classification algorithm (also known as the goodness of variance fit method) as a suitable technique to use to determine the class intervals for a choropleth map (Robinson et al., 1984: 363; Dent, 1993: 139), yet few GIS programs offer this option. The potential for creating a misleading choropleth map is particularly great when the user relies on the default classification method which may be entirely inappropriate for a given set of data.

The success with which other symbolization methods can be utilized within a particular GIS program is also dependent upon the flexibility allowed when selecting related parameters. Proportional symbol mapping requires that the length, area, or volume of symbols be scaled according to data values, but few programs offer all of these options. Atlas*GIS includes only proportional area scaling, while Hunter GIS can only scale symbols by length. Few programs handle volume scaling. Thus, the capabilities indicated in Table 13.8 can be misleading—just because a GIS program includes a particular thematic mapping technique does not necessarily mean that the cartographer will be able to apply the method in the manner desired.

Map Design and GIS

Creating an effective thematic map requires more than selecting and applying an appropriate data representation method. The graphic design and layout of the map must also facilitate effective communication. Components of graphic presentation and design that demand attention include the map's legibility, contrast, balance, figure-ground, hierarchical organization, patterns, colours, and typography (Robinson et al., 1984: 144). The degree to which GIS programs can modulate these elements is beyond the scope of this paper, but a few examples will highlight the nature of the problems that can be encountered.

One illustration of a graphic design problem concerns the treatment of overlapping map symbols. This is a common situation with multivariate mapping, but can occur even with a single distribution. Consider a proportional circle map where, because of the clustering of point locations, the symbols must overlap. Maps A and B seen in Figure 13.1 show the two design options available with Atlas*GIS—either the fill or the outline of the symbol can be specified. When a filled circle is used, as in Map A, information behind the circles is completely hidden. Specifying an open circle, as in Map B, solves this problem, but now legibility is sacrificed as all the lines on the map are reduced to the same visual level. Maps C and D were created with an illustration package (CorelDraw) and represent two preferable design solutions not possible with the GIS program. In Map C a "halo" effect is placed around each circle thus revealing any smaller circles underneath larger ones, while in Map D the map boundary layer is placed in front of the circles where it is visible.

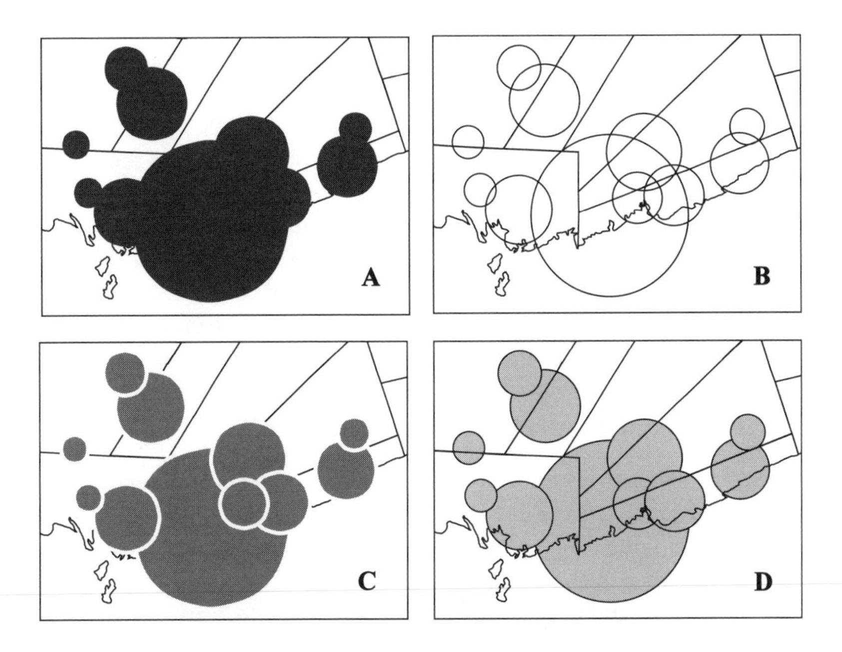

Figure 13.1 Design options for proportional circle mapping.

Because a map plate contains a variety of map and non-map elements, GIS programs must be able to handle multiple displays simultaneously. Windows based software tends to be more flexible in this regard as it allows images (such as graphs created with another application package) to be pasted into place via the clipboard. Most GIS programs do not provide complete freedom for positioning and resizing diverse elements within a single frame. One exception is TNTMips by MicroImages which contains a special map layout module that allows multiple vector, raster, and text objects (including graphs) to be selected, arranged, and ordered for large format printing.

The content analysis revealed that atlases rely heavily on graphs to display data. GIS programs vary widely in their capabilities for creating graphs, but most are not nearly as flexible in this regard as specialized graphing programs such as Freelance or SigmaPlot. Similarly, map plates requiring large amounts of text may find GIS programs inadequately equipped to deal with paragraph formatting and font selection.

STRATEGIES FOR IMPROVING GIS OUTPUT

Over the past several years GIS software developers have become increasingly concerned with the quality of their GIS output and are actively trying to improve the visualization component of their products (Buttenfield, 1992: 33). Strategies that have evolved for incorporating principles of cartographic design into GIS map production are outlined below.

Expansion and refinement of design features within the GIS program.

Most GIS programs are updated annually (Lang, 1993: 187), and each new version represents an increase in functionality and flexibility. A 1991 review of MapInfo, for example, criticized the lack of proportional symbol mapping (Kendall, 1991: 296); this option is included in a more recent version (Lang, 1993: 188). Many vendors have introduced a Windows version of their software (i.e, Atlas*GIS from Strategic Mapping) which allows more graphic control over the display. The GIS market is highly competitive, and as vendors become aware of user needs, or features of competing programs, they quickly strive to improve their own products.

Development of independent data display programs.

Another strategy that has been adopted, notably by several manufacturers of high-end GIS programs, is the development of independent "front-end" data presentation software which separate simple query and view commands from more computationally intensive analytical GIS operations. Compared to the parent GIS program, these tools tend to be less expensive, easier to use, provide a greater range of symbolization options, and allow considerable flexibility for manipulating graphic design elements such as colour and layout. ESRI's ArcView and Intera Tydac's SpansMap are examples of this approach. Both programs emphasize the production of high quality thematic maps and utilize an iconic toolbox for easy editing.

Utilization of integrated third party graphic programs.

Rather than developing new graphic design software, some GIS manufacturers have opted to bundle their programs with a third party package especially written to facilitate the manipulation of graphic images. As Green notes, this solution is easy, cost-effective, and users benefit from experience they may already have with illustration software (Green, 1993: 93). Examples of this approach include the pairing of Acorn Paint and Draw programs with ITC's GIS program, Alexander; and Dr. Halo and Image-Pro with Hunter GIS.

Post-processing with an illustration program.

Cartographers have been quick to recognize the potential of specialized graphic illustration programs, such as CorelDraw, Adobe Illustrator, Aldus Freehand and

Micrographix Designer, for creating and printing high quality maps. These programs are not intended specifically for cartographic purposes, but when coupled with a GIS or mapping package, can greatly increase the map maker's control over the graphic design of the map. Because illustration programs do not link map areas or points to a database, symbolization based on classifying or sizing symbols according to data values must be accomplished in a mapping or GIS program. Once the map is imported into the illustration program, however, significant enhancements can be made to the map's layout, colours, patterns, typography, etc.

To facilitate the subsequent importation of GIS output into an illustration program, many GIS programs now allow digital maps to be saved in one or more standard image file formats. Because illustration programs will import digital files in a variety of both bitmapped and vector formats, it is also possible to import many non-map elements such as graphs and scanned photographs from other graphing or imaging programs. These can be combined and arranged in a single frame with the maps, and text blocks can be added. It is often the case, however, that transferring image files between application packages is problematic, especially for vector-based graphics, and some experimentation with different formats may be required (Morber and Mersey, 1994: 3). One of the most powerful functions of illustration programs is the capability for printing the colour separations required for lithographic reproduction, which cannot be done with PC-based GIS programs. Idrisi is one GIS program that encourages its users to refine and print their maps with illustration software, and their latest manual provides detailed information about the file formats and procedures necessary for accomplishing this task.

CONCLUSIONS

The production of thematic maps with microcomputer-based GIS software offers significant advantages for cartographers, but certain constraints may exist as well. GIS programs are particularly ideal for the data processing stage of quantitative thematic mapping. Statistical operations can be readily carried out on the spreadsheet of data attributes georeferenced to the map features. Maps can be interactively modified, allowing the cartographer to vary such parameters as the classification interval on a choropleth map, and quickly assess the results. Also, GIS allows data to be displayed in new ways not possible or difficult to accomplish with manual techniques. Animated mapping and three dimensional views provide exciting alternative approaches for exploring geographic themes.

This paper has shown that, particularly for quantitative mapping, currently available GIS software may not provide the flexibility required for creating and manipulating some types of thematic map displays. Proportional symbol mapping, for example, is beyond the scope of many programs and, even when it is available, may not allow parameters to be varied in the manner required. Graphic design constraints may also exist, and incorporating text and graphs with maps may be problematic.

GIS technology continues to develop and improve rapidly. By becoming aware of the needs of the user community through studies such as this, vendors have the opportunity to respond by incorporating the desired features into their products. Several strategies may be necessary to address the shortcomings in available symbolization methods and graphic design flexibility identified in this report. Employing a specialized illustration program, either as an integrated module or via an export facility, offers tremendous potential for overcoming many design constraints. However, because map features are not linked to attributes in a database, quantitative symbolization cannot be performed with these programs. GIS vendors must continue to develop and expand the thematic mapping functionality of their own products, either by incorporating new symbolization modules or, as several manufacturers have done, by providing independent map display programs.

In his recent book, *Some Truth with Maps,* Alan MacEachren states "To use visualization tools effectively with GIS demands that we examine procedures for their use as carefully as we have incorporated the rules for statistical analysis in the past" (MacEachren, 1994: 122). Guidelines for the effective display of geographic data are well established in the cartographic literature; the increased focus on symbolization and design evident in recent years will help ensure that these same standards of cartographic excellence are brought to bear in the automated GIS environment.

ACKNOWLEDGEMENTS

The author gratefully acknowledges a grant from Natural Resources Canada and the Natural Sciences and Engineering Research Council of Canada which supported part of this study. Brian Morber's assistance with the content analysis is very much appreciated.

REFERENCES

Ahner, A.L. (1993). Modern cartography plays an essential role in GIS. *GIS World,* 6(10), 48-50.

Aronoff, S. (1989). *Geographic Information Systems: A management perspective.* Ottawa: WDL Publications.

Burrough, P.A. (1986). Principles of Geographical Information Systems for land resources assessment. *Monographs on Soil and Resources,* Survey No. 12. New York: Oxford University Press.

Buttenfield, B.P. (1992). Cartography year in review: Quality and availability of spatial information. *ACSM Bulletin,* 135, pp. 30-33.

Buttenfield, B.P., and Mark, D.M. (1991). Expert systems in cartographic design. In D.R.F. Taylor (Ed.), *Geographic Information Systems: The microcomputer and modern cartography* (pp. 129-150). Toronto: Pergamon Press.

Buttenfield, B.P., and Mackaness, W.A. (1991). Visualization. In D.J. Maguire, M.F. Goodchild, and D.W. Rhind (Eds.), *Geographical Information Systems, Vol. 1* (pp. 427-443). New York: John Wiley and Sons, Inc..

Buxton, T. (1992). Meeting the marketing challenge. In *1993 International GIS Sourcebook* (pp. 238-242). Fort Collins, Colorado: GIS World, Inc.

Dent, B.D. (1993). *Cartography: Thematic map design*. Dubuque, Iowa: Wm.C.Brown Publishers.

Donnelly, J. (1991). Geographic Information Systems in mapmaking. *ACSM Bulletin*, 134, 30-34.

Fisher, P., Dykes, J., andWood, J. (1993). Map design and visualization. *The Cartographic Journal*, 30, 136-142.

GIS World, Inc. (1992). *1993 International GIS Sourcebook*. Fort Collins, Colorado: GIS World, Inc.

Green, D.R. (1993). Map output from geographic information systems and digital image processing systems: A cartographic problem. *The Cartographic Journal*, 30, 91-96.

Hocking, D., and Keller, C.P. (1992). A user perspective on atlas content and design. *The Cartographic Journal*, 29, 109-117.

Keller, C.P., and Waters, N.M. (1991). Mapping software for microcomputers. In D.R.F. Taylor (Ed.), *Geographic information systems: The microcomputer and modern cartography* (pp. 97-128). Toronto: Pergamon Press.

Kendall, R. (1991). Mapping software: Analyzing a world of data. *PC Magazine*, July, 249-299.

Lang, L. (1993). Getting there with software maps. *PC Magazine*, March, 182-189.

MacEachren, A.M. (1994). *Some truth with maps: A primer on symbolization and design*. Washington, D.C.: Association of American Geographers.

MacEachren, A.M., Buttenfield, B.P., Campbell, J.C., and Monmonier, M.. (1992). Visualization. In R.F. Abler, M.G. Marcus, and J.M. Olson (Eds.), *Geography's inner worlds* (pp. 99-137). New Jersey: Rutgers University Press.

Monmonier, M. (1991). *How to lie with maps*. Chicago: The University of Chicago Press.

Morber, B., and Mersey, J.E. (1994). Thematic mapping with illustration software: Unravelling the mystery of graphic file formats. *Cartographic Perspectives*, 18, 3-16.

Muehrcke, P.C. (1990). Cartography and geographic information systems. *Cartography and Geographic Information Systems*, 17(1), 7-19.

Rittschof, K.A., Kulhavy, R.W., Stock, W.A., and Hatcher, J.W. (1993). Thematic maps and text—An analysis of `What happened there?'. *Cartographica*, 30(2/3), 87-93.

Robinson, A.H., Sale, R.D., Morrison, J.L., and Muehrcke, P.C.. (1984). *Elements of cartography*. New York: John Wiley and Sons.

Slocum, T.A., and Egbert, S.L. (1991). Cartographic data display. In D.R.F. Taylor (Ed.), *Geographic information systems: The microcomputer and modern cartography* (pp. 167-200). Toronto: Pergamon Press.

Star, J., and Estes, J. (1990). *Geographic information systems: An introduction*. Toronto: Prentice-Hall Canada.

Tufte, E.R. (1990). *Envisioning information*. Cheshire, Connecticut: Graphics Press.

Tufte, E.R. (1983). *The visual display of quantitative information*. Cheshire, Connecticut: Graphics Press.

Young, J.E. (1994). Reexamining the role of maps in geographic education: Images, analysis and evaluation. *Cartographic Perspectives*, 17, 10-20.

14

An Experiment with Choropleth Maps on a Monochrome LCD Panel

Matthew McGranaghan
Department of Geography, University of Hawaii

INTRODUCTION

This chapter is concerned with the design of choropleth maps for presentation on monochrome liquid crystal displays (LCDs). It joins two long-standing themes in cartographic research: the adoption of new technology; and the use of experimentation to improve cartographic design. The approach taken here assumes that past cartographic experience is generally applicable to new media, and that controlled experimental evaluation of map usability is the best way to improve map design.

Cartographers adopt new media as they become available. Technical pens, photomechanical materials, computer driven plotters, CRTs, laser printers, and image setters are examples. In each case, cartographers have experimented with the technology, rethought cartographic design and practice to accommodate its capabilities and limitations, and added it to the cartographic repertoire. Casual observation indicates that LCDs are making the same transition. They are increasingly used to display maps, and this use suggests that experience and some thought will determine their best use.

The chapter examines one way to use LCD technology to display choropleth maps, and experimentally compares several map design options. The chapter is organized as follows. The introduction describes the operation and capabilities of LCDs and relates these to research on using laser and dot matrix printers to display choropleth maps. That is followed by an experimental assessment of several designs for choropleth maps on LCDs. Finally, the implications of this experiment for our understanding of choropleth mapping are discussed.

Liquid Crystal Display Panels

Liquid crystal displays (LCDs) are the most common display technology on laptop computers, personal digital assistants, and specialized appliance computers for navigation, games, electronic books, and other uses. They excel in applications where low power consumption, thin form factor, flat display, high reliability, moderate to good image quality, high legibility, and low cost are necessary (Drzaic, 1990).

In addition, their non-emissivity, light weight, and small size make them attractive as display devices for desktop display applications as well (Werner, 1988). These properties suggest that LCDs are of interest to cartographers.

LCDs have been used for several years in automotive navigation systems (Ross, 1988), and in avionics displays (Litton Systems, 1992). They have been used in marine navigation devices, such as Trimble Navigation's NavGraphic II and Nav-Trac GPS (Trimble, 1991), which display simple maps on monochrome LCDs. These systems, together with the popularity of graphically capable laptop computers and the growing availability of spatial data and mapping software (included now in several spreadsheet programs), suggest that LCDs will increasingly be used to display maps, and that cartographers should study their application.

Capabilities of LCDs

An LCD operates by transmitting or absorbing light, rather than emitting it. Electro-optical materials such as amorphous silicon, cadmium selenide and poly-crystaline silicon (Brody, 1992: 7), transmit light when no electrical current is applied to them, but absorb (i.e., block) light when a current is applied. Molecules of these materials act like the slats of a venetian blind, closing and opening as current is applied and removed. An array of pixels of the material, deposited on a sheet of glass and driven by appropriate circuits, forms a display panel. Pixels to which a voltage is applied absorb light and appear dark, those to which no voltage is applied transmit light and appear bright. The light may come from in front of the panel, pass through it, and be reflected back through it again to the viewer (a reflective LCD panel), or may come from a source behind the glass (a back-lighted LCD panel). In either case, the principle is that light is either blocked or transmitted.

This principle can be extended. A simple, bi-level monochromatic LCD may have each pixel either dark or light. By spatially mixing dark and light pixels over an area, an intermediate tone could be produced. A grey scale LCD allows each individual pixel to appear to have intermediate tones, typically 16 or 64, by modulating the state of the pixel. Colour LCDs use both spatial mixing and temporal modulation simultaneously. Each pixel is split into three parts, one with a red-, one with a blue-, and one with a green-pass filter. The amount of light to pass through each part is modulated separately to mix the three additive primaries to produce a desired colour.

The cartographic capabilities of an LCD panel depend on its size, on the number and size of its pixels, and on the contrasts that can be achieved between pixels. Bi-level monochrome LCDs with 320 by 200 pixel (CGA resolution) have been available in laptops since the mid 1980s. Full colour LCDs have been on the market for several years (Gulick, 1991). LCDs with 3,072 columns by 2,048 rows of pixels have been demonstrated (Martin et al., 1993), but 640 pixel by 480 pixel (VGA resolution) arrays are in common use. Currently, the leading edge of commodity panels measure 11.3 inches diagonally, have 800 pixels by 600 pixels, and display 262,144 colours (Display Technology Report, 1995: 5). The capabilities of LCD panels improve steadily.

Casual observation of laptop users indicates that the transmissive state is usually used as the default background and that characters or graphics are then drawn with dark pixels on the light background, resulting in positive contrast like that found in most printing, and on "paper-white" CRTs. This is apparently by choice, as the contrast can usually be easily reversed through a hardware or software switch. The ease of switching the screen contrast might be a problem for cartographers.

Choropleth Mapping

Choropleth symbology is deceptively simple and incompletely understood. Many researchers (e.g., Williams, 1958; Jenks and Knos, 1961; Williamson, 1982; Kimerling, 1975; Cox, 1980; Leonard and Buttenfield, 1989; Kimerling, 1985; Monmonier, 1979, 1980; Groop and Smith, 1982; Plumb and Slocum, 1986; Castner and Robinson, 1969; Cuff, 1973, 1974; McGranaghan, 1986, 1989) have studied it. These studies cover two broad topics: producing choropleth symbology and understanding how it works.

How does choropleth symbology work? Assuming sufficient contrast in a display, understanding a choropleth map depends on recognizing that variously shaped areas on the map represent geographic areas, and that colour contrast among the shapes represents conceptualized differences among the geographic areas. Value contrast establishes order among area symbols, usually with darker symbols representing greater data magnitudes (Cuff and Mattson, 1982). This convention has been explained as a common sense association of more ink on the page with more of the mapped phenomenon (Cuff, 1974: 54), which makes sense in printing with dark ink on a light-coloured page, but not in media with light figure on a dark background, such as a blackboard (McGranaghan, 1989). Further, not all maps adhere to the convention, yet even non-adhering maps are useable.

Several studies have addressed why map readers do not always associate darker choropleth symbols with greater data magnitudes. Cuff (1973, 1974) reported experiments in which map readers were free to associate either darker or lighter symbols with greater data magnitudes (there was no map legend). He found that some map readers took lighter symbols to mean greater magnitudes, perhaps because of the thematic content of the map.

Experiments in which no particular thematic content was specified indicate that the association of darker symbols with greater magnitudes is independent of map content. McGranaghan (1988, 1989) found that, while the majority of map readers associate darker choropleth symbols with greater data magnitudes, and some map readers consistently choose lighter symbols to mean more, roughly a quarter of map readers take the symbol whose darkness is most different from that of the map's background to signify greater magnitudes, that is, they reverse the sense of the symbol order depending on the background. This led to the suggestion that contrast within the map display might influence the interpretation (ordering) of choropleth symbols. Unfortunately, the design of that experiment precluded determining whether response time differences were attributable to map designs, or to individuals' natural strategies for interpreting them.

Other studies have addressed the production of choropleth symbols on pixel addressable devices, such as dot matrix and laser printers, on which a grey area symbol must be produced by spatially mixing black and white pixels to produce a desired visual value. Perceptible texture affects the apparent value of area symbols (Castner and Robinson, 1969; Leonard and Buttenfield, 1989), and producing imperceptible fill patterns has been one goal in this line of research. Several studies suggest strategies that can be applied to bi-level LCDs.

Groop and Smith (1982) produced relatively coarse area symbols using a technique that simulated halftone screening with character cells on dot matrix line printers. Plumb and Slocum (1986) produced smoother, finer textured area symbols by taking better advantage of the dot matrix printer's ability to address pixels within a character cell. They compared uniform and random dot patterns and found that random dot patterns reduce apparent texture but require greater computation than uniform dot shading patterns. Leonard and Buttenfield (1989) produced even smoother tones using a laser printer.

Current LCDs have neither the contrast nor the resolution of laser printers; they are more comparable to dot matrix printers in these respects. The random dot shading and uniform dot shading strategies examined for dot matrix printers seem promising for bi-level monochrome LCDs. The computational overhead of random dot patterns is a concern with slow computers. The effects of apparent texture might be reduced by care in defining fill patterns and by the relatively lower contrast of the LCD. In the current work, uniform dot patterns were used.

From the foregoing, one might then ask how to design choropleth maps using uniform dot patterns for display on low resolution, bi-level monochrome LCDs. The following study attempts to determine the effects of three map design parameters: background value, symbol order, and legend design on map reading performance in this environment.

EXPERIMENT

Experimental Design

This experiment was intended to assess the effects of background value, symbol order, and legend design on reading choropleth maps rendered using uniform dot patterns on a bi-level monochrome LCD. A within-subjects, repeated measures, forced-choice, reaction time experiment was devised in which subjects extracted information from 32 five-class choropleth maps of the western United States, similar to those shown in Figure 14.1. Each map depicted a unique distribution and was presented on a reflective monochrome LCD using uniform dot shading fill patterns as choropleth symbols. On each map, subjects determined which of two enumeration districts was symbolized as having "more" of the (unspecified) mapped phenomenon. Response time (RT) and accuracy were recorded as measures of the cognitive difficulty of performing the task with each map. (See Shortridge and Welch, 1982: 72, for a justification of the use of RT in cartographic studies). It was

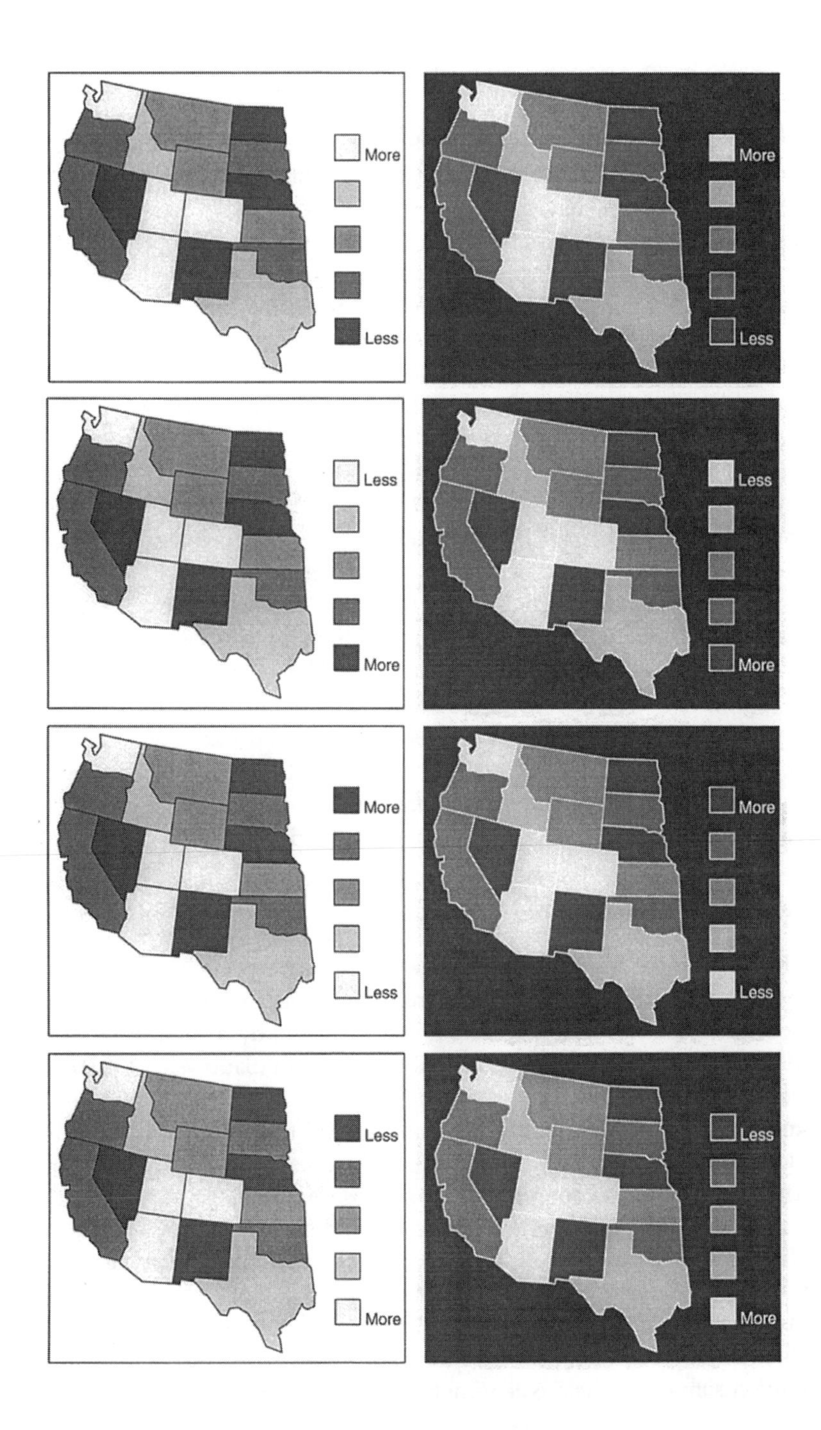

Figure 14.1 Examples of the maps used as stimuli in the experiment. Note the eight combinations of background, symbol order, and legend labels.

assumed that greater cognitive difficulty (longer RTs and lower accuracies) would be associated with poorer designs.

The experiment compared two-levels of each of three factors: background value (dark or light), symbol order (whether darker or lighter symbols represented greater magnitudes) and legend arrangement (whether the symbol for greater magnitudes was at the top or the bottom of the legend), giving rise to eight experimental conditions. Half of the maps were presented on the dark background and half of them on the light background. Each map's legend indicated how that map was to be interpreted. Half of the legends had the darkest symbol at the top and half had the lightest symbol at the top. In each of these halves, half of the trials had the top of the legend labelled "More" and the bottom "Less" ("standard format" in the terminology of DeLucia and Hiller, 1982), and the other half had these labels reversed. If the top of the legend had a dark symbol and was labelled "More", the dark symbols indicated greater data values. Similarly, if the bottom box of the legend were light and labelled "More", the lighter symbols represented greater data magnitudes. Thus, there were two legend designs that gave rise to a "darker is more" symbol order and two that gave rise to a "lighter is more" symbol order.

For each of the eight experimental conditions each subject made four comparisons: the darkest symbol to the lightest, and the second darkest symbol to the second lightest symbol, once with the darker symbol to the left, and once with it to the right. Responses were made by pressing the F- and J-keys on a 'qwerty' keyboard. F indicated a choice of the left symbol, and J the right. This provided a natural stimulus-response mapping for the task. Correct responses were balanced so that an equal number fell on the left and on the right.

Software and Equipment

The software for displaying maps and recording responses was written primarily in Borland's Turbo-C (TM) whose graphics routines were used for the display. Millisecond resolution reaction timing was provided by a slightly modified version of Richard S. Sadowsky's TCTIMER (version 1.0) (Sadowsky, 1988). This is a collection of C and assembler routines, based on TPTIME, by Brian Foley and Kim Kokkonen (of TurboPower Software), and released to the public domain. Similar timing programs are available on many public access software archives.

The experiment was conducted using a Toshiba T-1000 laptop microcomputer to present the map stimuli, record responses, and measure response times. The T-1000 has a CGA resolution (640 by 200 pixel) reflective monochrome LCD panel. Each pixel (0.37 millimetres on a side) can be set to one of two colours, nominally dark blue and light green. In CIE 1931 coordinates, these were Y=718.1 x=.309 y=.315 and Y=1174 x=.326 y=.356 respectively, as measured with a Minolta CL-100 reflective chromameter held 6 centimetres from the screen. On this panel, an area larger than a pixel can be given a colour perceived to be between these extremes by filling the area with a mixture of dark blue and light green pixels and counting on the observer's visual system to interpret the mixture as an intermediate colour.

Fill Patterns

Five uniform dot shading fill patterns (see Figure 14.2) were developed to approximate an equal value grey scale, subject to the constraints of the LCD screen, uniform dot shading, and the 8 by 8 pixel map used to define fill patterns in Borland's Turbo-C (Borland International, 1988). Each of the uniform dot fill patterns had to be distinct from the light and dark backgrounds (solid "off" and solid "on") as well as from each other. To accomplish this, five equally spaced perceptual values (10, 30, 50, 70, and 90 percent black) were chosen and the corresponding percent of physical area inked determined using Williams' graph in Robinson et al. 1984: 185). Assuming that "on" pixels were black and "off" pixels were white, an integer number of pixels (out of the 64 pixels in the fill pattern map) was determined to approximate the percent of area inked. Table 14.1 summarizes this transformation. Finally, uniform dot patterns of pixels were devised and coded into the display program for each of the grey-tones. Figure 14.2 shows an enlargement of these fill patterns.

Table 14.1 Grey scale to fill pattern pixels.

Desired Grey (%)	*Printing Screen (%)*	*LCD Area (%)*	*Pixels (of 64)*
10	4.3	6.25	4
30	12.1	12.5	8
50	34.8	31.25	20
70	55.3	50	32
90	83.3	75	48

Figure 14.2 The 8 by 8 pixel patterns used as area symbols in the experiment. The solid white and solid black patterns were used as the light and dark backgrounds, respectively.

Subjects and Procedure

Eighteen volunteers participated in the experiment. They ranged from 20 to 37 years old. Three were female and 15 were male. Each subject was told that they were "participating in an experimental study of map designs, and were free to stop the experiment and leave at any time," and to think of the experiment "as a primitive video game in which your task is to make comparisons on a series of maps as quickly and accurately as possible." The subject was familiarized with the computer, the maps, the map legends, and told to indicate whether Nevada or Colorado was symbolized as having a greater data magnitude on each map. Colorado and Nevada were pointed out to the subject, and the rationale for using 'F' and 'J' keys to indicate choices of Nevada or Colorado (respectively) was explained.

When the subject understood the task, 32 practice trials with maps like those used for the experiment were performed. All of the subjects quickly demonstrated competence with the task. At the end of the practice, the computer displayed the subject's accuracy rate and average response time, reinforcing the instructions to be both fast and accurate. Finally, the subject performed the experiment with the test stimuli. A different random presentation order was used for each subject to balance sequential learning and fatigue effects across stimuli.

Results and Conclusions

The subjects understood the task and could perform it very well. The overall accuracy rate was .969. The mean RT for the task was 1784 milliseconds (ms).

The raw data exhibited the outliers and skewness common to RT experiments. The extreme RTs for individual trials were 13509 ms and 1 ms, both, incidentally, for correct responses. To reduce the impact of rare and apparently anomalous RTs on subsequent statistical analyses, the data were smoothed. The 24 RTs greater than 4000 ms and the one RT less than 250 ms were replaced by the RT for the left-right replicate of the trial for the subject. In the single case where both the trial and the replicate RT exceeded 4000 ms (a map with lighter symbols for greater magnitudes, a dark ground, and "Less" at the top of the legend), a value of 4000 ms was substituted for both RTs. In all, 25 of the 576 data values were replaced, distributed over 11 of the subjects and 19 of the maps.

The experimental results are summarized in Table 14.2, which presents the means of the smoothed RTs for each of the eight experimental conditions, in Table 14.3, which presents the overall accuracy for each of them, and in Figure 14.3 which summarizes both tables.

Table 14.2 Mean reaction time (ms) for each of the eight experimental conditions.

Ground:	Light	Dark		
More:	Dark	Light	Dark	Light
Legend: "More" at Top	1491	1556	1658	1789
Legend: "Less" at Top	1512	1663	1696	1684

Considering the RTs in Table 14.2, maps on a light background, with darker symbols to represent greater magnitudes, and "More" at the top of the legend were read fastest (1491 ms). The second fastest design (1512 ms) was identical except that "More" was at the bottom of the legend. The slowest design (1789 ms) had a dark background, lighter symbols for greater magnitudes, and "More" at the top of the legend. Thus, the difference between the fastest and the slowest conditions was 298 ms; just longer than a single visual fixation.

Comparisons of the average RTs reveal the magnitude and direction of the RT differences associated with each of the design parameters. An analysis of variance of the RTs indicated that of the three main effects (background value, symbol order, legend label position) and their 2- and 3-way interactions, only the background value effect was statistically significant at the .05 alpha level. On average, maps presented on the light background (1556 ms) were read 151 ms faster than those on a dark background (1707 ms). The rest of the effects were not statistically significant, yet they are interesting. The second largest effect was associated with the symbol order. Maps using darker symbols for greater magnitudes (1589 ms) were read 86 ms faster than those with lighter symbols representing greater magnitudes (1673 ms). Maps in which the symbol for greater magnitude differed most from the background were read 24 ms faster than those in which they were similar (1619 ms versus 1643 ms). Less difference was attributed to the legend design. The maps whose legends had "More" at the top (1623 ms) were read 16 ms faster than those with "Less" at the top (1639 ms).

The high accuracy rates in Table 14.3 indicate that subjects succeeded in their attempts to read these maps. The proportion of trials with correct responses were high for all of the experimental conditions, with values ranging from .93 (67 of 72 trials correct) to .98 (71 of 72 correct). Chi-square analysis indicated that the differences in accuracy are not significant, but it is interesting to note several patterns. Two conditions tied for being most accurate. Subjects were most accurate when lighter symbols represented greater magnitudes, on either a dark background with "More" at the top of the legend or a light background with "Less" at the top of the legend. Subjects were least accurate when dark symbols represented greater magnitudes on the dark background with "More" at the top of the legend. Notice that four conditions tied as the second most accurate, .97 (or 70 out of 72) correct.

Table 14.3 Accuracy for each of the eight experimental conditions.

Ground:	Light	Dark		
More:	Dark	Light	Dark	Light
Legend: "More" at Top	.9722	.9722	.9306	.9861
Legend: "Less" at Top	.9722	.9861	.9722	.9583

The average RTs and aggregate accuracies are plotted together in Figure 14.3. The vertical axis is RT, the horizontal axis separates the two backgrounds, and the lines link experimental conditions with the same legend design. The larger diamonds indicate more accurate conditions; the largest diamonds mean 1 wrong, the next largest mean 2 wrong, the next 3 wrong, and the smallest mean 5 wrong. The figure suggests several generalizations. The most accurately read maps were also the slowest, but accuracy is not monotonically related to RT. The range of accuracies is greater for the dark background than for the light background. Maps on light backgrounds tend to be read faster, and on average more accurately. Figure 14.3 also reveals that maps with lighter symbols for greater magnitudes and "Less" at the top of the legends had unusually similar RTs across the two backgrounds.

Summarizing the results then, only the background value effect produced significantly different RTs and there were no significant differences among the aggregate accuracies. Maps on light backgrounds were read faster than those on dark backgrounds. Stretching the data perhaps further than it should be stretched, one might note the (insignificant) differences among these conditions and suggest that conventional looking choropleth maps (*i.e.*, those presented on light backgrounds, with darker symbols representing more, and with legends that put "More" at the top) are to be preferred.

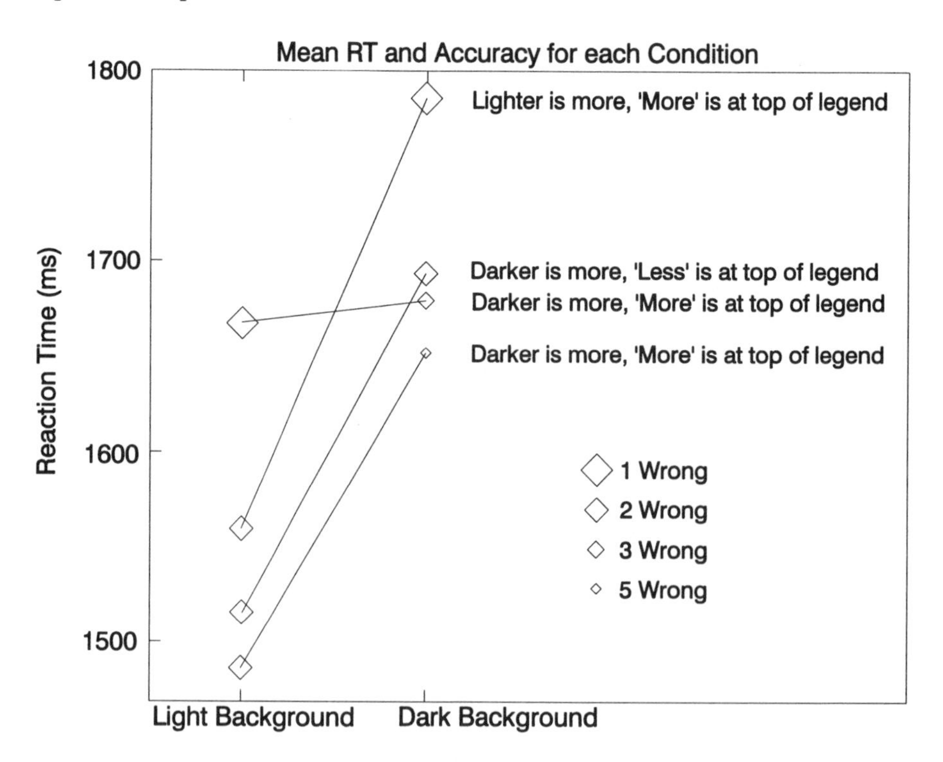

Figure 14.3 Summary of the reaction times and accuracies associated with the eight experimental conditions.

DISCUSSION

This section discusses the implications of these results for our understanding of choropleth symbology. On the practical (and not at all startling) side, the high accuracy rates for all conditions show that uniform dot shading can be used to display choropleth maps on bi-level monochrome LCD panels. The experiment provides evidence that choropleth maps on light backgrounds are read faster than those on dark backgrounds. Perhaps the main observation to be made is that people are very good at making sense out of even very odd maps.

This experiment was similar to earlier studies (McGranaghan, 1988, 1989), but differed from them in that subjects were asked to interpret maps according to a legend. Thus, data were collected that reflect the effects of using symbology in ways that may have been counter intuitive or unnatural to the subjects. The measured effects were small and for the most part not statistically significant at normal alpha levels. Interestingly, the effect of background was significant, indicating that maps on light backgrounds are read faster. Further, the directions of the main effects are consistent with the notion that conventional-looking maps (i.e., those on light grounds, with darker symbols for greater magnitudes, and "standard legends") are read faster than unconventional maps.

These observations should be accounted for by a theory of choropleth map reading. Why are lighter backgrounds faster? Several mechanisms have been suggested to explain how we make sense of choropleth symbols, but these have concentrated on symbol order rather than background. Associating darker symbols (more ink) with greater data magnitudes, and associating gloomier conditions with darker symbols were mentioned before. Neither is a satisfying explanation of the background effect. Following map legends, intra-image contrast, and convention are other possibilities. What can we make of these data?

If map readers followed map legends perfectly, one would expect that all maps with legends would be interpreted with equal ease. That is not the case. The high accuracy rates on all of the conditions indicates that map readers do follow legends, but the RT differences indicate that differently designed maps require more or less processing effort. Background value was associated with significant RT differences, leading one to conclude that some cognitive or perceptual effect involving background value influences map reading performance (in concert with the legend).

That again raises the notion of contrast within the map influencing interpretation of choropleth symbology. If contrast were the only factor, one would expect to find that maps in which the symbol for the greatest magnitudes is most different from the background are fastest, and, given equal contrast, maps with darker symbols for greater magnitudes on light backgrounds and maps with lighter symbols for greater magnitudes on a darker background should have equal RTs. Further, one would expect that the two cases in which the background and the symbol for greater magnitudes are similar should have equal RTs, and these should be greater than the RTs for the first two cases. The data in this experiment do not exhibit this pattern. The high contrast stimuli were both fastest (light background, darker symbols for greater magnitudes) and slowest (dark background, lighter symbols for greater magnitudes). This argues against contrast as the only driver of map interpretation and suggests that there is something superior about the design that uses darker symbols for greater magnitudes and presents this on a light background.

What about familiarity and convention? People seem to prefer working with positive contrast on LCDs. Familiarity with conventionally printed text and maps is a possible explanation. Once it is learned, applying the convention should require little cognitive effort. Conventional maps would be read quickly while unconventional maps may seem odd and take more time to decipher. In this study, the most conventional map was read faster than all of the others, but it was not read more accurately. On both backgrounds, the most accurately read maps used lighter symbols for greater magnitude. This is consistent with the notion that subjects took an extra glance to check an odd looking map, thus taking longer and at the stroke increasing accuracy for these conditions. This line of reasoning is intriguing, but it requires refinement and further experimentation.

Perhaps the most important observation that can be made from the current study is that people are very good pattern processors and very adaptive to map design. Given half a chance they will make sense out of even very odd maps. Giving people a legend and a monotonic value progression virtually ensures that they will make sense of the map. However, the rules and mechanisms by which that sense is made are not yet clear enough that we can predict how well or how rapidly a given individual will perform with a given map. These depend on a combination of learned familiarity with symbolic conventions and available contrast within a map display, but the functional relationships have not been determined. In all likelihood they vary from person to person.

Is it worth worrying about it? Maybe, maybe not. Monmonier (1979, 1980) argued persuasively that fine perceptual adjustments to maps are probably not warranted because they are likely to be swamped by uncontrollable circumstances in the map reading environment and by individual differences among map readers. A map reader reversing an LCD's contrast switch may be a gross example of uncontrollable circumstance. Practically, and thankfully, one would need to read a lot of maps for the RT advantages measured here to add up to a good coffee break.

ACKNOWLEDGEMENTS

Support from NSF Grant SES 88-10917 for the NCGIA is gratefully acknowledged, as is the thoughtful advice on data analysis provided by Dr. Lawrence Smith of the Psychology Department at the University of Maine - Orono, and the cooperation of the subjects who agreed to participate in this study.

REFERENCES

Borland International. (1988). *Turbo C reference guide.* Scotts Valley, CA: Borland International.

Brody, T.P. (1992). Active-matrix TFTs are in trouble. Cadmium selenide is the answer. *Information Display,* 8(2), 5-9.

Castner, H., and Robinson, A. (1969). Dot area symbols: The influence of pattern on their perception, pp. 20-28, Technical Monograph No. CA-4, American Congress on Surveying and Mapping, Washington, D.C.

Cox, C. (1980). The effects of background on the equal value grey scale. *Cartographica,* 17(1), 53-71.

Crawford, P. (1971). Perception of gray-tone symbols. *Annals of the Association of American Geographers,* 61(4), 721-735.

Cuff, D. (1973). Shading on choropleth maps: Some suspicions confirmed. *Proceedings of the Association of American Geographers,* 5, 50-54.

Cuff, D. (1974). Impending conflict in colour guidelines for maps of statistical surfaces. *The Canadian Cartographer,* 11(1), 54-58.

Cuff, D., and Mattson, M. (1982). *Thematic maps: Their design and production.* New York: Methuen.

DeLucia, A., and Hiller, D. (1982). Natural legend design for thematic maps. *The Cartographic Journal,* 19(1), 46-52.

Display Technology Report (1995). Sharp's 11.3" LCDs. *Display Technology Report,* 1(7), 5. Insight Communications Inc., Littleton, MA.

Drzaic, P. (1990). Liquid-crystal displays. *Information Display,* 6, n. 12, pp. 15-17.

Groop, R. and Smith, R. (1982). Matrix line printer maps. *The American Cartographer,* 9(1), 19-24.

Gulick, P. (1991). Liquid-crystal Displays, *Information Display,* 7(12), 16-18.

Jenks, G. and Knos, D. (1961). The use of shaded patterns in graded series. *Annals of the Association of American Geographers,* 51, 316-334.

Kimerling A. (1975). A cartographic study of equal value gray scales for use with screened areas. *The American Cartographer,* 2, 119-127.

Kimerling A. (1985). The comparison of equal value gray scales. *The American Cartographer,* 12, 132-142.

Lavin, S. (1986). Mapping continuous distributions using dot density shading. *The American Cartographer,* 13(2), 140-150.

Leonard, J., and Buttenfield, B. (1989). An equal value gray scale for laser printer mapping. *The American Cartographer,* 16(2), 97-107.

Litton Systems Canada Ltd. (1992). Advanced technology in action (advertisement). *Information Display,* 8(11), inside cover.

Martin, R. et al. (1993). A 6.3M pixel AMLCD. *Society for Information Display International Symposium Digest of Technical Papers,* pp. 704-707.

McGranaghan, M. (1986). Effective use of color in choropleth maps on CRTs. [Unpublished Ph.D. dissertation.] Buffalo, NY: State University of New York at Buffalo.

McGranaghan, M. (1988). *Symbolizing quantitative differences on color CRTs.* Report number 1115, Naval Submarine Medical Research Laboratory, Submarine Base Groton, CT.

McGranaghan, M. (1989). Ordering choropleth map symbols: The effect of background. *The American Cartographer,* 16(4), 279-285.

Monmonier, M. (1979). Modelling the effect of reproduction noise on continuous-tone area symbols. *The Cartographic Journal,* 16, 86-96.

Monmonier, M. (1980). The hopeless pursuit of purification in cartographic communication: A comparison of graphic-arts and perceptual distortions of graytone symbols. *Cartographica,* 17(1), 24-37.

Paslawski, J. (1983). Natural legend design for thematic maps. *The Cartographic Journal*, 20(1), 36-37.

Patton, J., and Crawford, P. (1977). The perception of hypsometric colours. *The Cartographic Journal*, 14(2), 115-127.

Plumb, G., and Slocum, T. (1986). Alternative designs for dot matrix printer maps. *The American Cartographer*, 13(2), 121-133.

Robinson, A.H., Sale, R.D., Morrison, J.L., and Muehrcke, P.C. (1984). *Elements of cartography*. 5th ed. New York: John Wiley and Sons.

Ross, D. (1988). Automotive LCDs. *Information Display*, 4(2), 12-16.

Sadowsky, R. (1988). TCTIMER (v1.0) - Routines for high-resolution timing of events for Turbo C, public domain software.

Shortridge, B., and Welch, R. (1982). The effect of stimulus redundancy on the discrimination of town size on maps. *The American Cartographer*, 9(1), 69-80.

Trimble Navigation Marine Division (1991). *Trimble Navigation 1991 Marine Navigation Products Catalog*, Trimble Navigation Marine Division, 645 N. Mary Ave., Sunnyvale CA 94086.

Werner, K. (1988). The flowering of liquid-crystal technology. *Information Display*, 4(2), 6-11.

Williams, R. (1958). Equal-appearing intervals for printed screens. *Annals of the Association of American Geographers*, 48(2), 132-139.

Williams, R. (1960). Map symbols: The curve of the grey spectrum - An answer. *Annals of the Association of American Geographers*, 50(4), 487-491.

Williamson, G. (1982). The equal contrast gray scale. *The American Cartographer*, 9(2), 131-139.

15

An Evaluation of Multivariate, Quantitative Point Symbols for Maps

Elisabeth S. Nelson

Department of Geography, San Diego State University

Patricia Gilmartin

Department of Geography, University of South Carolina

INTRODUCTION

> *Getting information from a table is like extracting sunlight from a cucumber.*
>
> Farquhar and Farquhar, 1891.

The use of visual displays to represent object locations in space has a long history in human development. The map is perhaps the earliest form of these spatial representations. More abstract than the map is the diagram and both have played important roles in graphic communication. By 1786, with the publication of *Playfair's Political Atlas*, most symbols representing univariate data had been invented (Wainer and Thissen, 1981). Attempts to represent three or more dimensions with symbols, however, lagged greatly due to the complex nature of the data. A renewed interest in the topic has occurred during the last 20 years or so, due, in part, to the advent of computers and computer mapping. A number of researchers from various fields have devised symbols to represent multivariate data during this time (see Figure 15.1 for illustrations of a few such symbols). For example, Anderson (1960) proposed a glyph symbol, in which the centre of the symbol consisted of a circle. Radiating from the circle are rays, the length of which represent the values of the various variables in question. Many variations on this theme have also been proposed. Seigal et al. (1971) suggested that the ends of the rays be connected and the rays themselves deleted, leaving only the outline of the polygon. Shapes of the resulting polygons could then be compared. More recently, Borg and Staufenbiel (1992) have suggested the use of factorial suns to represent multivariate data. Factorial suns combine the values of the variables and their correlational structure, supposedly making them easier to interpret than some of the other multivariate symbols. Perhaps the most creative of all these multivariate symbols, however, is the Chernoff face, developed in 1973 by Herman Chernoff (Chernoff, 1973). This symbol takes advantage of humans' ability to perceive and remember small changes in the structure of faces. A Chernoff face can be composed of up to 18 variables which are represented by various facial features such as the mouth, ears, and eyes.

Although numerous methods for representing multivariate data have been proposed, there has been little research on their effectiveness, especially in a map environment. How well do these symbols work, and what are their relative strengths and weaknesses? The purpose of the research reported here was to evaluate four types of multivariate, quantitative point symbols within a cartographic context. Of the four symbols examined, two were abstract geometric designs (crosses and circles), one was a variant of Chernoff faces, and one was a rectangular symbol containing graduated alphabetic characters which stood for the mapped variables. Communicative effectiveness was measured by presenting subjects with maps of socioeconomic data symbolized by the methods just described and asking them to interpret the mapped distributions. Cartographers and others still have much to learn about how to graphically represent multivariate data, and it is hoped that the research reported here will enhance our understanding of how humans perceive multivariate point symbols.

BACKGROUND: RESEARCH AND THEORY

Cartographers have paid scant attention to the problem of representing two or more variables on a single map, and most such efforts have focused on choropleth techniques rather than point symbols. Almost 30 years ago Board and Wilson (1966) proposed a method for symbolizing three quantitative variables on a black and white choropleth map. They employed line patterns printed in three orientations (vertical, horizontal, and at a 45° angle) in different spacings and line weights to represent the proportion of the population in Southern Africa in three racial categories. Similar bivariate choropleth maps consisting of variably-spaced lines in crossed-line shading patterns were studied by Carstensen (1982, 1984, 1986a, 1986b) and Lavin and Archer (1984).

Other authors have investigated subjects' ability to interpret spectrally-encoded bivariate choropleth maps designed using three primary colours (Olson, 1981; Wainer and Francolini, 1980). Later Eyton (1984) proposed an alternative colour system based on pairs of complementary colours rather than the primary colours employed on the earlier maps.

After reviewing several methods of multivariate choropleth mapping, Chang concluded that, "The future of multi-component maps ultimately depends upon their reception by the map reader. There is an urgent need to study the communicative effectiveness of multi- component maps from the standpoints of researchers and the general audience" (Chang, 1982: 103). His call for empirical research on such maps has been largely ignored by cartographers, however.

Cartographic research on multidimensional point symbols is virtually nonexistent, a fact also noted recently by MacEachren (1994). Although several authors have formulated designs for such symbols, most have not conducted any empirical research to evaluate how well they function. For example, Carlyle and Carlyle (1977) devised an ellipse which represented three variables: the number of sheep sold at various markets in Scotland (symbolized by the length of the semi-major axis of the ellipse), the distance the sheep had been transported to market (indicated by the length of the semi-minor axis), and the proportion of sales accounted for by various breeds (denoted by shading sectors of the ellipse, pie-graph style). However, they simply proposed the design and demonstrated the calculations required; they did not test its effectiveness with map readers.

Several other multivariate point symbols for maps also have been utilized. Turner designed a map using Chernoff faces to symbolize four socio-economic variables for Los Angeles (reproduced in Muehrcke and Muehrcke, 1992: 162). A similar map showing nine quality-of-life variables for the United States was published by Wainer (1979). Bi-variate ray-glyph point symbols were used by Carr to symbolize trends in sulphate and nitrate deposition in the eastern U.S. (Carr, 1991; Carr, et al., 1992). The symbol consisted of two line segments joined end to end, one ray pointing left, to represent sulphates, and one pointing right, to indicate nitrates. The angle of the lines away from vertical (up or down) symbolized the rate of increased or decreased deposition per year. In order to illustrate the number of course offerings in cartography and the relative importance of the cartography programs at U. S. colleges and universities, Dahlberg combined circle size and shading value in a bivariate point symbol (1981: 111). And Bertin (1983) discussed several ways in which multivariate data can be symbolized with point symbols. For example, he constructed a map to show the distribution of three anthropomorphic characteristics of Europeans using point symbols which varied in size (to indicate height), value (for predominant hair color) and shape and orientation (to represent the cephalic index of the population). He maintained, however, that it is not possible to construct multivariate point symbols which will "provide an immediate response to all types of question [sic]"(p. 154). Specifically, he stated that such a map can answer the question "In a given area, what is there?" but cannot deal efficiently with the question, "Where are the greatest values of X?" He provided no rationale or evidence for this claim, however.

In the only empirical study of multivariate point symbols we have found, Rhind et al. (1973) tested the effectiveness of a three-arm "wind-rose"-type symbol for summarizing geochemical data. The length of each arm represented the concentration of copper, lead, or zinc in stream sediments, and the symbols' locations on the map showed where the sediment samples had been collected. The researchers settled on this particular symbol after evaluating alternative designs in terms of how easily they could be computer-produced (given the computational and plotting technology of 20 years ago) and how much information the symbols could convey. An experiment involving several counting and estimation tasks was conducted to determine how the wind-rose symbol would function under experimental conditions of map scale and background "noise," consisting of a geological base map. Results of the study were troubling. The authors found that none of the experimental variables had much effect and that under all conditions, the subjects (many of whom were professional geologists) performed poorly. Subjects answered correctly about 42 percent of the estimation questions and only 18 percent of the counting questions. Not surprisingly, the authors stated the need for additional research on the design and perception of multivariate point symbols.

The lack of empirical research on multivariate point symbols represents a significant gap in the existing body of research on cartographic design. In recent years, increasingly powerful computers have facilitated the production and manipulation of large multivariate data sets as well as complex graphical displays. One outgrowth of these advances in computer technology and the accompanying proliferation of data has been a surge of interest in visualization methods for analytical, educational, and presentational purposes. Thus, some understanding of how best to represent multivariate data graphically would seem to be an issue of some importance to cartographers and others.

Researchers in other disciplines, notably psychology and statistics, have conducted a number of studies that are relevant to the cartographic questions addressed here. Of the several multivariate point symbols they have explored, Chernoff faces probably have received the most attention. Jacob et al. (1976) examined whether faces have an advantage for displaying multivariate data when compared to other graphic designs. He asked subjects to sort a number of data sets into groups, based on a prototypical symbol for each group. The data sets included faces, polygons, and digits. The results of this experiment showed that subjects were most accurate when sorting faces, followed by polygons, and digits. In a second experiment, the authors explored the "... generality of the face-superiority effect ..." (Jacob et al., 1976: 195) found in his first experiment. In this task, he exposed subjects to one of four display sets. The data sets used for the displays were faces, upside down faces, digits, glyphs, and polygons. Each symbol in a display set was paired with its corresponding Greek letter name. After memorizing their assigned display, subjects were shown the display without names and asked to name each symbol. The results, based on the number of trial sequences it took to correctly name all symbols in a display, showed that performance was significantly better with faces than with the other symbols. Subject introspection indicated that they remembered the faces by using holistic techniques labelling the face as happy, sad, etc. while the other symbols were remembered by focusing on individual features.

In another experiment, Huff and Black (1978) examined subjects' ability to perform a cluster analysis of data using Chernoff faces. They initially ran a computerized cluster analysis on data for thirteen cities, in an effort to determine the similarity of cities based on seven economic measures. The authors then asked subjects to manually sort face symbols for each of these cities into five distinct groups to determine how accurately they could match the computer's analysis. Results indicated that the highest correlations between the actual and expected groupings occurred only with those groups containing more than one city. Following this initial task, subjects were asked to rank the seven facial features used in order of importance in differentiating the faces from one another. In an effort to determine if a closer correspondence couldn't be obtained between the computer and subjects, Huff and Black conducted a second experiment using these rankings. Using an ANOVA, they determined the relative importance of the economic measures in explaining the variance associated with each measure, and then assigned each variable by rank to the ranked facial features. They then asked a new set of subjects to repeat the manual cluster analysis. Results showed that correlations increased dramatically, indicating that the assignment of variables to facial features is very important. This suggests that people gravitate towards specific features to perceive differences in facial structures, which is in direct opposition to perceived differences on the basis of some holistic criteria.

After reviewing numerous studies of comparative graphics, Carswell and Wickens (1988) concluded that Chernoff face displays are quite effective for making complex comparisons, especially when a large number of variables are involved (i.e., a greater number of facial features are varied). The faces' advantage tends to diminish in tasks requiring only a few comparisons, according to the reviewers, probably because, "... the overall distinctiveness of faces is a function of the number of dissimilar features" (p. 34). Schmid was quite critical of Chernoff faces for communicating statistical information (1983). Commenting on Wainer's map of the U.S. (1979), Schmid said that, "The deciphering of the facial characteristics of [Chernoff faces], along with an attempt to derive a clear overall understanding of what the map purportedly conveys is a time-consuming and frustrating experience. [The map] seems to possess more the characteristics of an esoteric puzzle than a clear, straightforward, readily understood, and reliable vehicle of visual communication" (pp. 188-9). And Chernoff, himself, stated that, "Faces are of little use to illustrate or communicate unless the audience is specially trained in which case they can be of limited use" (Chernoff, 1978: 2). Thus there has been considerable debate regarding the efficacy of Chernoff faces for communicating multivariate data, and the question is still far from resolved.

Researchers have evaluated a variety of other graphic methods for portraying multivariate data in addition to Chernoff faces. Of these, the ones that might serve as cartographic point symbols include circular Fourier plots, glyphs, petal charts, or four-fold circular displays, bar charts, and polygons (polar line profiles) (see Figure 15.1 for examples of some of these symbols). It is difficult to draw general conclusions from these investigations because they are so diverse in terms of the symbols employed, the kinds of questions asked, the subjects involved, and the experimental design. Nevertheless, it appears that for questions addressing complex comparisons and integrating several categories of information, the most effective symbols were those that, visually, were most integrative or holistic (i.e., polygons, faces, and Fourier plots) (Carswell and Wickens, 1988). Least effective were line graphs, petal charts, and bar charts. (Much of the comparative research on these and other symbols has been reviewed by Carswell and Wickens (1988), DeSanctis (1984), and MacDonald-Ross (1977). The interested reader should refer to these reviews and primary sources cited therein for more detailed information.)

The conclusions summarized in the preceding paragraph bring us to the theoretical dimension which underlies some of psychologists' interest in the perception of graphic forms: selective attention. Selective attention can be defined as the ability to attend to one dimension of a stimulus, while ignoring all others. If the dimensions of a figure can be attended to independently of one another, the figure is considered to be ungrouped or separable; if not, it is termed "integral." Selective attention is relevant to the design of multivariate symbols in that those symbols designed to facilitate selective attention are inherently different from multivariate symbols designed to be perceived holistically.

Garner and Felfoldy (1970) performed one of the first experiments to explore the idea of selective attention. Using a speeded classification task, they asked subjects to sort colour chips on the basis of either one relevant dimension, two perfectly correlated dimensions, or two uncorrelated dimensions. The dimensions used in testing were value and chroma. Results showed that it was easier to classify the stimuli with one relevant dimension or two correlated dimensions, suggesting that subjects were not able to selectively attend to either value or chroma in the stimulus while ignoring the other characteristic.

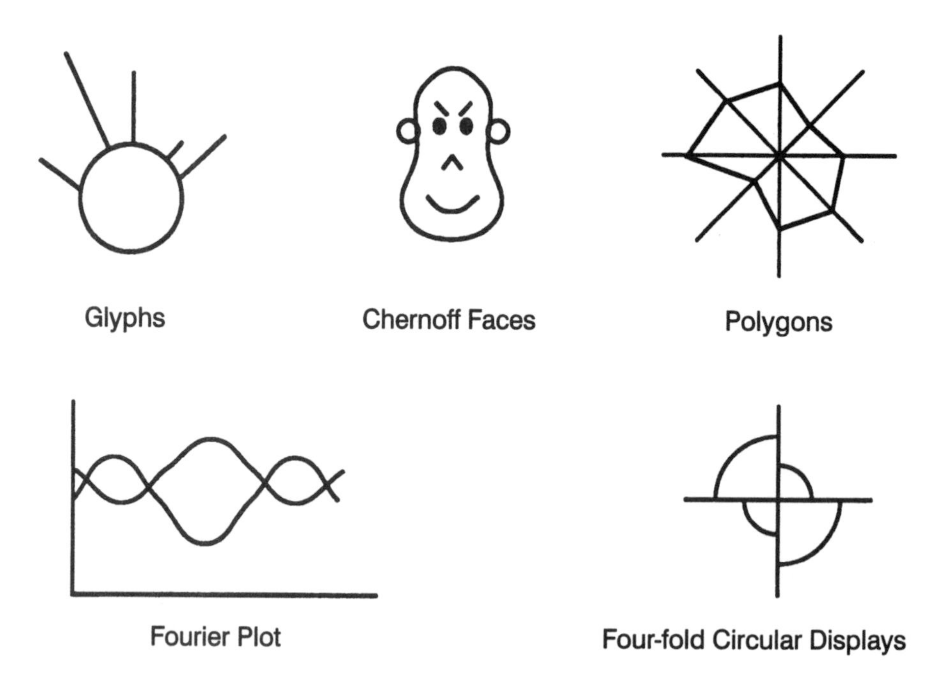

Figure 15.1 Examples of multivariate point symbols designed by various other researchers (after Carswell and Wickens, 1988: 13).

Pomerantz and Garner (1973) performed a similar experiment, testing vertical and horizontal orientation using sets of parenthesis-like symbols. Their results corroborated Garner and Felfoldy's study. Pomerantz and Schwaitzberg (1975) extended this work in 1975 by examining the role of proximity in the grouping process. The methodology for this study was similar to the 1973 work, but used different parenthesis-like symbols to emphasize the concept of proximity. Results showed that variable proximity does affect perception of symbol grouping.

Carswell and Wickens (1990) examined the selective attention issue by investigating the incidence of integrality and configurality in graphic displays. Integrality assumes that the multiple physical attributes of a symbol correspond to a single perceptual code, whereas configurality assumes that although symbol attributes maintain separate perceptual codes, emergent properties resulting from their combinations are also coded. Using a speeded classification task similar to the previous studies, they asked subjects to sort symbol sets composed of attributes common to graphical displays (orientation, colour, and linear extent). Their results indicated that none of the stimulus sets were integral in nature. Rather, a continuum of configurality existed for the stimulus sets, suggesting that this principle may be more accurate in describing the processing advantages and disadvantages of multivariate symbols.

In a cartographic context, both Shortridge (1982) and MacEachren (1994) have discussed the potential relevance of selective attention to map design. Shortridge, for example, suggested that separable graphic dimensions, such as hue and shape, would be good choices for emphasizing qualitative data differences, while integral dimensions (saturation and value) might be more appropriate for symbolizing quantitative variation (p. 165). In fact, several cartographic studies have been published in which the authors did employ two visual dimensions to symbolize one variable (redundant coding). Lavin et al. (1986) and Amedeo and Kramer (1991) represented the distribution of rainfall by combining a dot density background with an overlay of isolines. Their notion was that the continuously varying background shading would convey the impression of a continuous value transition, while the isolines would communicate values at specific locations. Dobson (1983) also examined the effect of redundant visual coding by adding grey-tone shading to proportional circles: the greater the quantity represented by the circle, the larger the circle in area and the darker the shading within the circle. He found that the redundant symbolization resulted in subjects responding more quickly and accurately.

Although none of the redundantly-coded maps mentioned above were created with the principles of selective attention in mind, they might well have been. It would seem logical that to emphasize and clarify a single graphic message utilizing more than one visual variable, the variables chosen should be integral. That is, they should be perceived holistically, as a single dimension, as are value and chroma and the height and width of rectangles, according to psychologists (Garner and Felfoldy, 1970; Dykes, 1979). Conversely, if the cartographer's goal is to represent two or more thematic variables with multidimensional symbols so that the variables' individual characteristics can be retrieved, then the symbols should consist of separable visual dimensions.

As is so often the case in cartography, however, design goals are seldom that discrete. As many other authors have observed, thematic maps must serve a variety of functions, from being a static storehouse of spatial information to facilitating sophisticated synthesis and analysis. We want to enable readers to make relatively accurate estimates of "what" and "how much" exists at specific locations on the map, but in addition, it should be possible for them to detect spatial patterns and correlations at a regional level. Things become much more complicated with multivariate maps. In addition to the local/regional distinction, multivariate point symbols, ideally, should convey information at two other levels: discretely, providing data about the variables individually, and holistically, about the variables as a group. Thus the design challenge can be summarized as a two-by-two matrix with spatial dimensions (local/regional) on one axis and data dimensions on the other (individual categories versus an integrated summary). It may be that these goals create conflicting demands on multivariate symbols within the framework of selective attention. Symbols such as the Chernoff face, which previous research has suggested is an integral figure, may function well in that role, but be quite difficult from which to obtain information about individual variables. In the following study we evaluate four different four-variable point symbols to see how well they function for each cell of the two-by-two matrix described above.

EXPERIMENTAL DESIGN

The Symbols

Four multivariate, quantitative point symbols were designed to represent the socioeconomic variables of crime rate, education level, income level, and health care availability for a nine-county region. The choice of symbols was guided by the research and theory outlined above, that some graphical symbols, such as Chernoff faces, are experienced holistically, while others seem to retain the individuality of their separate dimensions. The symbols used for this research included a variant of Chernoff faces, two abstract symbols—a circle divided into quadrants and a cross—and boxed letters representing the variable names (Figure 15.2). Each of the four variables represented by the symbols could have a value of low, medium, or high. These ordinal classes were represented graphically in a variety of ways, depending on the symbol in question. As illustrated in Figure 15.3, crime rate was symbolized by changing the head shape of the face symbol (Figure 3a,15), by increasing or decreasing the length of one of the arms of the cross (Figure 3b,15), by manipulating the grey tone in one quadrant of the circle (Figure 3c,15), or by changing the size of the letter "C" in Figure 3d,15. Based on the three possible levels of the four variables represented by each of the four symbols, a total of 81 possible combinations were available.

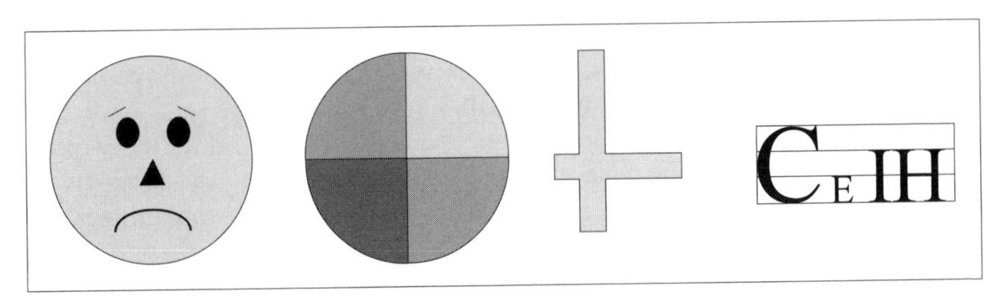

Figure 15.2 The four multivariate, quantitative point symbols used in the study.

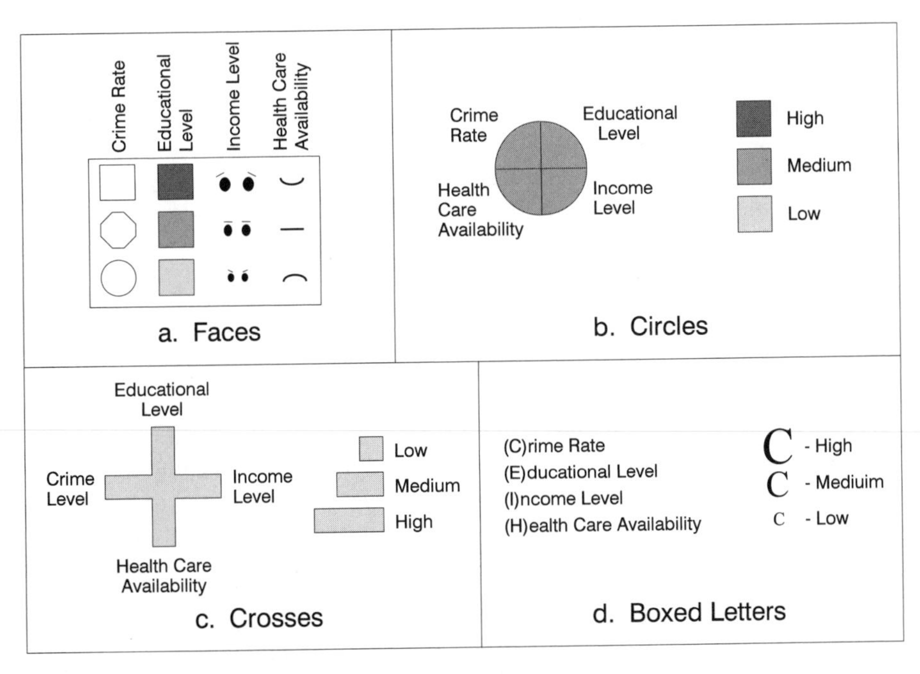

Figure 15.3 Legends for each symbol illustrating the attributes represented by the symbol parts and the graphic methods employed to represent ordinal levels of value.

Base Maps and Questions

A digital file of political boundaries from the state of Kentucky was used to create a base map consisting of nine counties. Thirty maps were randomly generated for each symbol-type by choosing among the 81 possible symbol variations for each of the nine counties on the map. The questions about the maps fell into one of the four categories resulting from the two-by-two matrix described earlier. Subjects were asked to focus on either an individual symbol (within a highlighted county) or a cluster of symbols (within a highlighted region). They also were queried about either the value of a single category within the symbol or the composite value of the four variables together. The latter question was explained to the subjects as being the county's or region's overall "quality of life," and for those cases, all four variables carried the connotation of "good," "poor," or "middling" living conditions. These questions were designed with the issues of symbol integrality and variable separability in mind. For example, the quality of life question should address the holistic quality of the symbols being tested, while the questions concerning individual variables should test how easily the symbol's parts could be accessed independently of one another.

Each variable—crime rate, education level, income level and health care availability—was the subject of a question six times, with the remaining six questions (for a total of 30 questions) concerning the overall quality of life for a county or region. These map combinations were randomized by computer. Each symbol set—face, circle, cross, letters—used the same 30 map/question combinations. The experiment was divided into four blocks, one for each symbol type. The block orders were randomized for each subject and within each block, subjects alternated between seeing the maps in a top-to-bottom order, or a bottom-to-top order.

Subjects and Methods

Thirty-two undergraduate volunteers from the University of South Carolina, 9 females and 23 males, took part in the study. Subjects were tested individually on an IBM 486 computer, and each subject participated in all four blocks of the experiment. The experiment began with some oral instructions summarizing the subject's task; the instructions were then repeated in text on the computer screen. Each block of symbol-types began with a practice session designed to familiarize the subject with the experimental procedure and the symbol which they would be seeing. Subjects could spend as much time as they wanted in practice trials before proceeding with the actual experiment. Each trial was structured so that subjects first saw a question on a blank screen. When they had studied the question and were ready to move on, they pressed a key and a map and legend appeared on the screen with the question (see Figure 15.4). Subjects had been instructed to answer the question as quickly and accurately as possible by pressing one of three keys (low, medium, high) to indicate their answer. This procedure was repeated for each of the blocks. Subjects thus answered a total of 120 questions—30 questions per symbol type. Reaction times and accuracy were recorded for each trial.

Hypotheses

The communicative effectiveness of the four symbols was tested by examining the idea that some multivariate symbols are more effective for representing variables in a holistic manner in which variable dimensions cannot be easily separated, while other symbols lack this property and are more efficient for data sets in which the separability of variables is important. It was hypothesized that the variant of

Chernoff's faces would be the most appropriate symbol choice for data perceived in a holistic fashion, such as quality of life for a place, whereas the more abstract symbols—circle, cross, and letters—would provide better performance for questions targeting individual variables.

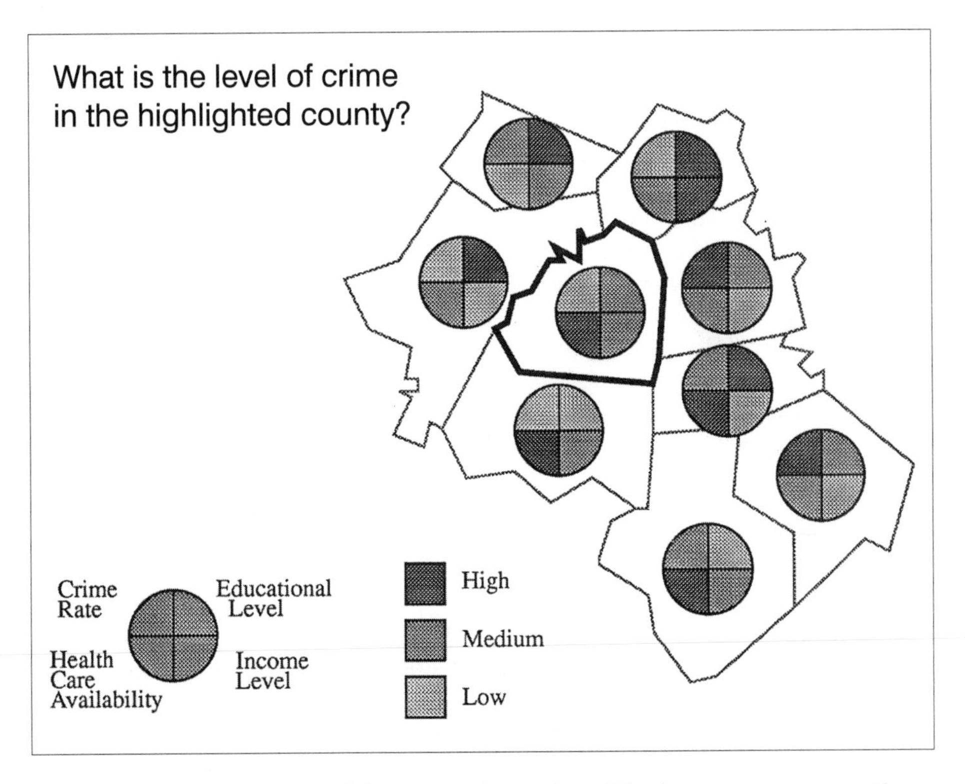

Figure 15.4 Presentation of the map and question. (On the computer screen the target county was outlined in red.)

ANALYSES AND RESULTS

The raw data files were edited to eliminate one negative reaction time (an instance in which a subject pressed a response key before the map appeared on the screen) and responses with extreme reaction times, as defined by Tukey's "outer fences" (Tukey, 1977: 43-44). The latter procedure identified 47 responses (1.2 percent of the total number of responses) as being extreme. Results of the study are summarized in the following paragraphs, figures, and tables.

There were two dependent variables in the research design, namely percent correct answers, and reaction times; and three independent variables: 1) symbol type, 2) whether the question referred to one part of the symbol or the entire symbol, and 3) whether the question was for a single county or a multi-county region.

For the dependent variable of percent correct, Table 15.1 shows that subjects' responses were relatively accurate in all cases, averaging about 95 percent correct overall. (High accuracy is normal and expected in the reaction time paradigm. In fact, some researchers ignore incorrect responses and analyse only reaction times associated with correct responses.) Accuracy was slightly lower for the Chernoff face symbols than for the other three types. In general, subjects were less accurate in reading individual parts of the symbols than the wholes and in focusing on a single county than in scanning a several-county region.

Table 15.1 Percent correct by symbol type and question type.

		Question Type			
	Overall	*Symbol*		*Target*	
Symbol Type	*Mean*	Part	Whole	Region	County
Faces	93	88	98	95	92
Circles	96	94	97	97	94
Crosses	95	92	98	97	94
Boxed Letters	95	92	97	99	91
Overall Mean		92	98	97	93

Analysis of variance (ANOVA) for percent correct responses showed that the slight differences due to symbol type were not significantly different from each other ($p < .31$), while the differences between attending to the whole symbol versus a part and a region versus a county were ($p < .001$) (see Table 15.2). However, a significant interaction existed between the county/region questions and the whole/part discriminations: on questions that referred to an individual part of the symbol, subject's answers were more accurate when the questions were about a region (96 percent correct) rather than about an individual county (87 percent correct).

Table 15.2 ANOVA results for percent correct responses.

Overall Model: Dependent Variable = Percent Correct	
Symbol	PR > F (1.20) = 0.31
Feature Type (Whole/Part)	PR > F (33.14) = 0.001
Question Type (County/Region)	PR > F (14.72) = 0.001
Symbol x Feature	PR > F (1.59) = 0.19
Symbol x Question	PR > F (1.84) = 0.14
Feature x Question	PR > F (19.16) = 0.001
Symbol x Feature x Question	PR > F (0.36) = 0.78

When the question required subjects to look at the whole symbol at once, or one part of a symbol for a region, the difference in accuracy almost disappeared, with only two points difference among the scores. Figure 15.5 illustrates this interaction. It can be seen from Figure 15.5 that all the statistically significant differences in percent correct responses found in the study are due to one factor: that subjects were less accurate in their answers to questions about an individual attribute in an individual county than any other kind of question. This pattern was consistent for all four symbol types.

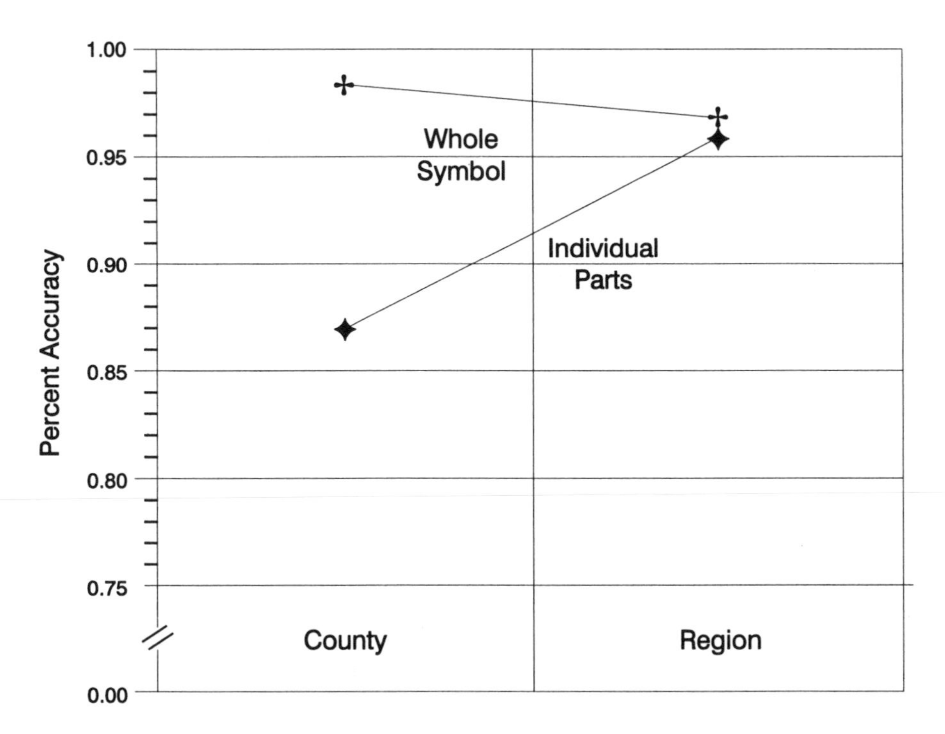

Figure 15.5 Interaction for percent correct—whole/part by county/region.

A summary of reaction times in the study is shown in Table 15.3. There were large differences in reaction times among the four symbol types, with the Chernoff faces having the longest times and the boxed letter symbols the shortest. An analysis of variance and Tukey's HSD (Honest Significant Difference) multiple comparisons test disclosed that the mean reaction times for all four symbols were significantly different from each other ($p < .001$) (Table 15.4). The time required to interpret several symbols within an area was significantly higher ($p < .001$) than the time needed to read the symbol for a single county, and a significant two-way interaction ($p < .015$) was found between symbol type and whether subjects had to look at the whole symbol or just one of its parts. As Figure 15.6 shows, with the shaded circles and cross symbols, subjects took longer to attend to their individual parts than to integrate the message of the whole symbol. The results for the other two symbols were just the opposite: individual parts of the faces and letters could be read faster then could the whole. No other significant interactions were found in the results for reaction times.

Table 15.3 Mean reaction time (msec) for symbol type and question type.

		Question Type			
		Symbol		*Target*	
Symbol Type	*Overall Mean*	Part	Whole	Region	County
Faces	3756	3686	3827	3922	3590
Circles	3091	3231	2951	3289	2893
Crosses	2729	2807	2651	2896	2562
Boxed Letters	2273	2126	2421	2419	2128
Overall Mean		2963	2962	3132	2793

Table 15.4 ANOVA results for mean reaction times.

Overall Model: Dependent Variable = Reaction Time	
Symbol	PR > F (77.521) = 0.001
Feature Type (Whole/Part)	PR > F (0.00) = 0.997
Question Type (County/Region)	PR > F (22.793) = 0.001
Symbol x Feature	PR > F (3.501) = 0.015
Symbol x Question	PR > F (0.092) = 0.965
Feature x Question	PR > F (1.512) = 0.219
Symbol x Feature x Question	PR > F (0.786) = 0.502

DISCUSSION

The symbols and maps used in this study were rather elementary. The maps contained only nine point symbols, and the questions required subjects to make ordinal judgments, certainly an easier task than estimating numerical quantities. The rudimentary nature of the study can be justified from several perspectives. First, as in all empirical research such as this, it is necessary to control every experimental condition as much as possible, a constraint which tends to produce a somewhat abstract set of conditions. Second, since there is virtually no prior cartographic

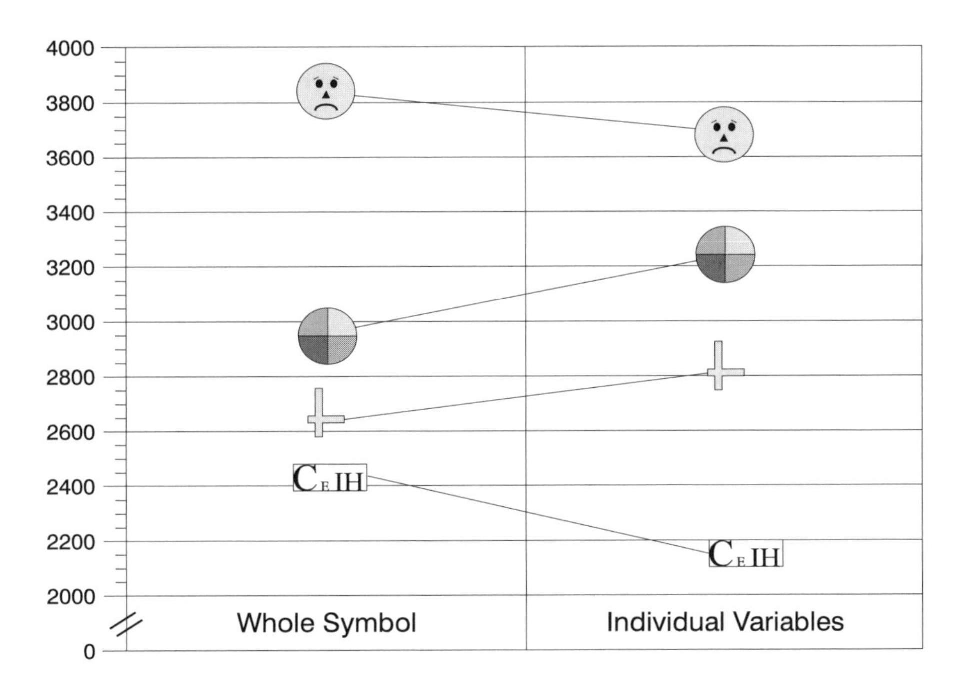

Figure 15.6 Interaction for reaction times—whole/part by symbol type.

research on multivariate point symbols which we could use as a precedent for methods, symbol designs, and hypotheses, we believed that a complex experimental design could have lead to unintelligible results. One thing we did know from previous cartographic studies was that numerical estimation tasks are fraught with problems inherent in the task itself, and these problems could have confounded the results of our research. Thus we chose to avoid numerical estimations altogether. Finally, we believed we could obtain useful information from an initial, basic study which could serve as a foundation for more complex research designs in the future.

The findings of this study provided valuable insights—some expected and some unexpected—into how well people are able to interpret certain types of multivariate point symbols. The accuracy measures demonstrated that if people took whatever time they needed to study the symbols, they could interpret all four types quite accurately. It is understandable that their answers were more accurate for "quality of life" questions, where they could simply look at the appearance of an entire symbol, than for questions about individual variables, for which they had to locate and focus on just one segment of the symbol. In addition, most subjects had to refer to the map legend before answering each question about a single variable, except for the boxed letters symbol. It was apparent throughout the experiment and confirmed by post-test interviews that subjects were not able to memorize most of the symbols during the 30 trials. (That is, they could not remember which variable was symbolized by which arm of the cross, which quadrant of the circle, or which facial feature. The boxed letters required no memorization, of course; the letters served as abbreviations for their referents, much as a pictorial symbol would.)[1]

What was surprising was the interaction shown in Figure 15.5: subjects were considerably less accurate in answering questions about an individual part of a symbol in an individual county than for corresponding parts of several symbols in a region (87 percent versus 96 percent correct). Perhaps some of the errors were due to perceptual difficulties in distinguishing between three levels of the visual variables of which the symbols were composed (grey tones or symbol-component sizes), even though we consciously restricted the number of levels in an attempt to avoid such problems. If this explanation does account for some or most of the difference in accuracy, it would simply underscore an additional problem associated with designing multivariate point symbols. That is, that the component parts of such symbols are necessarily rather small (since the size of the overall symbol is limited by the display dimensions and cartographic design considerations), and perceptual limitations make it difficult for people to compare small shapes, areal shadings, and the like. Cartographers also encounter this problem in designing choropleth shadings for maps having small areal units.

Another possible explanation for why people made more errors in answering questions about one part of one symbol might be that the symbols were, to some degree, integral, in selective attention terms, making it difficult to focus on just one portion of them. If this reasoning is correct, however, how were subjects able to answer regional questions about parts of symbols about as accurately as they answered questions about the symbol as a whole? Previous research on the theory of selective attention does not provide an answer to this question.

Reaction times provided further information about the efficacy of multivariate point symbol designs. This dependent variable established a clear hierarchy in how difficult the four symbols were to process. Although subjects could provide about the same proportion of correct answers for all the symbols if they studied them long enough, it was much easier to reach the correct answer for the boxed letters, followed by crosses, divided circles, and Chernoff faces. Part of the reaction time advantage for boxed letters could have been the fact, discussed earlier, that participants did not have to refer to the legend before answering each question as they did to determine, say, which arm of the cross represented income rates. If this element of the research design were controlled by having subjects memorize the assigned position of each variable within each symbol before taking the experiment, some of the reaction time differences might disappear. On the other hand, as Figure 15.6 illustrated, there were significant differences among the symbols even for "quality of life" questions, and subjects did not need to refer to the legend on any of the maps to answer these questions; they simply had to note the value represented by all the variables in the symbol, high, medium, or low. Thus, the low reaction times for the boxed letters cannot be dismissed as merely an artifact of the experimental design.

Based on the earlier research conducted by psychologists, we hypothesized that the reaction times for Chernoff faces would be lower than they were relative to the other symbols—especially when subjects were called upon to evaluate the symbol holistically for quality of life questions. Recall, however, that some researchers have found that Chernoff faces are most effective for portraying a large number of variables—up to 18—and that they are less effective when just three or four visual elements vary (Carswell and Wickens, 1988: 34). Our results seem to be consistent with the latter effect.

The fact that the region questions took significantly longer to answer than questions about an individual county is neither surprising nor particularly interesting. It simply took more time to scan several symbols than to look at one symbol. This would not imply that looking at several symbols is less efficient than looking at one symbol unless one found a non-linear relationship between the number of symbols scanned and the amount of time required to do so. We did not address that question in this study, however.

The interaction between symbol type and the part/whole questions is of interest for what it implies about selective attention theory and the separable-ness and integrality of the symbols in this study. The higher reaction times on quality of life (whole symbol) questions compared to the individual part questions for Chernoff faces and boxed letters suggest that they are more separable than integral; that is, that subjects could focus more quickly on an individual component of those symbols than on their holistic image. Conversely, the higher reaction times for questions about individual variables for the crosses and circles may indicate that they are more visually integral than separable. Among the particular set of symbols used in this study, the more abstract designs (circles and crosses) appear to have formed a relatively integral image, while the faces and letters, which were more pictorial (in the case of the faces) or direct abbreviations of their referents (for the letters) were more separable. Whether this result is a coincidence, or whether it represents a more general model would need to be determined through additional research.

A number of interesting questions come to mind in connection with this study. For example, what if we had designed the multivariate symbols using colours instead of black and white? Colour extends the range of symbolization options considerably, of course, and it is likely that more efficient symbols could be created by employing colour. If our divided circle symbols had been designed using three values of four different hues, say, the visual separability of the four quadrants would probably have been quite different. In addition, subjects would have had the cue of hue in addition to quadrant position with which to associate a specific variable, thus facilitating their map-reading task. The effect of colour in multivariate point symbol design provides a wealth of research opportunities. One issue that inevitably arises in connection with research on multivariate maps is whether a set of n univariate maps is more or less effective that a single multivariate map. We did not address that question in our study, but it would be a valuable follow-up for the future.

CONCLUSION

Based on the results of the study reported here, it appears that Chernoff faces which represent just a few variables do not function very well as cartographic symbols, at least if ease and accuracy of reading the symbols are primary considerations. There may be offsetting advantages to the faces, however, such as their uniqueness and attention-getting quality. The boxed letters appeared to function best overall and to offer several advantages such as transparency of meaning and ease of design and production. However, they should be evaluated further from the standpoint of aesthetics and legibility in a realistic map context.

Most importantly, we need to develop a robust theoretical basis for understanding and predicting how people perceive multiple-cue graphics. With a better grasp of integral, separable, and congruent visual dimensions, cartographers should be able to design map symbols from which readers can retrieve information about either discrete or integrated values or distributions. The ever-increasing flood of information which is available to researchers in all fields today compels us to find more efficient ways of analysing and communicating such data. With this study we contribute to the small amount of empirical knowledge which currently exists on the topic of quantitative multivariate point symbols for maps.

ENDNOTE

[1] Subjects were not asked to memorize the meaning of individual symbol segments, and, thus, had to refer to the legend continuously. Although it would be interesting to see how the results of the study would be affected if subjects had been required to memorize the legend, that was not the aim of our study. We assumed that in an actual map-reading situation, readers would be unfamiliar with such map symbols and that our findings would be most useful if they were based on a realistic set of conditions. Some insight into this issue might be available from the boxed letter symbol in this study, since its component parts, alphabetic characters, directly tied the symbol to its referents. No memorization—no legend, in fact—was required to interpret the symbol. This set of conditions made no apparent difference in results for the specific symbol other than to facilitate faster responses.

REFERENCES

Amedeo, D., and Kramer, P. (1991). User perception of bi-symbol maps. *Cartographica*, 28(1), 28-53.

Anderson, E. (1960). A semi-graphical method for the analysis of complex problems. *Technometrics*, 2, 387-392.

Bertin, J. (1983). *Semiology of graphics: diagrams, networks, maps*. Madison, WI: University of Wisconsin Press. Translated by William Berg from *Semiologie Graphique* (1967). Editions Gauthier-Villars.

Board, C., and Wilson, Mrs. E. (1966). The compilation of a three-component map of Southern Africa. *The Cartographic Journal*, 3(2), 83-86.

Borg, I., and Staufenbiel, T. (1992). Performance of snowflakes, suns, and factorial suns in the graphical representation of data. *Multivariate Behavioral Research*, 27(1), 43-55.

Carlyle, I.P., and Carlyle, W.J. (1977). The ellipse/A useful cartographic symbol. *The Canadian Cartographer*, 14(1), 48-58.

Carr, D. (1991). Looking at large data sets using binned data plots. In A. Buja and P. Tukey (Eds.), *Computing graphics in statistics* (pp. 7-39). New York: Springer-Verlag.

Carr, D., Olson, A., and White, D. (1992). Hexagon mosaic maps for display of univariate and bivariate geographical data. *Cartography and GIS*, 19(4), 228- 236, 271.

Carstensen, L. (1982). A continuous shading scheme for two-variable mapping. *Cartographica*, 19(3), 53-70

Carstensen, L. (1984). Perceptions of the variable similarity of bivariate choroplethic maps. *The Cartographic Journal*, 21 (2), 23-29.

Carstensen, L. (1986a). Bivariate choropleth mapping: The effects of axis scaling. *The American Cartographer*, 13(1), 27-42.

Carstensen, L. (1986b). Hypothesis testing using univariate and bivariate choropleth maps. *The American Cartographer*, 13(3), 231-251.

Carswell, C.M., and Wickens, C.D. (1988). Comparative graphics: History and applications of perceptual integrality theory and the proximity compatibility hypothesis. *Technical memorandum 8-88*. U. S. Army Engineering Laboratory.

Carswell, C.M., and Wickens, C.D. (1990). The perceptual interaction of graphical attributes: Configurality, stimulus homogeneity, and object integration. *Perception and Psychophysics*, 47, 157-168.

Chang, K. (1982). Multi-component quantitative mapping. *The Cartographic Journal*, 19 (2), 95-104.

Chernoff, H. (1973). The use of faces to represent points in k-dimensional space graphically. *Journal of the American Statistical Association*, 68(342), 361-368.

Chernoff, H. (1978). Graphical representation as a discipline. In P. Wang (Ed.), *Graphical representation of multivariate data* (pp. 1-11). New York: Academic Press.

Dahlberg, R. (1981). Educational needs and problems within the National Cartographic System. *The American Cartographer*, 8(2), 105-114.

DeSanctis, G. (1984). Computer graphics as decision aids: Directions for research. *Decision Science*, 15, 463-487.

Dobson, M. (1983). Visual information processing and cartographic communication: The utility of redundant stimulus dimensions. In D.R. Fraser Taylor (Ed.), *Graphic communication and design in contemporary cartography* (pp. 149-176). London: John Wiley & Sons.

Dykes, J.R. (1979). A demonstration of selection of analyzers for integral dimensions. *Journal of Experimental Psychology: Human Perception and Performance*, 5, 734-745.

Eyton, J.R. (1984). Complementary-color two-variable maps. *Annals of the Association of American Geographers*, 74(3), 477-490.

Farquhar, A.B., and Farquhar, H. (1891). *Economic and industrial illusions: A discourse of the case for protection*. New York: Putnam.

Garner, W.R., and Felfoldy, G.L. (1970). Integrality of stimulus dimensions in various types of information processing. *Cognitive Psychology*, 1, 225-241.

Huff, D.L., and Black, W. (1978). A multivariate graphic display for regional analysis. In P. Wang (Ed.), *Graphical representation of multivariate data* (pp. 199-218). New York: Academic Press.

Jacob, R.J.K., Egeth, H.E., and Bevan, W. (1976). The face as a data display. *Human Factors*, 18 (2), 189-200.

Lavin, S., and Archer, J.C. (1984). Computer-produced unclassed bivariate choropleth maps. *The American Cartographer*, 11(1), 49-57.

Lavin, S., Hobgood, J., and Kramer, P. (1986). Dot-density shading: A technique for mapping continuous climatic data. *Journal of Climate and Applied Meteorology*, 25(5), 679-690.

MacDonald-Ross, M. (1977). How numbers are shown: A review of research on the presentation of quantitative data in texts. *AV Communication Review*, 25, 359-409.

MacEachren, A. (1994). *How maps work: issues in representation, visualization, and design*. Chapter 3. New York: Guilford Press.

Muehrcke, P., and Muehrcke, J. (1992). *Map use: Reading, analysis and interpretation* (3rd ed.). Madison, WI: JP Publications.

Olson, J. (1981). Spectrally encoded two variable maps. *Annals of the Association of American Geographers*, 71, 259-76.

Pomerantz, J.R., and Garner, W.R. (1973). Stimulus configuration in selective attention tasks. *Perception and Psychophysics*, 18(5), 355-361.

Pomerantz, J.R., and Schwaitzberg, S.D. (1975). Grouping by proximity: Selective attention measures. *Perception and Psychophysics*, 14(3), 565-569.

Rhind, D., Shaw, M.A., and Howarth, R.J. (1973). Experimental geochemical maps—A case study in cartographic techniques for scientific research. *The Cartographic Journal*, 10(2), 112-118.

Schmid, C. (1983). *Statistical graphics: Design principles and practices*. New York: John Wiley and Sons Ltd.

Seigal, J.H., Goldwyn, R.M., and Friedman, H.P. (1971). Pattern and process in the evolution of human septic shock. *Surgery*, 70, 232-245.

Shortridge, B. (1982). Stimulus processing models from psychology: Can we use them in cartography? *The American Cartographer*, 9(2), 155-167.

Tukey, J. (1977). Exploratory data analysis. Reading, MA: Addison-Wesley Publishing Company.

Wainer, H. (1979). Graphic experiment in display of nine variables uses faces to show properties of states. *Newsletter of the Bureau of Social Science Research*, XIII, Fall 1979.

Wainer, H., and Francolini, C. (1980). An empirical inquiry concerning human understanding of two-variable color maps. *The American Statistician*, 34, 81-93.

Wainer, H., and Thissen, D. (1981). Graphical data analysis. *Annual Review of Psychology*, 32, 191-241.

16

Feature Matching and the Similarity of Maps

Robert Lloyd

Department of Geography, University of South Carolina

Elzbieta Rostkowska-Covington

School of Public Health, University of South Carolina

Theodore Steinke

Department of Geography, University of South Carolina

What knowledge enables successful map design? Knowledge about the map construction process is obviously important, but map designers also can make use of knowledge on how maps are processed by map readers (Figure 16.1). Examples of processes frequently used by map readers are: (1) searching for a target (Brennan and Lloyd, 1993); (2) identifying objects (Lloyd and Steinke 1986); (3) classifying objects (Lloyd and Steinke 1985); and (4) judging similarity (Lloyd and Steinke 1977). How do map readers learn such processes? Nigrin (1993) argued there are three basic types of learning processes. Map readers can come to know how to process maps from any of these three types of learning processes. In some cases *supervised learning* could involve the map designer as educator (Figure 16.1). The designer might teach a course on how to read maps or write a text accompanying the maps on how they should be read. It is relatively rare that maps are learned by individuals whose learning process is directly or indirectly supervised by the designer of the maps. It is more generally true that designers of maps construct them with little knowledge of who will be reading them or how they will be processed.

Most map readers have relatively little formal training in how to process spatial information. These map readers rely on common sense and use generalized cognitive processes developed for processing visual information. Two types of learning experiences can be identified that have no direct supervision by an expert (Nigrin, 1993). The first type, *unsupervised learning*, results in knowledge being acquired without the map reader knowing if the processing has produced a proper result. *Unsupervised learning* has neither an expert to tell map readers they are making correct interpretations, nor does it have feedback from the consequences of the map readers' interpretations. This type of learning involves experiences with a

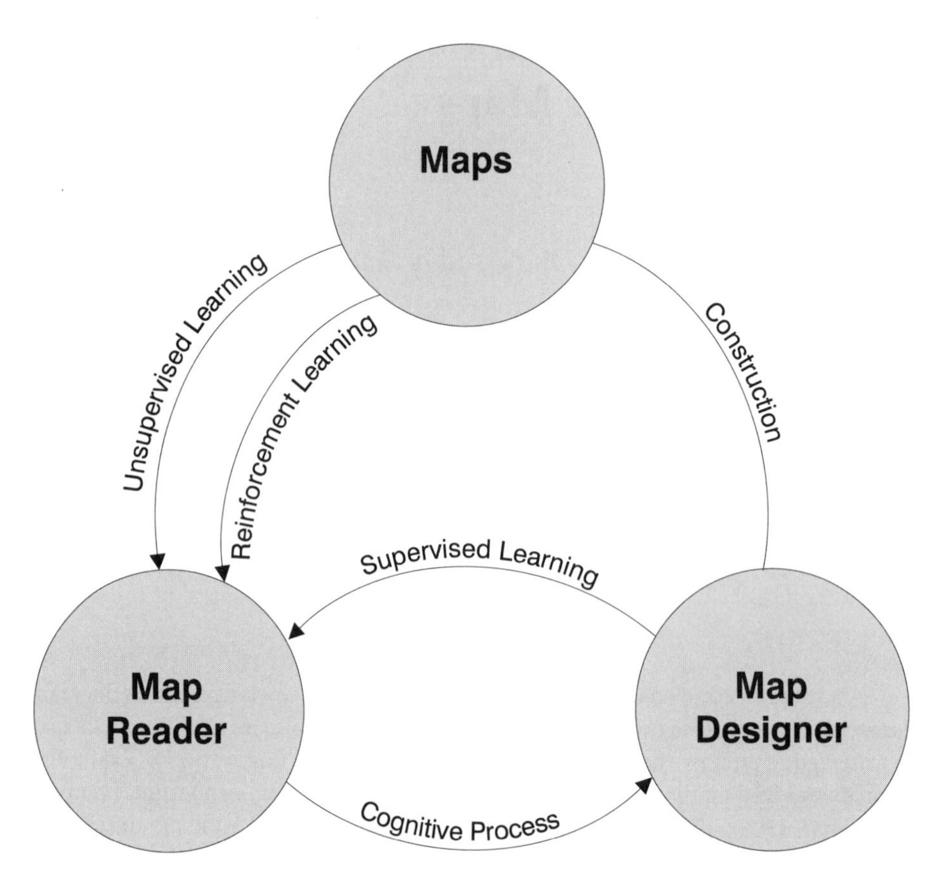

Figure 16.1 Relationships among maps, map readers, and map designers illustrating the knowledge needed for map design.

map that acquire information, but no actual decisions are made by the map reader that could provide feedback on the success of the learning experience. Casual experiences with maps that are not motivated by imminent decisions are examples of *unsupervised learning*.

The second type of learning without expert supervision, *reinforcement learning*, involves indirect feedback from the consequences of decisions made by the map reader (Nigrin, 1993). For example, a person might encode information from a map to learn the locations of some landmarks in an environment. Finding the landmarks in the environment would indicate the map reading process was a success and failing to find the landmarks would indicate it was a failure. Success positively reinforces the learning process used, while failure provides negative reinforcement for that process. Failure, however, does not provide the solution to a successful learning process. After a number of experiences that resulted in failure, the map reader may find a successful process. This sort of learning is clearly not supervised by an expert, but based on the trials and errors of an individual learning by making mistakes.

Both *unsupervised learning* and *reinforcement learning* are not assisted by an expert, yet they both may be at least partially successful, if not optimal, solutions for learning. The basic difference between them is the map reader's awareness of the quality of the learning process. *Unsupervised learning* may produce very good or very bad map-reading results, but the map reader will not be aware of the outcome. Although *reinforcement learning* may eventually lead to optimal success, it may take persistence and a little luck to achieve this end. Given that most individuals who read maps are not using skills acquired through *supervised learning*, one important goal for map designers should be to understand the cognitive processes being used to process spatial information. Since they are, for the most part, a product of *unsupervised* and *reinforcement learning*, they may not conform to theoretical expectations implicit in map construction and *supervised learning* processes. This chapter argues that map designers can acquire knowledge of the *cognitive processes* used by map readers and use this knowledge to improve their designs (Figure 16.1).

BACKGROUND

Consider this common map reading task. Look at any two maps side by side and compare them for the purpose of stating how similar they are on a scale of 1 to 10. Can you describe the process you used to decide the appropriate rating? Most people do not describe a comparison process that involves numerical processing. No mental equations or sums of squared differences are reported as being computed. Visual comparisons of maps appear to be somewhat intuitive. They are likely to be the same processes used to compare everyday objects in the environment and be the product of *unsupervised learning* and *reinforcement learning*. A number of studies have reported that the visual comparisons of maps do not produce the same results as the formal statistical comparisons (Lloyd and Steinke, 1976, 1977; Steinke and Lloyd, 1981, 1983). For choropleth maps, factors unrelated to the patterns represented by the data can have a large impact on visual similarity. Two maps that are generally dark or light appear to be more similar than expectations based on the spatial patterns generated by their data. Designing maps with equal average blackness (lightness) has been recommended as one solution to this map comparison problem (Lloyd and Steinke, 1977).

For reference maps that display distributions of features, similarity cannot be computed with standard correlation coefficients. There is still, however, a definite visual impression of similarity if reference maps for different areas are considered side by side. If both maps have the same type of features symbolized, for example, a lake, that might make them more similar looking. This might be true even if the lakes were different sizes and located in different relative locations on their respective maps. If one map has a unique feature not shared by the other map, such as a volcano, this would contribute to the two maps' dissimilarity.

In general any two maps have some information on them that is shared and some that is unique. This chapter argues that the visual similarity of maps is a function of both the common information and the distinctive information found by the viewer on the maps. This theoretical notion was first proposed by Amos Tversky (1977) as a general model for making comparisons. He presented what he called the *contrast model* and argued that similarity is determined by matching the features of the objects being compared.

Although the current research focuses on the similarity of maps, this concern is only a part of geographers' larger interest in the similarity and differences of places.

> *When, however, we study more complex integrations in geography, we find a much smaller number of essentially similar specimens. As in many other sciences, we attempt to overcome this difficulty by recognizing categories or types within which differences in cases appear less marked than the similarities. But in thus classifying objects, or phenomena, each of which involves a host of independent or semi-independent elements, we do not have specimens that are similar in all essential elements. They are similar only in terms of the particular categories we have chosen, and may differ notably in other respects which research may demonstrate to be less important.*
>
> Hartshorne (1959: 150)

Hartshorne's quotation expressed the common problem geographers face when trying to determine the similarity of places. The same issues are important when someone is visually comparing two maps and determining their similarity.

THE SIMILARITY OF MAPS

Maps can be thought of as distributions of features, for example, colours, shapes, textures, sizes, and spatial frequencies. Choropleth maps with N classes also can be represented geometrically as normalized vectors in an N dimensional space. The cosine of the angle between pairs of vectors would then represent a measure of the similarity, that is, the statistical correlation for two maps (Johnston, 1978). Since many distribution patterns for pairs of maps can result in the same statistical correlation coefficient and features other than pattern may impact visual similarity, geometric models may not represent visual similarity well. We will argue that the visual similarity of two maps is dependent on the matching of their features. The apparent visual similarity of some maps may be described better by models based on matching features than those based on statistical correlation (Tversky, 1977). This is particularly true for maps representing nominal classes of information, such as land use patterns, because it may not be appropriate to compute standard statistical correlations for such maps. It may also be at least partially true for maps like choropleth maps that can be compared by computing statistical correlation coefficients.

THE CONTRAST MODEL

Amos Tversky and his associates have performed a series of experiments that considered how features are used to make similarity judgments (Ben-Shakhar and Gati, 1987; Gati and Tversky, 1982,1984,1987; Sattath and Tversky, 1987; Tversky 1977). Tversky (1977) argued that geometric models of similarity, that is, those that compute similarity as the distance between objects in a space of N dimensions make dimensional and metric assumptions that may not be true for all situations. He made a distinction between quantitative and qualitative dimensions. For geometric models to be appropriate one must be able to compute the distance between points. On qualitative dimensions, which indicate the presence or absence of a feature, for example, oriented with north at the top or not, this many not be possible. Although some features, such as colour, might be considered as values on quantitative dimensions, they may also function as qualitative features in making visual comparisons. The fact that two choropleth maps have legend classes defined as shades of blue might have a positive impact on their visual similarity. If one map has legend classes defined as shades of a single colour and another uses a variety of different colours, this may contribute to their visual dissimilarity. Tversky (1977) also argued that the metric assumptions of:

Minimality:	$d(a,b) \geq d(a,a) = 0,$
Symmetry:	$d(a,b) = d(b,a)$
Triangular Inequality:	$d(a,b) + d(b,c) \geq d(a,c)$

must be true for distance computations to be valid. Minimality violations in map reading might occur if there is not sufficient contrast between symbols or colours. If county X on a choropleth map was classified as belonging to class A and assigned the appropriate colour, more people might remember county X as belonging to class B than to class A if the visual contrast between the two classes was not sufficient. Tversky provides two very geographic examples of symmetry violations. "Similarity judgments can be regarded as extensions of similarity statements, that is, statements of the form 'a is like b' . . . . We say 'an ellipse is like a circle,' not 'a circle is like an ellipse,' and we say 'North Korea is like Red China,' rather than 'Red China is like North Korea'" (Tversky, 1977: 328). He also provided a geographic example for a triangular inequality violation that applied to 1977 political alignments:

> *However, the triangle inequality implies that if a is quite similar to b, and b is quite similar to c, than a and c cannot be very dissimilar from each other. The following example (based on William James) casts some doubts on the psychological validity of this assumption. Consider the similarity between countries: Jamaica is similar to Cuba (because of geographical proximity); Cuba is similar to Russia (because of their political affinity); but Jamaica and Russia are not similar at all.*

Distinctive features are considered to be important for discrimination tasks such as searching for something on a map (Brennan and Lloyd, 1993; Lloyd, 1988). "The detection of a distinctive feature establishes a difference between stimuli, regardless of the number of common features" (Gati and Tversky, 1984: 342). Common features are thought to be more important for classification tasks. The similarity of objects, however, is determined by both their common and distinctive features and "the relative weight varies with the nature of the task" (Gati and Tversky, 1984: 342). A methodology has been developed that can determine the importance of a specific feature as either a common or distinctive feature when judging similarity. Gati and Tversky (1984) made a distinction between characteristics of objects (maps) defined as additive attributes and those defined as substitutive attributes. An additive attribute defines the presence or absence of a particular feature and a substitutive attribute defines the presence of exactly one element from a possible set. To provide an explicit example, let **b** represent a common background for a map display and say it is a base map of South Carolina. Further define two substitutive attributes, **p** and **q**, that define the style of the map. It may seem unusual to compare maps produced in different styles. Since some spatial data may be appropriately displayed in various styles of maps and since map readers may need to compare maps produced in different styles, it is important to know how style affects the communication process. MacEachren (1982) has discussed the effectiveness of different map styles for communicating general spatial patterns. Style **p** could be a choropleth map and style **q** an isopleth map. Also define two additive attributes **x** and **y** that represent elements that may or may not be in the display. Let attribute **x** be colour and attribute **y** be a title for the map. The following displays could now be created and presented in pairs for a map reader to evaluate:

bp is a choropleth map of South Carolina with no colour and no title.

bq is an isopleth map of South Carolina with no colour and no title.

bpx is a choropleth map of South Carolina in colour and with no title.

bqx is an isopleth map of South Carolina in colour and with no title.

bpy is a choropleth map of South Carolina with no colour and with a title.

bqy is an isopleth map of South Carolina with no colour and with a title.

To assess the effect of additive component **x** (colour) as a common feature, denoted C(**x**), map readers would be asked to compare displays **bp** and **bq** and rate their similarity and to do the same for displays **bpx** and **bqx.** The difference between these similarities is an estimate of C(**x**). In equation form then:

$$C(\mathbf{x}) = S(\mathbf{bpx},\mathbf{bqx}) - S(\mathbf{bp},\mathbf{bq}) \qquad (1)$$

To assess the effect of additive component **x** (colour) as a distinctive feature, denoted D(**x**), map readers would be asked to compare displays **bpx** and **bpy** and rate their similarity and to do the same for displays **bpx** and **bpy**. The difference between these similarities is an estimate of D(**x**). In equation form then:

$$D(\mathbf{x}) = S(\mathbf{bp},\mathbf{bpy}) - S(\mathbf{bpx},\mathbf{bpy}) \qquad (2)$$

The effect of **x** as a common feature relative to its effect as a distinctive feature is defined as W(**x**) and computed as follows:

$$W(\mathbf{x}) = C(\mathbf{x})/[C(\mathbf{x}) + D(\mathbf{x})] \qquad (3)$$

The value of $W(x)$ ranges from 0, when $C(x) = 0$, to 1 when $D(x) = 0$, and equals 0.5 when $C(x) = D(x)$. The value of $W(x)$ can be used to determine the relative strength of feature **x** as a common and distinctive feature.

The methodology has been thoroughly tested on both verbal and pictorial stimuli. Verbal stimuli were descriptions of people or meals, and pictorial stimuli were schematic faces or landscapes. Common features were more important than distinctive features when verbal descriptions were used, and distinctive features were more important than common features when pictorial displays were used. Verbal descriptions of pictorial displays were also compared with the results for pictorial displays. They appeared to be evaluated like other verbal stimuli, that is, common features were more important than distinctive features. Gati and Tversky (1987) had subjects rate the similarity of both verbal and pictorial stimuli. The verbal description and pictorial displays were presented again with one common and one distinctive feature missing. "The analysis of both correct recall and misclassified recall suggest that components that were encoded as common are relatively more salient in verbal than in pictorial stimuli, whereas components that were encoded as distinctive exhibit the opposite pattern" (Gati and Tversky, 1987: 99). These experiments suggest that judgements of the similarity of maps should be dependent on both their common and distinctive features, but that distinctive features would dominate. If one were to read verbal descriptions of two maps, however, one would expect similarity judgements to depend on both common and distinctive features, but common features would dominate.

PURPOSE AND HYPOTHESES

The purpose of this chapter is to investigate processes humans use to determine the visual similarity of maps and also to investigate how such judgements are affected by categorical information known about the maps. Two experiments are used to test two basic hypotheses. The first experiment considered hypothesis one: *Maps representing categorical information are judged to be more similar if they share common features and less similar if they have distinctive features* (Tversky, 1977). It used reference maps with simple features and considered similarity judgements and reaction times as dependent variables. The second experiment again tests hypothesis one, but also considers hypothesis two: *Categorical information not directly related to the patterns on maps also affects how similar maps are thought to be.* For example, knowledge of the locations of the environments represented on two maps or the two time periods represented by two maps of the same place could influence how similar the maps are thought to be. If a person is told that two maps are from the same location, for example, two counties in Iowa, they might be influenced to think they are more similar than if they are told one map is from Iowa and the other from Bulgaria. In a similar fashion, two maps labelled as representing 1990 and 1991 agricultural patterns in Southeastern Australia might be considered more similar than if one was labelled as representing 1890 patterns and the other 1991 patterns.

Information that is related to the maps, but not represented directly in the visual patterns on the maps, can be referred to as top-down information (Treisman and Gelade, 1980). This is knowledge someone has in memory that is related to the maps being compared. Bottom-up information is the information visually represented on the maps and reflects data used to construct the maps. Top-down information, therefore, is *a priori* information about the places represented on the maps that a person brings to the map comparison process. Bottom-up information is the information carried on the maps and visually experienced by the map reader during a comparison process.

EXPERIMENT ONE

The first experiment had subjects compare both reference maps (visual process), and descriptions of reference maps (verbal process). Important results are summarized here. Complete details can be found in Covington (1992).

Materials

Experimental maps were constructed as simple rectangular spaces containing up to four features. Possible features on a trial map (e.g., lakes, rivers, highways, forests, railroads, corn fields, contour lines, and urban areas) represented typical information found on reference maps (Figure 16.2). A total of 114 pairs of maps were created as stimuli for the experiment by combining features in different ways.

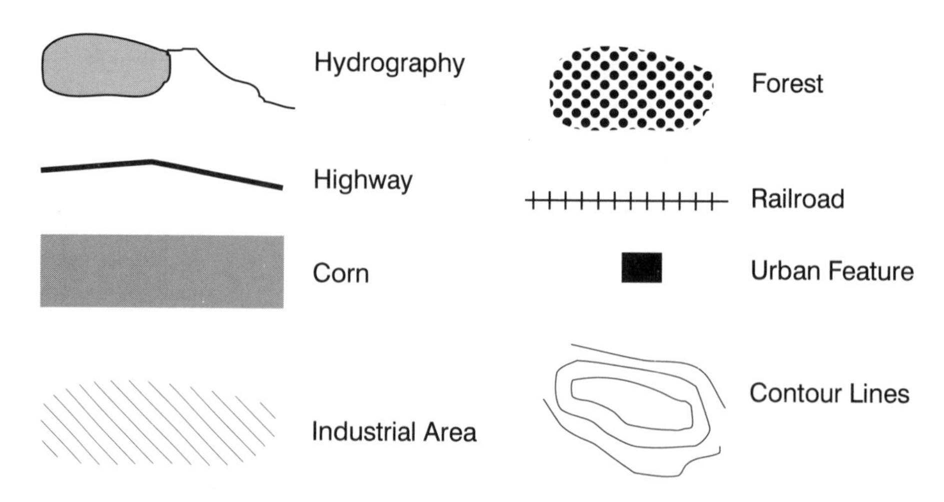

Figure 16.2 Combinations of these features were used to construct map pairs for Experiment One.

Each pair represented a particular situation that was needed to compute C(**x**) or D(**x**). The top pair of maps in Figure 16.3 illustrates a **p** map with a water feature and a **q** map with a highway feature that was one of the trials. The bottom pair has an urban feature added to both maps to make a **px** map and a **qx** map. The similarity of this pair was also judged among the 114 trials. Equation 1 can use the similarity judgements for these two pairs to compute the effect of **x**, for example, urban area, as a common feature. In other words C(**x**) measures how much more similar maps **p** and **q** are if **x** is added to both of them.

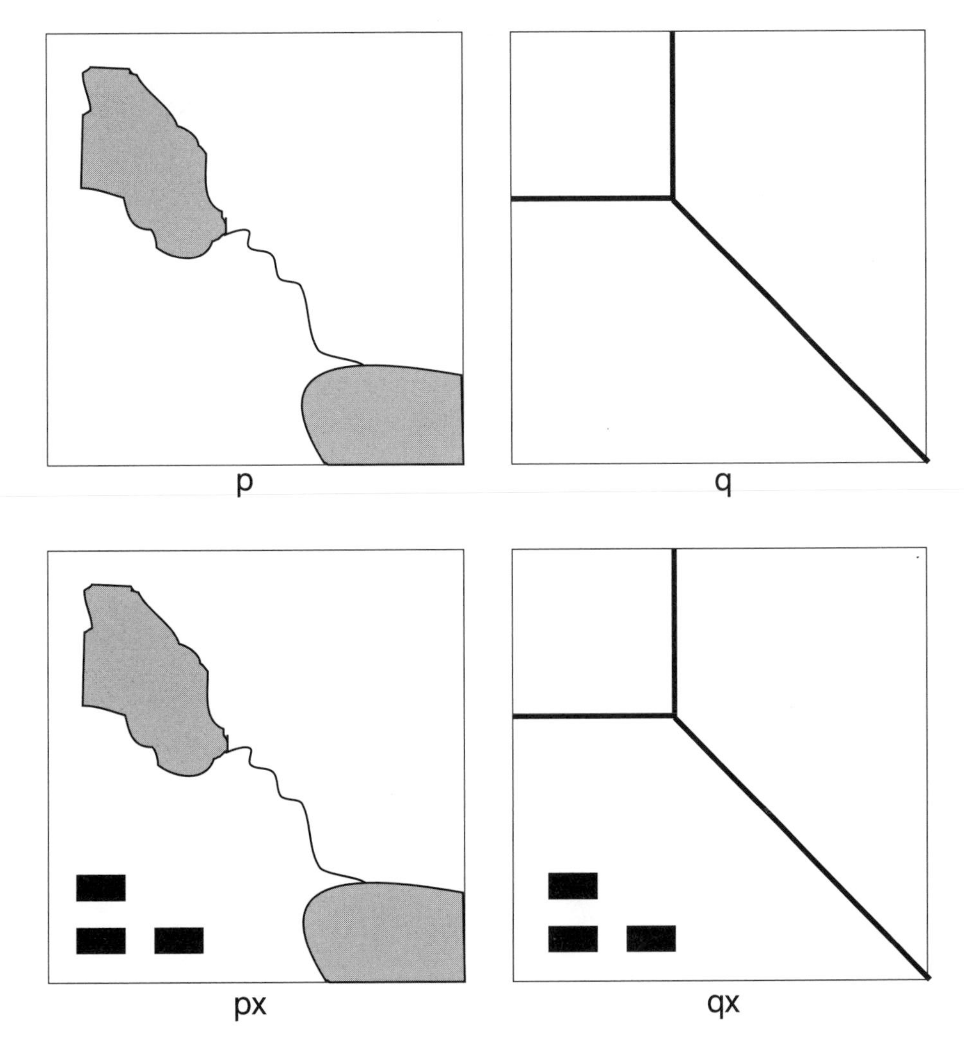

Figure 16.3 Sample of pairs of maps that were used to estimate the effect of adding a single common feature on subjects' judgement of map similarity.

The top pair of maps in Figure 16.4 illustrates a **p** map with a water feature and a **py** map with water and forest features. This was another of the 114 trials. The bottom pair has an urban area added to the left map making a **px** map and repeats the **py** map with the water and forest features. Equation 2 can use the similarity judgements for these two pairs to compute the effect of **x**, for example, urban area, as a distinctive feature. In other words D(**x**) measures how less similar maps **p** and **py** are if **x** is added as a distinctive feature to map **p**. For the experimental trials **p** maps and **q** maps were defined using the features represented in Figure 16.2.

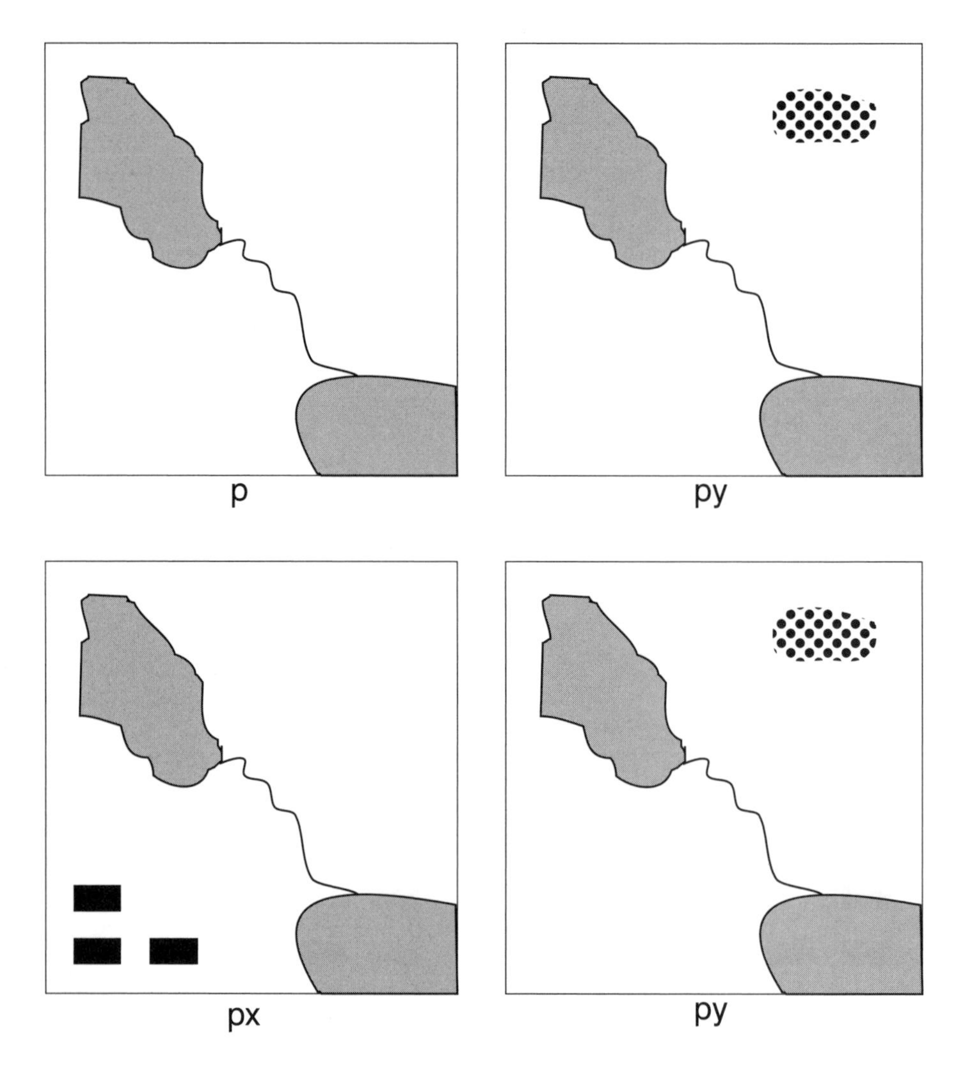

Figure 16.4 Sample of pairs of maps that were used to estimate the effect of adding a distinctive feature on subjects' judgements of map similarity.

For any given **p** and **q**, the remaining features (Figure 16.2) were used to define **x** and **y** features. Some trials were structured so that the effect could be measured for adding **x** as a second common feature, $C_1(x)$, and as a common feature, but in different locations on maps **p** and **q**, $C_2(x)$. Other trials were structured so that the effect of **x** as a second distinctive feature, $D_1(x)$, could be measured. All the equations for measuring feature **x**'s influence on both similarity and reaction time are listed in Appendix A. Each map was also described in terms of its features and 114 corresponding pairs of verbal descriptions were created as stimuli.

Subjects

Subjects were student volunteers from the University of South Carolina. Half of the 60 subjects were assigned to make visual comparisons of maps and the other half compared verbal descriptions of maps. Subjects were paid $5 for participating in the experiment which took less than 1 hour to complete.

Procedures

Each subject was tested separately using a Macintosh computer to control the experiment. The subjects who visually compared pairs of maps were presented all 114 pairs in a random order and made similarity judgements by using a mouse to click on a linear scale from 0 (not very similar) to 100 (very similar). They were instructed to answer as quickly and accurately as possible. Both similarity judgements and reaction times were recorded for each map pair. The subjects who read verbal descriptions of pairs of maps also made similarity judgements following similar procedures. All subjects were given practice trials to enable them to become familiar with the procedures before doing the experimental trials.

Results

Similarity judgements and reaction times were aggregated for eight classes of comparisons: 1. (**p**, **q**), 2. (**px**, **qx**), 3. (**px**, $\mathbf{qx_a}$), 4. (**py**, **qy**), 5. (**pxy**, **qxy**), 6. (**p**, **qy**), 7. (**px**, **qy**), and 8. (**p**, **qxy**). These were used to compute the five similarity indices: 1. $C(x)$, 2. $C_1(x)$, 3. $C_2(x)$, 4. $D(x)$, and 5. $D_1(x)$; and five corresponding reaction time indices: 1. $TC(x)$, 2. $TC_1(x)$, 3. $TC_2(x)$, 4. $TD(x)$, and 5. $TD_1(x)$ (Appendix A). Subjects who compared maps and subjects who compared verbal descriptions were aggregated separately.

The graphical representation (Figures 16.5) of this aggregate information can be used to assess the theoretical arguments advanced by Tversky's (1977) *contrast model*. For the six indices that assess the impact of **x** as a common feature, similarity should increase if a common **x** is added to either maps or map descriptions. Change vectors for maps, M_1, M_2, and M_3, and map descriptions, V_1, V_2, and V_3, are represented as the hypotenuses of right triangles (Figure 16.5). The vectors represent the change in similarity along the horizontal axis and reaction time along the vertical axis. The lengths of the legs of the right triangles are equal to the similarity and reaction time indices specified in Appendix A. For example, the horizontal leg of the M_1 triangle equals C(x) and the vertical leg equals TC(x). Similar changes result when a common **x** is added to both maps (M_1) or map descriptions (V_1), an **x** is added to both maps (M_2) or map descriptions (V_2) as a second common feature, and a common **x** is added to both maps (M_3) or map descriptions (V_3), but in different locations (Figure 16.5).

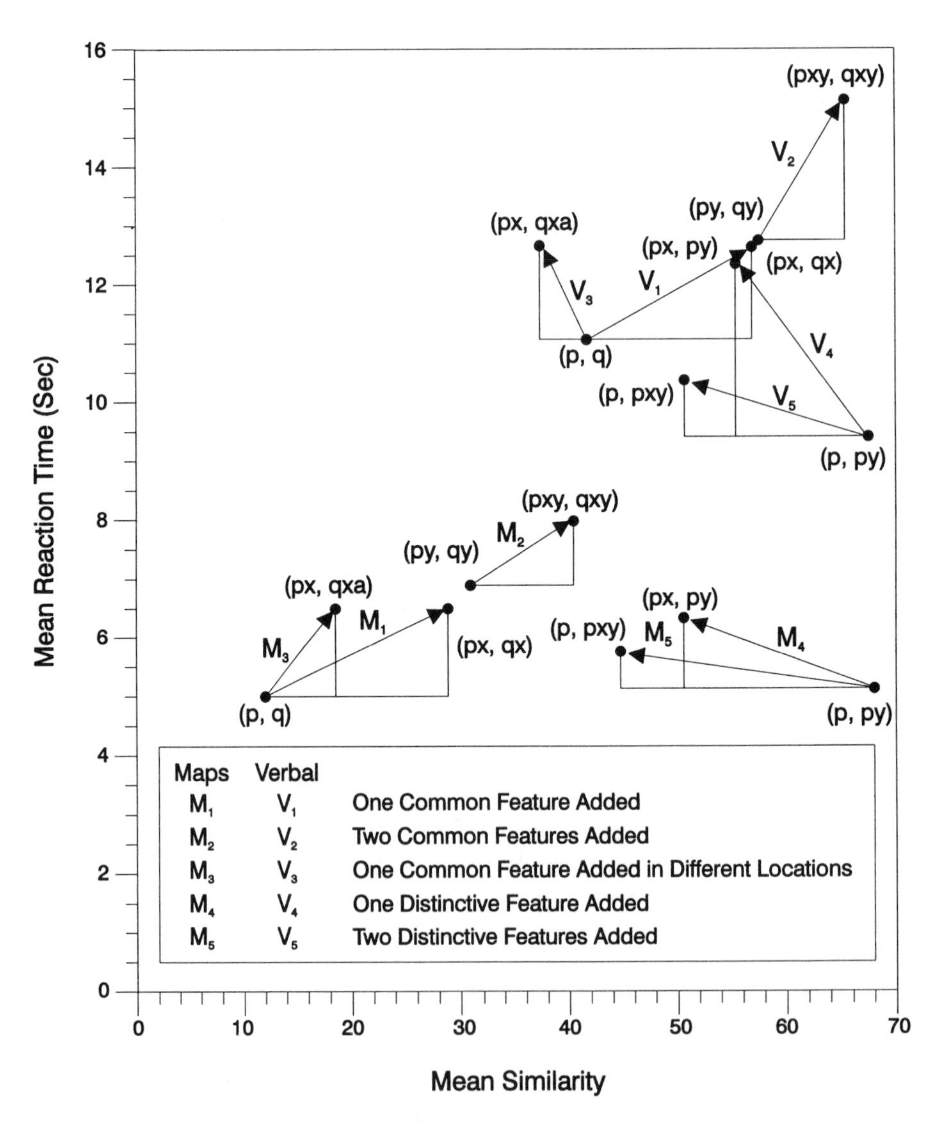

Figure 16.5 Change vectors for various comparisons of maps and verbal descriptions. Vectors pointing up indicate an increase in reaction time. Those pointing to the right indicate increasing similarity, and those pointing to the left indicate decreasing similarity.

Similarity should decrease if a distinctive **x** is added to one map for the two indices that assess the impact of **x** as a distinctive feature. These change vectors, M_4 and M_5, are also represented as the hypotenuses of right triangles (Figure 16.5). The vectors represent the change in similarity and reaction time that occurs when a distinctive **x** is added to one map (M_4), or a distinctive **x** is added to one map as a second distinctive feature (M_5). Corresponding change vectors (V_4 and V_5) represent the same effects for comparisons of map descriptions (Figure 16.5).

Discussion

One noticeable pattern for the change vectors is that verbally processing descriptions of maps takes longer than visually processing maps, that is, all the V vectors are higher in the graph than the M vectors (Figure 16.5). This should be expected because verbal processing is a serial process and visual processing is a parallel process (Lloyd, 1988; 1989). Another noticeable pattern for the change vectors is that they all point upward on the graph. It is obvious that adding features to maps generally increases the time it takes to make comparisons. This is true when the feature being added is a common or distinctive feature. It is true when maps are being compared and when verbal descriptions of maps are being compared. This general increase in time might suggest that features are processed serially with each feature taking some additional amount of time to process. Note, however, that adding a second distinctive feature to a map or a description of a map (M_5 and V_5 in Figure 16.5) results in faster reaction times than adding it as a first distinctive feature (M_4 and V_4). In fact, the addition of a second distinctive feature did not increase the reaction time a significant amount for either map comparisons ($t=1.32$, $P>t=0.1869$), or map description comparisons ($t=1.32$, $P>t=0.1912$) while adding the first feature was significant for maps ($t=2.73$, $P>t=0.0074$) and map descriptions ($t=3.94$, $P>t=0.0001$). It is obviously easier to decide similarity when the maps or descriptions of maps are distinctly different.

Adding one (M_1) or two (M_2) common features to maps increased reaction times a significant amount ($t=3.26$, $P>t=0.0015$ and $t=2.54$, $P>t=0.0125$) (Figure 16.5). Adding one common feature to map descriptions (V_1) did not significantly increase reaction time ($t=1.68$, $P>t=0.0963$), but adding a second common feature (V_2) to map descriptions did significantly increase reaction time ($t=2.78$, $P>t=0.0065$) (Figure 16.5). Adding a common feature at different locations on two maps (M_3) significantly increased reaction time ($t=3.12$, $P>t=0.0023$), while adding the common feature at different locations in descriptions of maps (V_3) did not significantly increase reaction time ($t=-1.68$, $P>t=0.0966$) (Figure 16.5).

Reaction times were always significantly increased when common features were added to maps, but not when distinctive features were added to maps. This lends some support to Tversky's argument that visual processing attends to distinctive features more than common features. The evidence suggests that adding more distinctive features to a map makes processing similarity faster (M_5 in Figure

16.5), while adding more common features to maps makes processing slower (M_2 in Figure 16.5). The evidence is less clear for map descriptions. Adding more common or distinctive features did not significantly increase reaction time (V_2 and V_5 in Figure 16.5).

Another obvious pattern for the change vectors is that the addition of distinctive features always decreased the similarity for both maps (M_4: t=6.26, P>t=0.0001 and M_5: t=9.83, P>t=0.0001) and the descriptions of maps (V_4: t=4.80, P>t=0.0001 and V_5: t=6.77, P>t=0.0001), that is, change vectors point significantly to the left (Figure 16.5). Adding common features in the same location on both maps always increased similarity (M_1: t=6.26, P>t=0.0001, M_2: t=4.11, P>t=0.0001, V_1: t=5.30, P>t=0.0001, and V_2: t=31.8, P>t=0.0019), that is, change vectors point significantly to the right (Figure 16.5). Gati and Tversky (1984) argued that the impact of adding a common feature is strongest if the feature is identical on the two displays, however that similarity is still increased if only the same type of feature is added. It is interesting that adding a common feature, but in a different location, (M_3), increased similarity for map comparisons a significant amount (t=2.43, P>t=0.0166), but not nearly as much as when the feature was added in exactly the same location, M_1. Subjects apparently considered the specific location of objects to be of importance when processing the maps. Adding a common feature in different locations on descriptions of maps (V_3) had an even more interesting impact. Changing the location of features in the descriptions actually decreased similarity, but not a significant amount (t=1.38, P>t=0.1693), that is, the vector points to the left instead of to the right. Differences in the locations of features may have become more important to subjects when it was explicitly pointed out in a short description. The differences in the locations of common features should be noticeable to those comparing maps, but the importance may not be made as explicit as when it is pointed out in map descriptions, because map readers are focusing more on distinctive features.

Gati and Tversky (1984) argued that individuals processing pictorial information like maps should focus more on the distinctive features and subjects processing verbal information like descriptions of maps should focus more on the common features. The theory predicts that the W(x) index (Equation 3) should be below 0.5 for map comparisons, indicating that distinctive features are having a stronger impact on similarity judgements. Conversely, the theory predicts that the W(x) index should be above 0.5 for comparisons of verbal descriptions, indicating common features are having a stronger impact. The W(x) computed for map comparisons was 0.448, and for map description comparisons it was 0.555. This supports the notion that distinctive features were given more weight by subjects who visually compared maps, while common features were given more weight by subjects who verbally compared descriptions of maps. The *contrast model* seemed to account for subjects' comparisons of simple reference maps. Experiment Two considers the model's ability to explain the similarity of simple qualitative land use maps.

EXPERIMENT TWO

Tversky's (1977) contrast model can be applied to a variety of situations requiring the measurement of the similarity of two objects through a simple equation (DeSarbo et al., 1992; Goldstone et al., 1991; Tversky, 1977):

$$Similarity\ (A,B) = \alpha * f\,(A \cap B) - B * f\,(A\text{-}B) - \gamma * f\,(B\text{-}A) \qquad (4)$$

For example, the similarity of Map A and Map B would be a function of three components:

1. The features that Map A shares with Map B defined by:
   $(A \cap B)$
2. The features that Map A has that Map B does not have:
   $(A\text{-}B)$
3. The features that Map B has that Map A does not have:
   $(B\text{-}A)$

The coefficients alpha, beta, and gamma are weights for their respective components. If alpha is set to 1 and both beta and gamma are set to 0, then only common features are being considered to affect similarity. If alpha is set to 0 and beta and gamma are set of 1, then only distinctive features are being considered to affect similarity. For similarity judgements, all coefficients are non-zero because similarity is positively influenced by common features and negatively influenced by distinctive features. If alpha is considered as a negative coefficient and beta and gamma as positive coefficients, then dissimilarity is being measured rather than similarity. This form of the equation was used by DeSarbo et al. (1992) in their TSCALE multidimensional scaling program.

Materials

An experiment was designed that had subjects compare pairs of visual displays on monitors that represented qualitative maps of land use. All maps consisted of nine square cells organized in a 3 by 3 grid-like display. These isolated windows represented locations randomly picked from larger raster displays of land use classified using remote sensing techniques. Each of the nine cells in a map had one of four colours (red, blue, green, or yellow) assigned to it. Each colour represented a different land use. Pairs of maps were generated that represented eight different levels of similarity (low to high) as measured by Equation 4.

The first component in Equation 4 was determined by counting the number of cells on the two maps having a matching colour. If both maps had three red cells, three blue cells, and three green cells, the two maps would match completely, and according to Tversky (1977) should be considered similar. This is because they have all common features and no distinctive features. The second component in Equation 4 was determined by counting the number of cells that were unique to the first map. If the first map had five red cells and the second maps had two red cells, the first map would have two common and three distinctive red cells. The third component in Equation 4 was determined by counting the number of cells that were unique to the second map. If the second map had four yellow cells and the first map had zero yellow cells, the second map would have no common yellow cells and four distinctive yellow cells. The coefficients alpha, beta, and gamma were all set to 1, making the components equally weighted. Ten map pairs were generated for each of the eight categories of similarity making a total of 80 map pairs available for the experiment.

Subjects

The 30 subjects who participated in the experiment were student volunteers from the University of South Carolina who were paid $5 for their time. All experiments were completed in less than one hour. Subjects were not required to have any previous experience in map reading, but did have to be able to distinguish among the colours being used in the experiment.

Procedures

Subjects were arbitrarily assigned to one of two groups based on the instructions they were to be given for the experiment (Figures 16.6 and 16.7). The instructions given to half the subjects (Premise I) showed a large raster display on the monitor that was identified as a map of land use for a particular date. Two isolated 3 X 3 windows were marked in separate parts of the larger display (Figure 16.6). The Premise I subjects were told they would be comparing locations like these examples that represented a common time, but different spatial locations. They were simply to consider how similar the land use maps were for the two areas represented and, using a mouse, mark a similarity scale which appeared on the screen below the maps. The scale varied from 0 on one end (not very similar) to 100 on the other end (very similar).

The other half of the subjects (Premise II) were shown two larger displays that represented land use for a particular location at two different times (Figure 16.7). The same location (3 X 3 window) was marked in both larger displays. The Premise II subjects were told they would be comparing maps like these examples that represented a common location, but different times. They were simply to consider how similar the land use maps were for this location for these two times and mark the similarity scale accordingly.

All subjects were instructed to make their decision as quickly as possible. Both groups of subjects saw the exact same 80 pairs of maps repeated in four blocks. The maps were presented in a random order each time. The maps were repeated four times to determine if the subjects' decisions were consistent over the four blocks of trials and to determine if the subjects' were becoming more skilled at making decisions with practice.

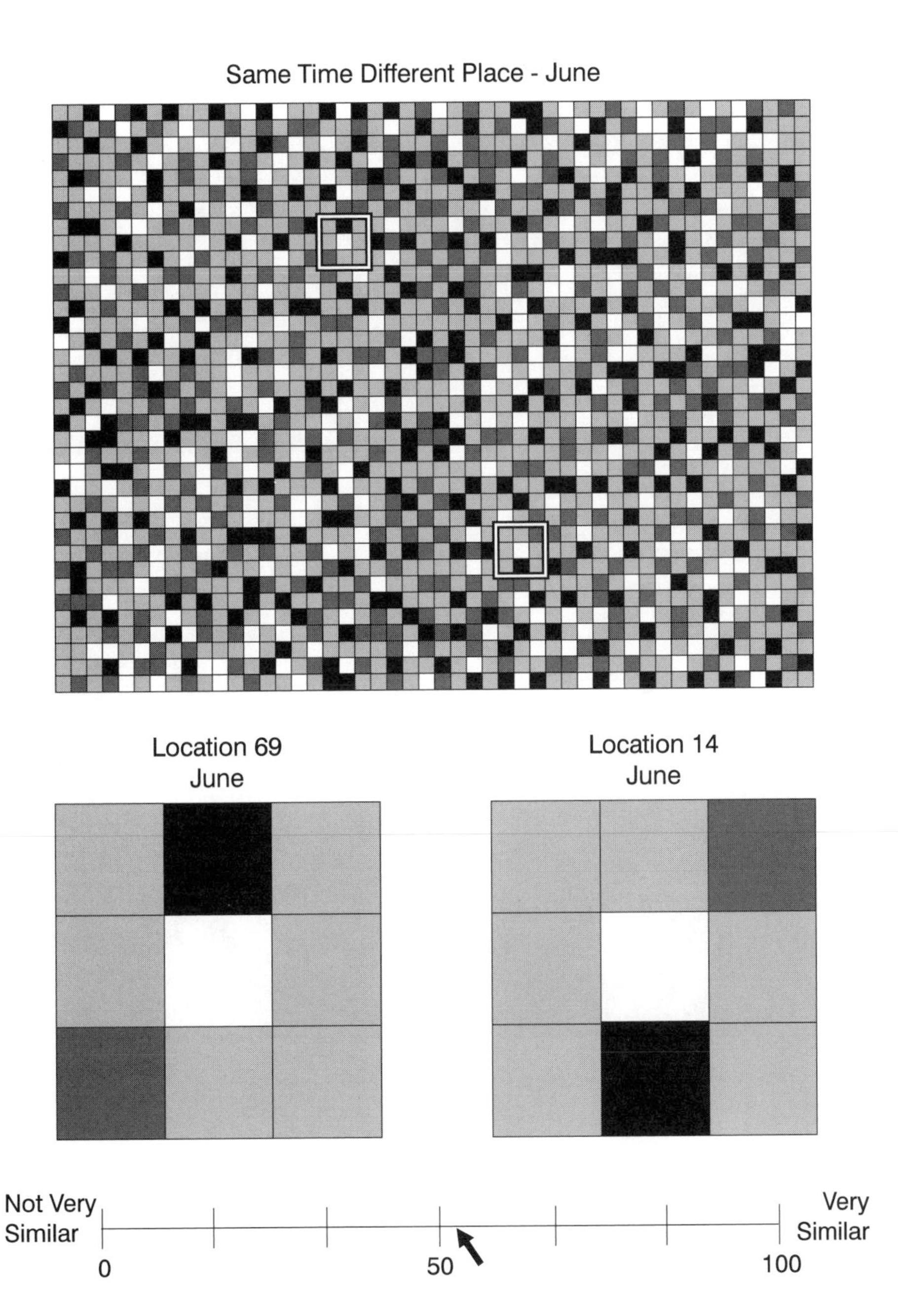

Figure 16.6 The land use display at the top illustrates two different locations at the same time. The isolated windows below were viewed by subjects who rated their similarity using the graphic scale at the bottom. The actual displays viewed by the Premise I subjects consisted of red, blue, green, and yellow cells.

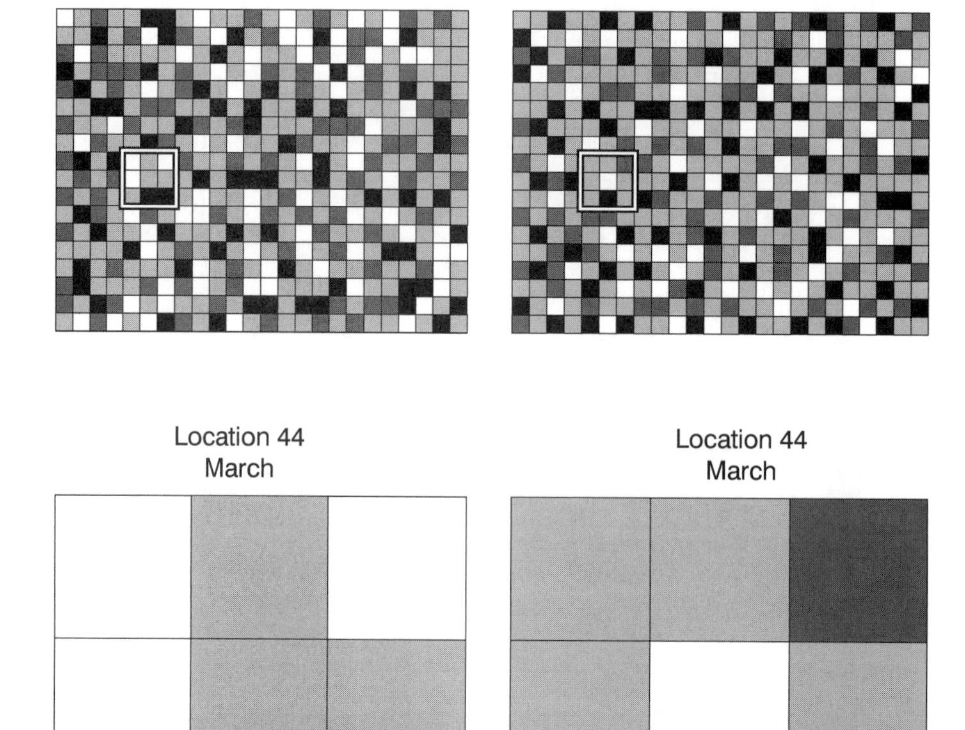

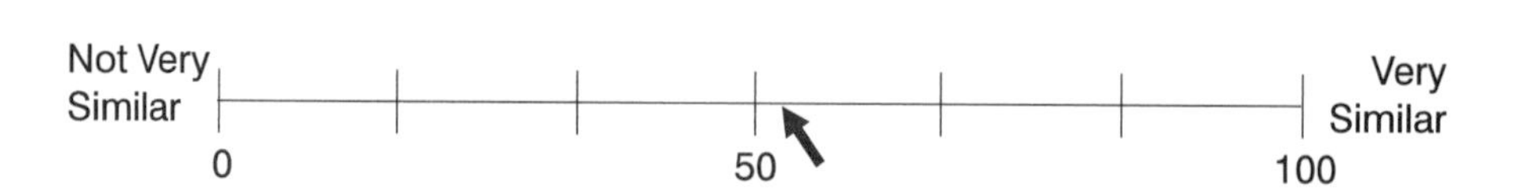

Figure 16.7 The land use display at the top illustrates the same location at two different times. The isolated windows below were viewed by subjects who rated their similarity using the graphic scale at the bottom. The actual displays viewed by the Premise II subjects consisted of red, blue, green, and yellow cells.

Results

The data were aggregated for the two premises over the individual maps and subjects. A repeated measures analysis of variance was computed using **Similarity Rating** and **Reaction Time** as dependent variables. Main effects were the **Tversky Index** (low to high), **Premise** (Same Time and Same Place), and repeated variable **Set** (block 1 to 4). Two-way interaction effects were also included in the analyses. Individual analyses of variance for the dependent variables **Similarity Rating** (F=170.54, P>F=0.0001, R^2=0.982) and **Reaction Time** (F=4.22, P>F=0.0001, R^2=0.569) were successful at explaining the variation of the dependent variables.

The ordinal categories of the **Tversky Index** generally showed a significant positive relationship both with **Similarity Rating** (F=307.9, P>F=0.0001) (Figure 16.8a) and **Reaction Time** (F=7.57,P>F=0.0001) (Figure 16.8b). The instructions given to the subjects that provided top-down information about the maps being considered (**Premise**) also was a significant main effect for both the similarity analysis (F=81.0, P>F=0.0001) (Figures 16.8c) and the reaction time analysis (F=9.2, P>F=0.0039) (Figure 16.8d). Two maps that were thought to be for the same place and a different time (mean = 35.7) were rated significantly higher than maps thought to be different places at the same time (mean = 31.7). The decisions were made faster when the pairs were thought to be the same place and a different time (mean = 3598 msec) than when the pairs were thought to be different places for the same time (mean = 3944 msec).

The **Tversky Index * Premise** interaction effect was different for the two analyses (Figures 16.9a and 16.9b). The interaction effect was significant for **Similarity Rating** (F=46.0, P>F=0.0001). The subjects' **Similarity Rating** increased at an increasing rate as the **Tversky Index** increased from low to high for the subjects who believed they were judging the same place at different times. Subjects who thought they were comparing different places for the same time had a different relationship between **Similarity Rating** and the **Tversky Index.** They appeared to make little distinction among the pairs representing the three highest similarity classes as measured by the **Tversky Index.** The **Tversky Index * Premise** interaction effect was not significant for the **Reaction Time** data (F=0.16, P>F=0.9924). Both groups of subjects had their **Reaction Time** increase as the **Tversky Index** increased, with the subjects who thought they were rating the same place at different times always performing the task faster.

Results for the four repeated blocks of trials (**Set**) were different for the two analyses (Figure 16.8e and 16.8f). The mean **Similarity Rating** was virtually identical over the four blocks of trials (means = 33.8, 33.8, 33.3, and 33.8), while **Reaction Time** decreased steadily from the first block (mean = 4271 ms) to the last block (mean = 3323 ms).

Repeated measures between subject effects, which ignore within **Set** effects, were significant for the **Tversky Index** (F=7.88, P>F=0.001) and **Premise** (F=8.98, P> F=0.0043). The **Tversky Index * Premise** interaction effect, however, was not significant (F=0.16, P>F=0.9925). Both **Similarity Rating** (Figure 16.9c) and **Reaction**

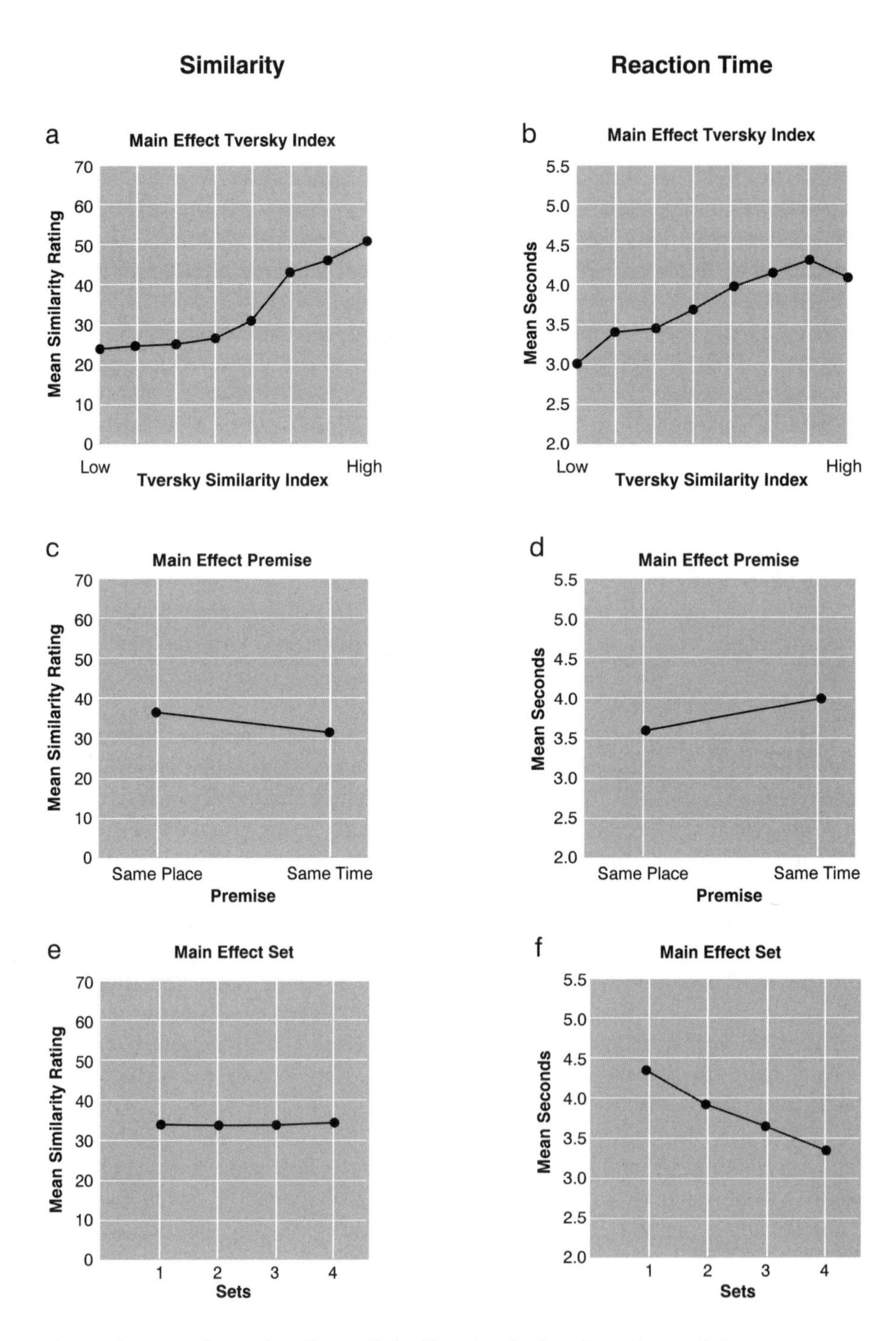

Figure 16.8 The main effects of the Tversky Index, Premise, and Set illustrated for both similarity and reaction time.

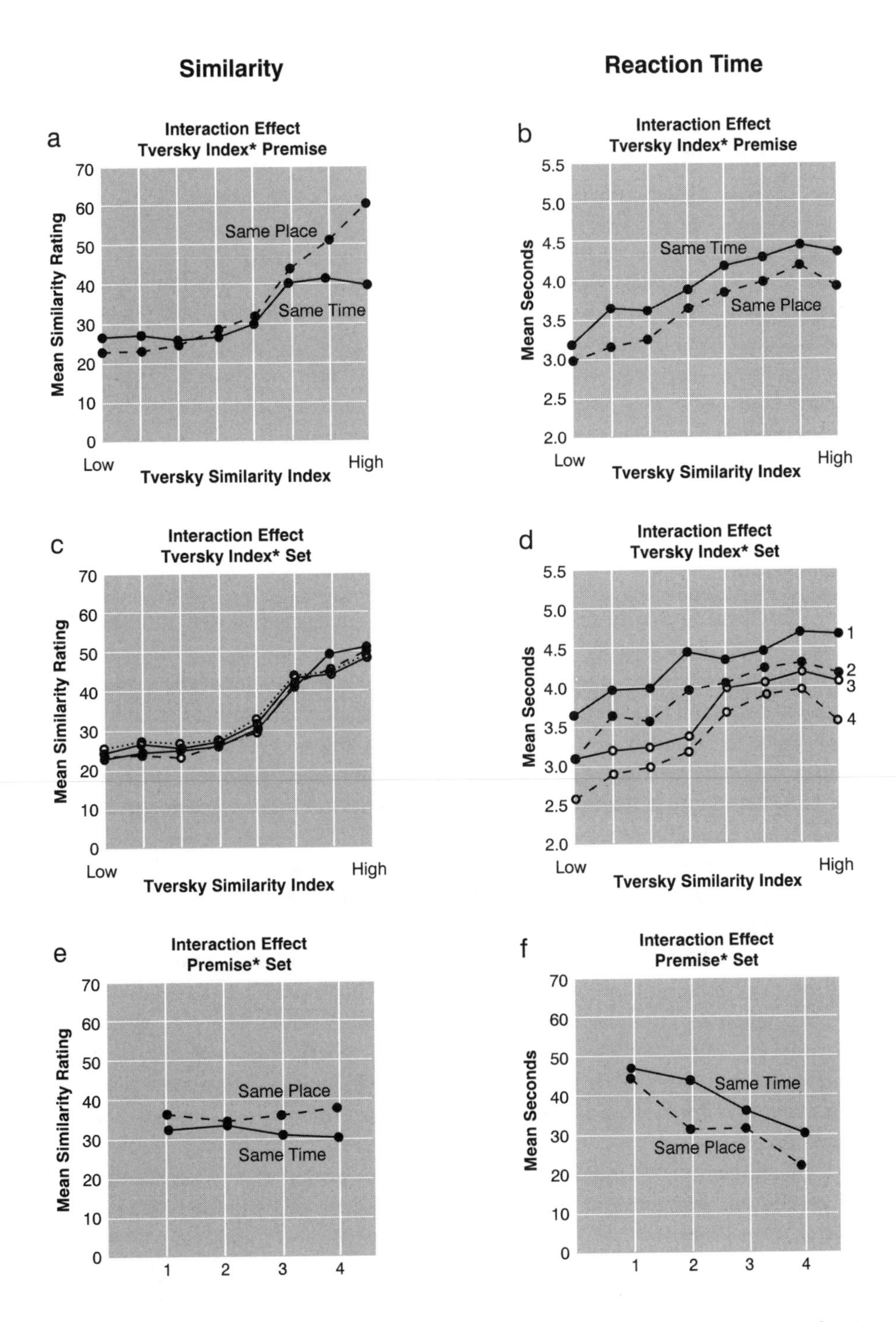

Figure 16.9 The interaction effects for Tversky Index * Premise, Tversky Index * Set, and Premise*Set illustrated for both similarity and reaction time.

Time (Figure 16.9d) increased with an increase in the **Tversky Index** for all four **Sets** of trials.

Univariate tests of within subjects effects indicated that **Set** was significant (F=4306.5, P>F=0.0001), as was the interaction effect **Tversky Index * Set** (F=7.27, P>F=0.0001). The interaction effect **Premise * Set** was also significant (F=9.42, P>F=0.0035). For **Similarity Rating** this was caused by an increase in the ratings for the last two blocks for the group of subjects who thought they were rating the same place for different times (Figure 16.9e). For **Reaction Time** the significant interaction effect appears to be related to a faster decrease (particularly between the first and second block) in mean decision time for the group of subjects who thought they were comparing the same place for different times (Figure 16.9f). Multivariate tests using **Similarity Rating** and **Reaction Time** as dependent variables produced very similar results. **Set** was significant (F=4306.5, P>F=0.0001) as was **Tversky Index * SET** (F=7.27, P>F=0.0001) and **Premise * Set** (F=9.42, P>F=0.0035).

Discussion

The visual similarity ratings provided by the subjects had a significant relationship with the similarity index values predicted by Tversky's contrast model. This supports the first hypothesis and leads to the conclusion that the visual similarity of maps is a function of both common and distinctive features. The instructions given to the subjects made them believe that they were either comparing the same location for two times, or different locations for the same time. This categorical top-down knowledge interacted with the identical bottom-up information on the maps in a significant way. The **Premise** given to the subjects created significant differences in both **Similarity Rating** and **Reaction Time**. Subjects who thought both maps were the same location were quicker to make their decisions. This suggests that the decisions were easier for them to make. Perhaps this top-down knowledge enabled these subjects to make their decisions with more confidence. If the two maps are for the same location are they not more likely to be similar? If one is predisposed to think that this is true, then one is more likely to see what is expected and not doubt it for a millisecond. The reverse is true, of course, for maps that are known to be for different locations. The differences in the means for the three highest similarity categories in Figure 16.9a provide good support for this argument.

The differences in **Similarity Rating** for the two groups of subjects appeared to be getting stronger with practice. This notion is supported by the significant interaction effect **Premise * Set**. The trend for the "same place" subjects was toward higher similarity ratings, and the trend for the "same time" subjects was toward lower similarity ratings (Figure 16.9e). Practice also enabled both groups to be more efficient and perhaps more confident in making their decisions. **Reaction Time** decreased for both groups over the four sets of trials (Figure 16.9f).

CONCLUSIONS

A number of important conclusion can be made from the results of the two experiments. Our first conclusion is that the first hypothesis, that is, *maps representing categorical information are judged to be more similar if they share common features and less similar if they have distinctive features,* is strongly supported. The results for both Experiment One, using reference maps, and Experiment Two, using qualitative land use maps, converge to suggest that map readers judge the similarity of qualitative maps consistent with the theoretical arguments at the core of Tversky's *contrast model*. The similarity of both types of categorical maps appears to be a function of both their common and distinctive features. The first experiment clearly illustrated that adding common features to maps increased similarity and adding distinctive features decreased similarity (Figure 16.5). This was also generally true for descriptions of maps except for adding common features described as being in different location on the maps.

The results of Experiment One also support Tversky's argument that people focus more on distinctive features when processing visual information like the type found on maps, and focus more on common features when processing verbal information like the type found in descriptions of maps. There is also some evidence that the visual similarity of the experimental maps was not only dependent on *what* features were on the maps, but also on *where* the features were on the maps (Figure 16.5). Tversky's original model did not consider the effect of location on similarity judgements. This means that two maps having exactly the same features are perfectly matched even if the features are not in the same locations. Others have suggested that this may not be a good assumption (Goldstone et al., 1991). *What* is on two maps being compared is, of course, important in determining their similarity, but *where* the features are located on the maps may also influence judgements of similarity. This is apparently true even for qualitative maps.

The second hypothesis, that is, *categorical information not directly related to the patterns on maps also affects how similar maps are thought to be,* is also strongly supported by the results of the second experiment. The top-down information (time and location) used in Experiment Two acted as a bias that significantly influenced the similarity judgements made by subjects. Although this may not be a total surprise, it is bad news for map designers who find bottom-up information on maps easier to control.

Future studies should consider similarity judgements for more complex maps and with additional features. We have known that the visual similarity of quantitative maps, for example, choropleth maps, is not reflected by standard correlation coefficients for some time. It is possible that the basic tenets of the *contrast model* could be used to explain the visual similarity of a range of quantitative maps as well. Other types of processes might be studied using the *contrast model*. Equation 4 could be used to study any number of map reading processes. Some promising research directions include investigating the suggestions that distinctive features are focused on during search processes and that common features are used more when making classifications. The current study set the alpha, beta, and gamma coefficients for Equation 4 to one, but they could be calibrated more precisely and adapted to modelling different types of processes. Finally, top-down information may affect decision-making in many other map reading contexts. This notion needs to be investigated over a range of map reading situations.

ACKNOWLEDGEMENTS

Experiment One was partially funded by a grant from the Cartography Specialty Group of the Association of American Geographers to Elzbieta Covington. Experiment Two was partially funded by a grant from the Carolina Venture Fund to Robert Lloyd.

REFERENCES

Ben-Shakhar, G., and Gati, I. (1987). Common and distinctive features of verbal and pictorial stimuli as determinants of psychophysiological responsivity. *Journal of Experimental Psychology: General*, 116, 91-105.

Brennan, N., and Lloyd, R. (1993). Searching for boundaries on maps: The cognitive process. *Cartography and Geographic Information Systems*, 20, 222-236

Covington, E. (1992). Similarity judgments of maps. Unpublished Master's Thesis, Department of Geography, University of South Carolina.

DeSarbo, W., Johnson, M., Manrai, A., Manrai, L., and Edwards, E. (1992). Tscale: A new multidimensional scaling procedure based on Tversky's contrast model. *Psychometrika*, 57, 43-70.

Gati, I., and Tversky, A. (1982). Representations of qualitative and quantitative dimensions. *Journal of Experimental Psychology: Perception and Performance*, 8, 325-40.

Gati, I., and Tversky, A. (1984). Weighting common and distinctive features in perceptual and conceptual judgments. *Cognitive Psychology*, 16, 341-70.

Gati, I., and Tversky, A. (1987). Recall of common and distinctive features of verbal and pictorial stimuli. *Memory and Cognition*, 15, 97-100.

Goldstone, R., Medin, D., and Gentner, D. (1991). Relational similarity and the nonindependence of features in similarity judgments. *Cognitive Psychology*, 23, 222-62.

Hartshorne, R. (1959). *Perspective on the nature of geography*. Chicago: Rand McNally Co.

Johnston, R.J. (1978). *Multivariate statistical analysis in geography*. London; New York: Longman.

Lloyd, R. (1988). The search for map symbols. *The American Cartographer*, 15, 363-78.

Lloyd, R. (1989). The estimation of distance and direction from cognitive maps. *American Cartographer*, 16, 109-122.

Lloyd, R, and Steinke,T. (1976). The decision-making process for judging the similarity of choropleth maps. *American Cartographer*, 3, 177-184.

Lloyd, R., and Steinke T. (1977). Visual and statistical comparison of choropleth maps. *Annals of the Association of American Geographers*, 67, 429-36.

Lloyd, R., and Steinke, T. (1985). Comparison of quantitative point symbols/The cognitive process. *Cartographica*, 22, 59-77.

Lloyd, R., and Steinke, T. (1986). The identification of regional boundaries on cognitive maps. *The Professional Geographer*, 38, 149-159.

MacEachren, A. (1982). The role of complexity and symbolization method in thematic map effectiveness. *Annals of the Association of American Geographers*, 72, 495-513.

Nigrin, A. (1993). *Neural networks for pattern recognition*. Cambridge, MA: MIT press.

Sattath, S., and Tversky, A. (1987). On the relation between common and distinctive feature models. *Psychological Review*, 94, 16-22.

Steinke, T., and Lloyd, R. (1981). Cognitive integration of objective map attribute information. *Cartographica*, 18, 13-23.

Steinke, T., and Lloyd, R. (1983). Judging the similarity of choropleth map images. *Cartographica*, 20, 35-42.

Treisman, A., and Gelade, G. (1980). A feature integration theory of attention. *Cognitive Psychology*, 12, 97-136.

Tversky, A. (1977). Features of similarity. *Psychological Review*, 84, 327-52.

APPENDIX A: EQUATIONS USED TO MEASURE THE EFFECT OF X FOR THE FIRST EXPERIMENT

s - similarity.

RT - reaction time.

p, **q** - base maps (verbal descriptions).

$\mathbf{x}$, $\mathbf{x_1}$, $\mathbf{y}$ - additive attributes (urban area, railroad, contour lines, forest, industrial area, and corn field).

Common feature - attribute added to base maps (verbal descriptions) **p** and **q**.

Distinctive feature - attribute added to one map (verbal description).

$C(\mathbf{x}) = s(\mathbf{px},\mathbf{qx})-s(\mathbf{p},\mathbf{q})$ - influence of **x** as a common feature on similarity judgements.

$C_1(\mathbf{x}) = s(\mathbf{pxy},\mathbf{qxy})-s(\mathbf{py},\mathbf{qy})$ - influence of **x** as a second common feature on similarity judgements.

$C_2(\mathbf{x}) = s(\mathbf{px},\mathbf{qx_1})-s(\mathbf{p},\mathbf{q})$ - influence of **x**, as a common feature at different locations, on similarity judgments.

$D(\mathbf{x}) = s(\mathbf{p},\mathbf{py})-s(\mathbf{px},\mathbf{py})$ - influence of **x** as a distinctive feature on similarity judgements.

$D_1(\mathbf{x}) = s(\mathbf{p},\mathbf{py})-s(\mathbf{p},\mathbf{pxy})$ - influence of **x** as a second distinctive feature on similarity judgements.

$W(\mathbf{x}) = C(\mathbf{x}) \ / \ C(\mathbf{x}) + D(\mathbf{x})$ - weight of common to distinctive features.

$TC(\mathbf{x}) = s(\mathbf{px},\mathbf{qx})-s(\mathbf{p},\mathbf{q})$ - influence of **x** as a common feature on reaction time.

$TC_1(\mathbf{x}) = RT(\mathbf{pxy},\mathbf{qxy})-RT(\mathbf{py},\mathbf{qy})$ - influence of **x** as a second common feature on reaction time.

$TC_2(\mathbf{x}) = RT(\mathbf{px},\mathbf{qx_1})-RT(\mathbf{p},\mathbf{q})$ - influence of **x**, as a common feature at different locations, on reaction time.

$TD(\mathbf{x}) = RT(\mathbf{p},\mathbf{py})-RT(\mathbf{px},\mathbf{py})$ - influence of **x** as a distinctive feature on reaction time.

$TD_1(\mathbf{x}) = RT(\mathbf{p},\mathbf{py})-RT(\mathbf{p},\mathbf{pxy})$ - influence of **x** as a second distinctive feature on reaction time.

17

An Examination of the Effects of Task Type and Map Complexity on Sequenced and Static Choropleth Maps

David K. Patton

Department of Geography and Environmental Studies, Slippery Rock University

Rex G. Cammack

Department of Political Science and Geography, Old Dominion University

INTRODUCTION

The growing dominance of and fascination with computers in the field of cartography has generated a host of questions for researchers to pursue. Some of these questions fit within existing cartographic paradigms such as: how can the dynamic capabilities of the digital environment be utilized to *design* more effective displays than were possible in the static mapping environment and, how effective are these new displays in *communicating* cartographic information? It is suggested by Olson (1983) and reiterated by Taylor (1987) that cartographic research must be concerned with both of the above questions. Taylor states that, "the development of computer-assisted cartography is posing new design responses based on a better understanding of communication processes" (Taylor, 1987: 594). This statement clearly indicates an important interaction of viewing design innovations within the larger framework of communication theory and allowing knowledge of cognitive processes to guide in the creation of new design methods. Examples of research addressing elements of both design and communication issues include Campbell and Egbert (1990), Slocum et al. (1990), DiBiase et al. (1992), Dorling (1992), Monmonier (1992), and Cammack (1994). In addition, it has been suggested that the design possibilities that exist within the digital framework should be utilized to move cartography into a new research paradigm: *visualization* (MacEachren and Ganter, 1990; MacEachren et al., 1992). Cartographic research in the area of visualization concentrates on designing tools that facilitate data exploration as opposed to data communication. While some proponents of a visualization paradigm attempt to distance themselves from a narrowly defined concept of the communication paradigm (search for the optimal map), it should be clear that the development of effective data exploration tools can only be aided by an understanding of perception and cognition. Therefore, while the goals of cognitive studies in cartography may shift with a movement from a strict communication paradigm to a visualization paradigm, cognitive research will continue to be of great value for evaluating and guiding the design of effective visualization tools.

This study attempts to further research within the design/communication framework, and at the same time, evaluate a design tool that might be useful for both cartographic communication and visualization. The use of "sequencing" in a choropleth map design appears to be a natural combination, particularly within the computer mapping environment. In fact, sequencing as a cartographic design method has been put into practice (Monmonier, 1992), and its effectiveness has been researched (Taylor, 1987; Slocum et al., 1990). Despite the attention given to sequencing, the previous research revealed somewhat "lukewarm" results in terms of overall effectiveness. The purpose of this study was to examine sequencing in more specific roles in order to determine if sequencing might be situationally effective. Towards this goal, the current study examined two different methods of sequencing along with a static presentation and isolated two basic map reading tasks within a framework of varying map complexity. It is hoped that the results of this research will more clearly indicate the benefits and detriments of sequencing as a map design method.

LITERATURE REVIEW AND BACKGROUND THEORY

The theoretical basis of this study is brought together from three areas of previous research. The first area of research, and the most critical in terms of map designs utilizing sequencing, is the theory of *chunking* as a means by which people hold information in working memory. Secondly, chunking is viewed as a situational method of rehearsal, therefore, different tasks may benefit differently from a given chunking scheme. The current study investigates the distinction of *"what"* tasks and *"where"* tasks. Finally, *complexity* is considered as an additional element that may influence the success of a particular chunking scheme given different tasks.

Chunking

As was noted in two prior studies looking specifically at the sequencing of choropleth maps (Taylor, 1987; Slocum et al., 1990), sequencing is basically a rehearsal mechanism that relies on the chunking of information. The term chunk was proposed as a means of defining units of information that are held in working memory (Miller, 1956). The vagueness of this term was Miller's solution to the problem of determining how many "bits" or items of information could be held in working memory. Miller determined that what constitutes a chunk of information is dependent upon the situation in which the information is processed and the existing knowledge, or level of expertise, that the person processing the information brings to the task. Howard (1983) demonstrates this effect with the example of a list of words, for example: Nebraska, Ohio, Michigan, South Carolina, Arizona. To an English speaking literate adult, the above list would be processed as six words, possibly five places. To a non-literate person, the same list might consist of 18 syllables, or even 40 letter symbols.

Once information has been chunked, the number of chunks that a person can hold in working memory appears to be seven plus or minus two (Miller, 1956). Miller's holding capacity of working memory has been supported in psychological research by deGroot (1966), Simon (1974), and Simon and Chase (1973). This group of studies compared novice and expert chess players for their ability to remember chess piece locations. It was found that the difference between novice and expert players was not the number of chunks that could be held, but the definition of a chunk. For the novice, each chess piece and its location constituted a chunk of information. For the expert, a chunk consisted of a meaningful *group* of chess pieces on the board. Obviously, there is a direct correlation between the two-class *map* of the chess pieces (black and white) and any choropleth map that a cartographer might create.

If experts utilize chunks that contain more bits of information, is it possible, through the use of a rehearsal mechanism, to increase the bits of information in novices' chunks? This basic question has generated several attempts by cartographic researchers to examine various display options testing if a significant increase in information retention would result. One group of cartographic studies examined the effects of hue on map communication at various levels of complexity (Mersey, 1990; Gilmartin and Shelton, 1989). In both of these studies, complexity was equated with the number of classes, and the results showed that effectiveness was acceptable in a range of five plus or minus two classes. Simply varying hue combinations does not constitute a rehearsal mechanism, but it could lead to more efficient chunking by the map reader.

In another set of cartographic experiments, artificially guided chunking or sequencing was tested as a rehearsal mechanism by comparing subject performance on chunked or sequenced maps versus static maps (Shimron, 1978; Taylor, 1987; Slocum et al., 1990). If it is accepted that people chunk information naturally, then presenting information to map users in a particular sequence could be considered artificially guided chunking. Shimron's experiment was the first to apply artificially guided chunking to a map presentation. Shimron experimented with two presentation types, a general purpose map displayed sequentially by category (natural features, roads, cities) and the same complete map displayed in three sections (left third, middle third, right third). Shimron found that subjects performed worse on positional tasks when shown the map sequenced by category. He concluded that map users have problems fitting themes together spatially when all the themes aren't present at the same time. It could be concluded that people do not chunk map information by categories.

Taylor (1987) made an important contribution by being the first to look at the effects of sequencing on a choropleth map. He compared a sequenced, five-class choropleth map to an identical, static, five-class choropleth map. Taylor's results showed that the subjects used map information presented in a sequenced fashion more accurately and efficiently than with the static map, but the subjects also appeared to forget the information more quickly when tested with memory tasks. These latter results, which were based on a decline in the accuracy of subjects' responses when asked to identify the correct category for each of the sequenced classes, may be explained by the lack of experimental control in Taylor's experiment. Subjects were tested in groups, not individually. It may also be that the spatial distribution of the data on the map made the latter questions more difficult. The paper does not indicate that the questions were asked in reverse order for some of the subjects. There is no reason to believe that information activated in working memory through artificial chunking should be less resilient than information activated by natural schema, particularly when the former results indicated stronger activation for the sequenced information.

Slocum et al. (1990) attempted to eliminate the methodological problems found in Taylor's study by creating a controlled experimental environment. In the formal part of the experiment, the authors, like Taylor, compared the effectiveness of sequenced five-class choropleth maps to the same five-class choropleth maps in a static presentation; however, in Slocum et al., subjects were tested individually to ensure experimental control. This study investigated both the impressions of the subjects towards the sequenced map and the subjects' performance in terms of both information acquirement and retention. The results indicated that subjects preferred the sequenced presentation, however, the authors point out that this could be based largely on the novelty of the presentation. In terms of performance, the results showed no significant difference between presentation types. An admitted flaw in this experiment was that subjects were allowed to dictate length of exposure time when viewing the sequenced maps. The authors allowed the "sequence" subjects to individually select viewing times based on Slocum et al. (1988) in which subjects indicated a preference for unrestricted viewing times. Unfortunately, unrestricted viewing times creates a difficult covariate in the analysis. Another important factor of the Slocum et al. (1990) research is the intensive study of the maps that occurred during the information acquirement phase of the experiment. "With respect to the memory portion of the experiment, there is no way of knowing what the results would be if the map were examined less intensively . . . it would be interesting to analyse the memory portion of the experiment as the number of information questions is reduced" (Slocum et al., 1990: 83).

What and Where Tasks

As important as experience or the development of expertise may be in the development of a rich chunking schema, another equally important element may be the task involved. In fact, in the realm of map reading, task may be considerably more important since it is questionable that there are "map reading experts" (Steinke, 1994). Two basic tasks in map reading are the identification of "what" and "where" objects are on a map. Initially, these tasks appear to be embedded within each other; however, psychologists have determined that these tasks are different and distinct and performed by separate systems within the brain (Rueckl et al., 1988; Kosslyn et al., 1989; Kosslyn et al., 1992). Although many of the cartographic studies cited above incorporated both "what" and "where" questions (Taylor, 1987; Mersey, 1990; Slocum et al., 1990), none of them specifically compared the two types of tasks within their research designs.

It is significant, for map design purposes, to acknowledge that these two tasks are not processed in the same manner by map readers. An important question, however, is, "Are both of these tasks important within a map reading environment?" While it is clear that the primary purpose of a map presentation is to convey spatial information, it is also clear that maps are used for a variety of purposes, including the communication of non-spatial information (or at least not spatial in terms of the patterns on a given map) exclusive of spatial context. For example, a presentation or investigation of a set of land use maps of the 50 United States might be approached simply as an inventory task. The map user might not be concerned, at this stage, with *where* corn is grown in Ohio, but simply with the information that corn *is* grown in Ohio. If it is accepted that "what" and "where" tasks are both important and different, it follows that a better understanding of the processes involved with these tasks will aid in the creation of more effective communication and exploratory tools.

Map Complexity

In addition to task type, another factor that may play an important role in the evaluation of an effective sequencing design is map complexity. Map complexity has received extensive examination in previous studies. The definition of map complexity has been addressed by Monmonier (1974), Muller (1975,1976), and MacEachren (1982a). Lavin (1979) investigated 12 different measures of complexity, including complexity based on the number of classes and the number of faces, and found them to be highly redundant. In addition, Lavin examined the connection between map complexity and the consistency in which map readers identified homogenous regions. He found that there was a direct relation between complexity and map reader agreement of regional boundaries.

Muehrcke (1973), and Steinke and Lloyd (1983) investigated the effects of map complexity on map comparison judgements. The first two studies determined that map complexity did have a direct effect on map readers' abilities to compare maps accurately. Steinke and Lloyd, however, found that both the level of blackness and the degree of cross-correlation were more important than complexity.

MacEachren (1982b) and Mersey (1990) both investigated the effects of map complexity within the framework of different map reading tasks. MacEachren's tasks included the extraction of detailed information at specific map locations, and the recognition of overall map patterns. His results showed that there was a difference in the effect of map complexity dependent upon the type of task being performed. When class was held constant and the pattern complexity was increased, the resulting interpretation of general map patterns was affected negatively, but not the extraction of specific map information. These results concerning identification of general patterns is supported by the Lavin (1979) results. In addition, when the number of classes was increased the extraction of specific map details was negatively affected, but not the identification of general map patterns. Mersey found that complexity did not affect the ability to discern spatial patterns in a direct acquisition situation. She used number of classes as the sole variation in complexity and held pattern constant, thus, her results are partially in agreement with MacEachren's latter findings. However, Mersey's results also showed that there was a significant effect on discernment of general spatial patterns when the tasks were done from memory. These findings are not in agreement with MacEachren.

In addition to the cartographic literature, an important theory has emerged from psychology which has direct bearing on the definition of complexity. Kahneman et al. (1992) have developed a theory of *object files*. Based on this theory, within a temporal presentation or natural occurrence, the focus of attention is on objects. Object *files* are "opened" for each *object* being attended and attribute information is stored. The authors found that object files can be repeatedly updated based on their spatial proximity and/or attribute change. The contribution of object file theory, for this study, is the affirmation that spatial definitions of complexity are as salient as attribute definitions. Certainly, measures of pattern have been used in definitions of cartographic complexity prior to object file research, but, the affirmation of the importance of spatial association serves as a signal to address pattern in future research.

Based on the literature cited above, it was determined that important questions concerning the design of computer-aided choropleth map displays could be addressed by examining the interactions of sequencing, task type, and complexity. We felt, by addressing the issue of sequencing within the framework of task and complexity, a more accurate picture of the possible uses of this technique can be achieved. In addition we hoped to highlight another important connection that exists between measures of complexity and task type, thereby furthering the research trend established by Lavin (1979), MacEachren (1982b), and Mersey (1990). The goals of this study were: (1) to examine the effects of two different sequenced choropleth map presentations compared to each other and a third static presentation in terms of the ability of subjects to perform memory tasks; (2) to evaluate the effectiveness of artificially guided "chunking" for "what" tasks and "where" tasks using accuracy and reaction time as a means of evaluation; and (3) to consider the effectiveness of sequencing for both task types within the framework of varying map complexity.

METHODOLOGY

A series of one hundred choropleth maps were designed for this experiment. The maps were all created from a digital file of political boundaries of 12 counties from Kentucky: flipped and rotated. The choice of this group of counties was based on their irregular but compact shapes and the overall compactness of the group. The maps were systematically varied based on the number of classes and the number of objects. Within the framework of this experiment, number of objects represented the number of faces of contiguous colour present on a given map. The total map set consisted of 10 maps for each of the following class and object combinations: 3*3, 3*5, 3*7, 3*9, 5*5, 5*7, 5*9, 7*7, 9*9 (Figure 17.1). The result was a map set that comprised only half of a matrix; however, based on these elements of complexity, this could not be helped. A nine class map can not have fewer than nine faces. The maps were coloured using a range of 10 hues that might appear on any nominally classed land use map: brown, orange, olive, blue, grey, cyan, green, red, yellow, and pink. Each map, within each group of 10 maps, was coloured using different hues from the set of total hues listed above. This was done in an attempt to eliminate the influence of colour within the experiment.

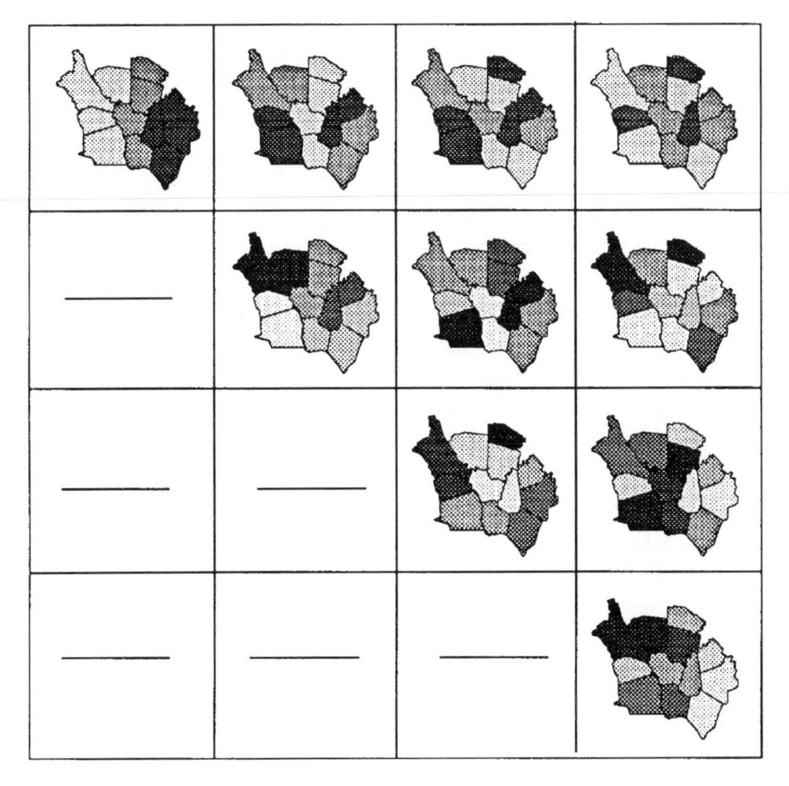

Figure 17.1 Matrix showing map types based on number of objects and number of classes.

Subjects were presented either: (1) the maps in a static presentation (static); (2) the maps shown complete on the screen with the boundaries of each class flashing in a given order (flashing); and (3) the maps revealed one class at a time, flashing each class as it was shown, until all of the classes were on the screen (building) (Figure 17.2). The static presentation represented a traditional map display. The flashing and building maps represented two different approaches to sequencing within a computer mapping environment.

Forty-five student volunteers from the University of South Carolina took part in the study. The subjects were partitioned into three groups of 15; one group for each of the three presentation types: static, flashing, and building. The method of presentation was the only varying factor between the three groups. The maps were presented to subjects, individually, on a NEC 486 computer with a NEC multisync 3fgx screen. Prior to beginning the experiment, each subject was asked whether or not they suffered from colour-blindness. If the subject was not sure, they were tested using the "pseudo-isochromatic plates for testing colour perception" created by Richmond Products, Boca Raton, Florida. No subjects were eliminated based on colour-blindness. The subjects were then instructed that they would be tested with a series of choropleth maps on the computer. A brief definition of choropleth maps was given. The subjects were told that the experiment would consist of two sections each lasting approximately 15 minutes in length.

Section one of the experiment consisted of the "what" task. For this task, the subjects were told that they would view a map for six seconds, the screen would clear, and a task screen would appear with a 1" by 1" square in the middle of the viewing area filled with one of the 10 hues described above. The subjects were told to answer the following question as quickly as possible, "was the colour in the square on the map you just studied?" They responded by pressing either the left (yes) or right (no) arrow keys. The correct response was "no" 50 percent of the time. The subjects performed this task 100 times in succession. The question was not repeated after the experiment began. A practice session, using a different base map, was administered prior to starting the main experiment. Reaction times and accuracy were recorded.

Section two of the experiment consisted of the "where" task. It was administered exactly as the "what" task with the following exceptions. For the "where" task, the question screen consisted of the same base map that was viewed during the study period, except only one of the counties was filled. The subjects were told to answer the following question as quickly as possible, "Is the filled county the correct colour based on the map you just studied?" The order of the sections (one and two) was reversed for half of the subjects.

The effectiveness of the three presentation types, in terms of information recall, was tested, the effects of the tasks "what" and "where" were tested, and the effects of object complexity and class complexity were tested. In addition, interaction effects were tested between presentation type and task type, task type and number of objects, and task type and number of classes. It was hypothesized that the two sequenced presentations would be more effective for the "where" task. This was

Figure 17.2 Static, flashing, and building presentation types.

based on the notion that sequencing is inherently spatial, therefore, it will highlight the spatial character of the maps more effectively than the static maps. It was hypothesized that the static presentation and the flashing presentation would be more effective for the "what" task. This hypothesis is based on the notion that *all* of the classes are viewed for the entire presentation on the static and flashing maps, therefore, "what" is on the static or flashing map would be learned more effectively than on the building map. It was also hypothesized that complexity, based on the number of classes, would affect the "what" task and complexity, based on the number of objects, would affect the "where" task. These two hypotheses were based on the premise that the number of classes defines "what" is on the map and that the number of objects defines "where" things are on the map.

RESULTS AND DISCUSSION

Analysis of Variance (ANOVA) based on percent correct and reaction time showed that there was a significant difference between presentation types (Figure 17.3a-b). The group means for percent correct ($PR > F(7.15) = 0.001$) showed an advantage for the building presentation (83 percent correct) over the flashing (80 percent correct) and the static presentations (79 percent correct). There was not a significant difference between the flashing and static presentations based on percent correct. The group means for reaction time ($PR > F(22.79) = 0.001$), showed an advantage for the building presentation (1070 ms) over the flashing (1121 ms) which was faster than the static presentation (1217 ms).

The presentation type results indicate that a building style of sequencing is more effective in terms of percent correct and reaction time. While these results differ from Slocum et al. (1990), this could be explained by the changes in experimental design. As stated in the literature review, Slocum et al. incorporated an information acquirement set of tasks in which the subjects studied the map intensively prior to completing the recall set of tasks. This intensive study of all the presentation types may have eliminated any benefits that the sequenced presentation would have had for the recall tasks if the study periods had been shorter.

The ANOVA revealed a significant difference between "what" and "where" tasks based on both percent correct and reaction time (Figure 17.3c-d). The group means for percent correct ($PR > F(323.79) = 0.001$) illustrated that "what" tasks (88 percent correct) are less difficult than "where" tasks (73 percent correct). The group means for reaction time ($PR > F(152.48) = 0.001$) showed that "what" tasks (1025 ms) also require less processing time than "where" tasks (1247 ms). The relationship between "what" and "where" tasks found in the present study are supported by the study of Taylor and Tversky (1992) in which subjects were tested with "what" and "where" tasks based on spatial relationships learned from both descriptions and maps.

As measures of complexity, both the number of objects ($PR > F(21.71) = 0.001$) and the number of classes ($PR > F(27.95) = 0.001$) were found to affect significantly

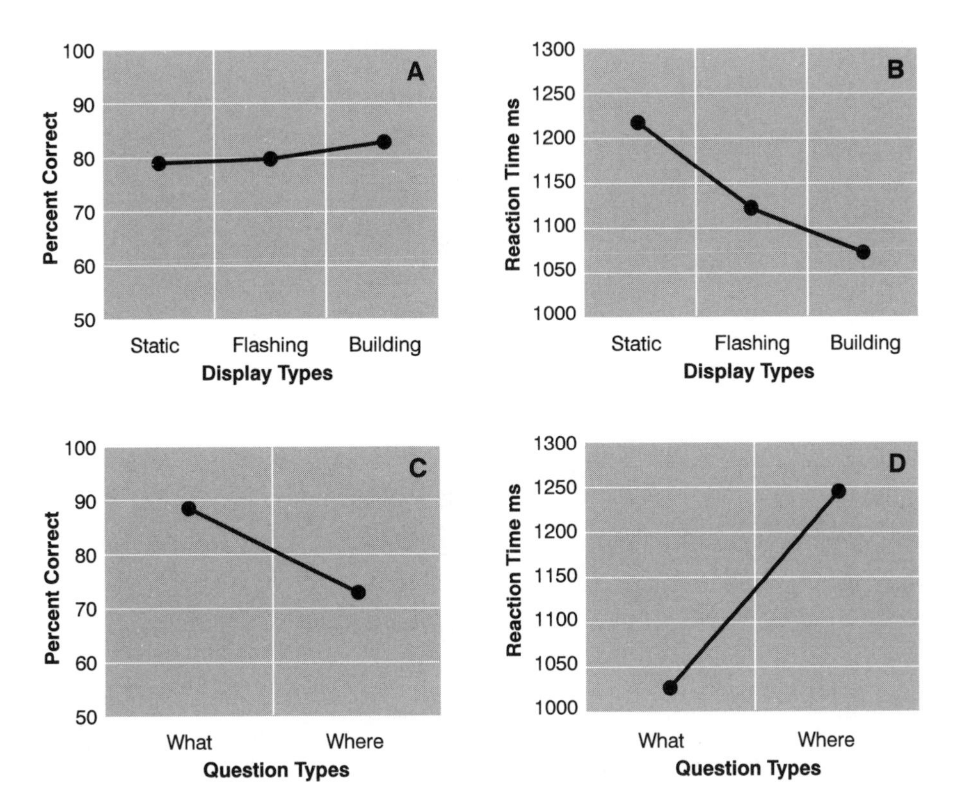

Figure 17.3 (a) presentation type by percent correct; (b) presentation type by reaction time; (c) task type by percent correct; (d) task type by reaction time.

the percentage of correct responses. However, neither of these measures of complexity significantly affected the reaction time of responses (Figure 17.4a-d). As illustrated in the figures, there is an inverse relationship between accuracy and the number of classes or objects. This inverse relationship appears to also exist for reaction time, but the range is actually less than 100 milliseconds for both number of objects and number of classes.

There was a significant interaction between task type and the number of objects based on the percent of correct responses (PR > F(4.87) = 0.002). The relationship of the group means for percent correct indicate that the number of objects has little effect on the ability of subjects to recall "what" was on a map; however, the number of objects does have an inverse effect on the ability of subjects to recall the spatial distribution of information on a map (Figure 17.5a). There was not a significant interaction between these variables by reaction time (Figure 17.5b). The significant interaction between task type and number of objects (PR > F(22.76) = 0.001) shows the opposite effect (Figure 17.5c-d). The "where" task is less affected by class, but the "what" task is dramatically affected by task. Unexpectedly, there was

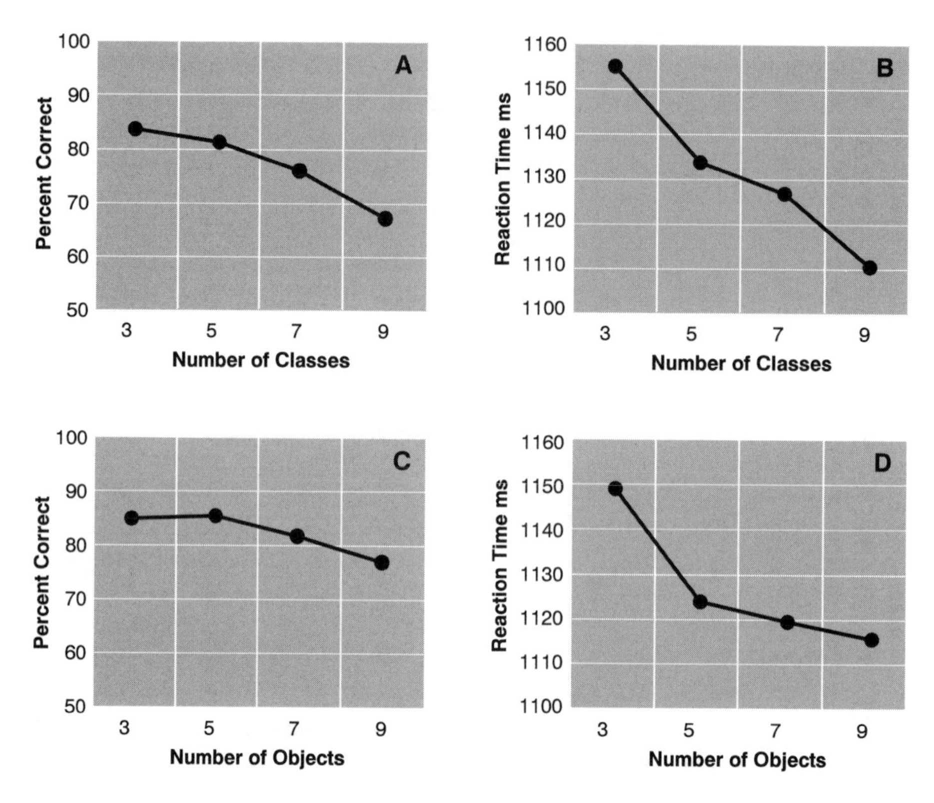

Figure 17.4 (a) number of classes by percent correct; (b) number of classes by reaction time; (c) number of objects by percent correct; (d) number of objects by reaction time.

a significant interaction between task type and number of classes based on reaction time (PR > F(10.43) = 0.001). Based on the pattern of the group means, this significant interaction appears to be due largely to a change in strategy for the "where" questions. It is possible that the subjects used verbal coding when there were fewer classes and imagery when there were more classes.

Finally, there was a significant interaction between presentation type and task based on both percent correct responses (PR > F(3.19) = 0.042) and reaction time of responses (PR > F(5.15) = 0.006). An examination of the interaction plot for percent correct illustrates two points: (1) presentation type did not appear to effect the "what" task; and (2) presentation type did appear to effect the "where" task (Figure 17.6a). These results only one of the two hypotheses concerning the presentation and task interaction. The building and flashing sequence presentations do improve accuracy for the spatial question, however, the static and flashing map did not effect higher accuracy for the "what" task.

In terms of reaction time, the pattern is more consistent with the hypotheses (Figure 17.6b). The sequenced presentations are more time efficient than the static

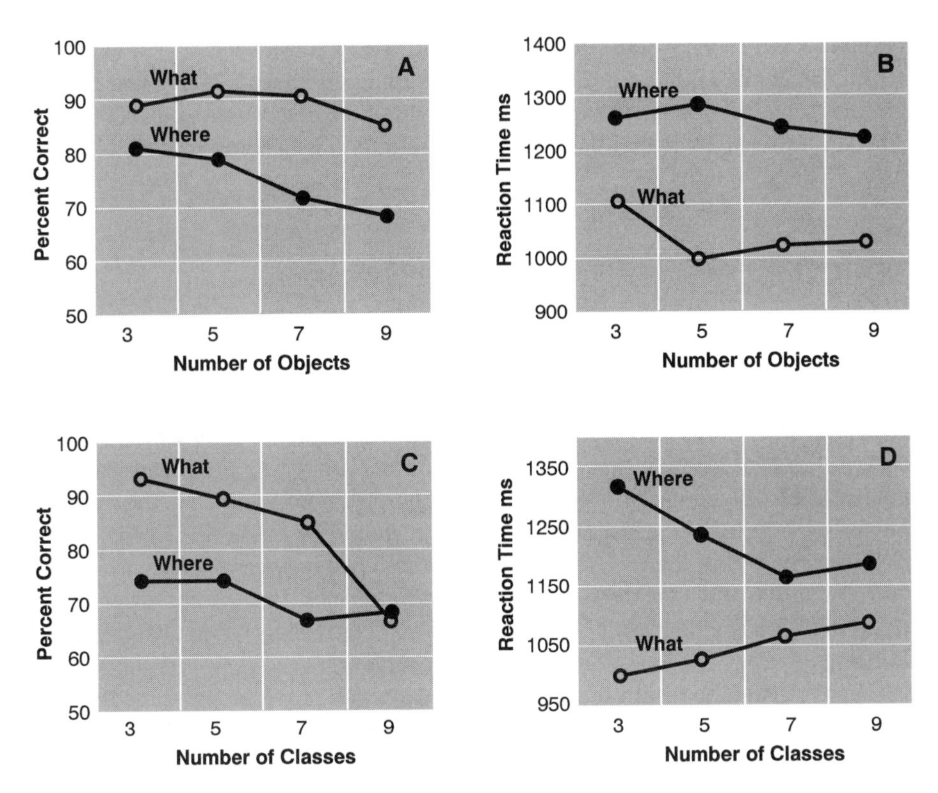

Figure 17.5 (a) interaction of task and number of classes by percent correct; (b) interaction of task and number of classes by reaction time; (c) interaction of task and number of objects by percent correct; (d) interaction of task and number of objects by reaction time.

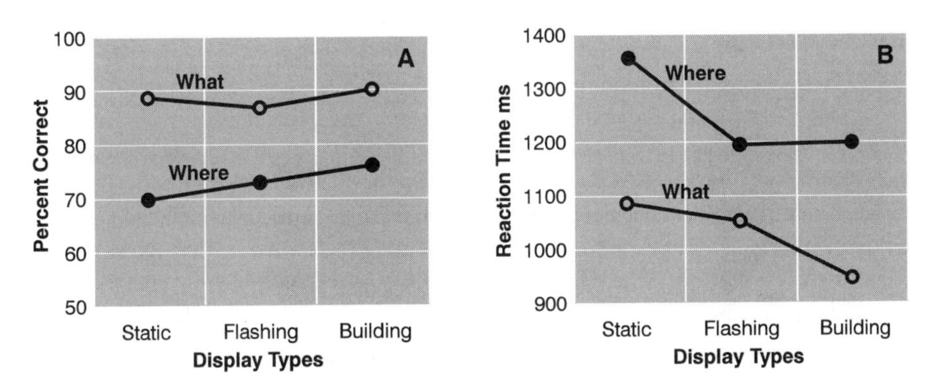

Figure 17.6 (a) interaction between task and presentation by percent correct; (b) interaction between task and presentation by reaction time.

presentation for the "where" question, and the static and flashing presentations, in which the entire map is in view throughout the presentation, both have similar reaction times for the "what" question. The variation from the hypotheses for the reaction time results is that the building presentation is faster than the static and flashing presentations.

CONCLUSIONS

The results of this study confirm several important theories concerning map design, task type, and map complexity. It has been shown that in a recall situation, with limited study time, a building form of sequencing is more effective in terms of accuracy. The results also indicate that this form of sequencing may be more efficient in terms of reaction time, particularly for "what" tasks. However, since the main effect for presentation type and reaction time was not significant, the assumption of superiority of the building presentation should be made cautiously. It was also confirmed that there was a significant difference between "what" tasks and "where" tasks, and that these tasks are in turn affected in different ways by an object definition of the complexity and definition of complexity based on the number of classes.

Considering the results of this experiment and the previous experiments done with sequencing, there are several issues that need to be examined in further research. First, it is still not clear as to the effect that sequencing will have on long term storage of map information. It is also unclear how effective sequencing will be within a much larger, more complex map display. Secondly, it is evident from this study that complexity based on the number of objects directly affects the ability of map readers to handle spatial information. Therefore, it might be appropriate to rethink the sequencing script. Instead of sequencing the classes of a choropleth map, it might be more effective to sequence the objects within classes.

REFERENCES

Cammack, R.G. (1994). Object de cart. [Unpublished Ph.D. dissertation.] Columbia, SC: University of South Carolina.

Campbell, C.S., and Egbert, S.L. (1990). Animated cartography: Thirty years of scratching the surface. *Cartographica*, 27(2), 24-46.

deGroot, A.D. (1966). Perception and memory versus thinking. In B. Kleinmuntz (Ed.), *Problem solving* (pp. 19-50). New York: John Wiley & Sons.

DiBiase, D., MacEachren, A.M., Krygier, J.B., and Reeves, C. (1992). Animation and the role of map design in scientific visualization. *Cartography and Geographic Information Systems*, 19(4), 201-214, 265-266.

Dorling, D. (1992). Stretching space and splicing time: From cartographic animation to interactive visualization. *Cartography and Geographic Information Systems*, 19(4), 215-227, 267-270.

Gilmartin, P.P., and Shelton, E. (1989). Choropleth maps on high resolution CRTs: The effects of number of classes and hue on communication. *Cartographica*, 26(2), 40-52.

Howard, D.V. (1983). *Cognitive psychology: Memory, language, and thought.* New York: Macmillan Publishing Co., Inc.

Kahneman, D., Treisman, A., and Gibb, B.J. (1992). The reviewing of object files: Object-specific integration of information. *Cognitive Psychology,* 24, 175-219.

Kosslyn, S.M., Chablis, C.F., Marsolek, C.J., and Koenig, O. (1992). Categorical versus coordinate spatial representations: Computational analysis and computer simulations. *Journal of Experimental Psychology: Human Perception and Performance,* 18, 562-77.

Kosslyn, S.M., Koenig, O., Barret, A., Cave, C.B., Tang, J., and Gabrieli, J.D.E. (1989). Evidence for two types of spatial representations: Hemispheric specialization for categorical and coordinate relations. *Journal of Experimental Psychology: Human Perception and Performance,* 15, 723-35.

Lavin, S.J. (1979). Region perception variability on choropleth maps: Pattern complexity effects. [Unpublished Ph.D. dissertation]. Kansas: University of Kansas.

MacEachren, A.M. (1982a). Map complexity: Comparison and measurement. *The American Cartographer,* 9(1), 31-46.

MacEachren, A.M. (1982b). The role of complexity and symbolization method in thematic map effectiveness. *Annals of the Association of American Geographers,* 72(4), 495-513.

MacEachren, A.M., and Ganter, J.H. (1990). A pattern identification approach to cartographic visualization. *Cartographica,* 27(2), 64-81.

MacEachren, A.M., Buttenfield, B.P., Campbell, J.B., DiBiase, D.W., and Monmonier, M.S. (1992). Visualization. In R.F. Abler, M.G. Marcus and J.M. Olson (Eds.), *Geography's inner worlds: Pervasive themes in contemporary American geography* (pp. 99-137). New Brunswick, NJ: Rutgers University Press.

Mersey, J.E. (1990). Colour and thematic map design: The role of colour scheme and map complexity in choropleth map communication. *Cartographica,* 27(3), Monograph 41.

Miller, G.A. (1956). The magical number seven, plus or minus two: Some limits on our capacity for processing information. *Psychological Review,* 63, 81-97.

Monmonier, M.S. (1974). Measures of pattern complexity for choropleth maps. *The American Cartographer,* 1(2), 159-169.

Monmonier, M.S. (1992). Authoring graphic scripts: Experiences and principles. *Cartography and Geographic Information Systems,* 19(4), 247-260, 272.

Muehrcke, P.C. (1973). The influence of spatial autocorrelation and cross correlation on visual map comparison. *Proceedings* of the ACSM. 33rd Annual Meeting.

Muller, J-C. (1975). Associations in choropleth map comparisons. *Annals of the Association of American Geographers,* 65(3), 403-413.

Muller, J-C. (1976). Numbers of classes and choropleth pattern characteristics. *The American Cartographer,* 3(2), 169-175.

Olson, J. (1983). Future research directions in cartographic communication and design. In D.R.F. Taylor (Ed.), *Graphic communication and design in contemporary cartography* (pp. 257-284). Chichester: John Wiley & Sons.

Rueckl, J., Cave, K., and Kosslyn, S. (1988). Why are "what" and "where" processed by separate cortical visual systems? A computation investigation. *Journal of Cognitive Neuroscience,* (2), 171-86.

Shimron, J. (1978). Learning positional information from maps. *The American Cartographer,* 5(1), 9-19.

Simon, H.A. (1974). How big is a chunk? *Science,* 183, 482-488.

Simon, H.A., and Chase, W.G. (1973). Skill in chess. *American Scientist,* 61, 394-403.

Slocum, T.A., Egbert, S.L., Prante, M.C., and Robeson, S.H. (1988). Developing an information system for choropleth maps. *Proceedings,* Third International Symposium of Spatial Data Handling, Sydney, Australia, 293-305.

Slocum, T.A., Robeson, S.H., and Egbert, S.L. (1990). Traditional versus sequenced choropleth maps: An experimental investigation. *Cartographica*, 27(1), 67-88.

Steinke, T.R. (1994). The development of expertise in three map reading tasks. Paper presentation. Association of American Geographers 90th Annual Meeting, San Francisco, CA.

Steinke, T., and Lloyd, R. (1983). Judging the similarity of choropleth map images. *Cartographica*, 20(4), 35-42.

Taylor, D.R.F. (1987). Cartographic communication on computer screens: The effect of sequential presentation of map information. *Proceedings* of the 13th International Cartographic Conference: Morelia, Mexico. Vol 1: 591-611.

Taylor, H., and Tversky, B. (1992). Spatial mental models derived from survey and route descriptions. *Journal of Memory and Language*, 31, 261-292.

18

Gestalt Theory Applied to Cartographic Text

John A. Belbin

College of Geographic Sciences, Lawrencetown, Nova Scotia

To anyone who decides to spend part of their career training others to become working cartographers, it rapidly becomes obvious that we have two different groups concerned with cartography and very little contact between them. Most available published material on the field is produced by the academic group, who seem largely preoccupied with developing cartography as a science. They have produced a large number of theories, formulas, tests, and studies of all kinds. In the graphic and design areas of cartography, there are extensive perception and psychological studies and large numbers of tests on simple graphic images, often using university students as the subjects. Relatively little of this large volume of material ever seems to be used in a production environment.

By contrast, most workers in production cartographic agencies produce their published material using a variety of guides, rules of thumb, standardized procedures, and the experience of observing what seems to work. These aspects can certainly be transmitted to students and even enforced, if you are so inclined, but it is difficult to explain to them the logic behind what is considered to be "normal" or "acceptable". Many people are suspicious that some traditional approaches could be greatly improved, while other approaches may be suspect and even illogical.

The two groups could obviously benefit one another as they are both concerned with the development of better cartographic communication and abilities. It seemed appropriate to attempt the link-up using the well known Gestalt theories, as they are enjoying somewhat of a renaissance with the current interest in desktop publication techniques, which has so enhanced our text manipulation capabilities.

Gestalt theorists were responsible for many advances in the fields of learning and memory. Many modern educators criticize the old-style rote learning and attempt to get their students to think creatively, to achieve insight. This so-called modern and revolutionary set of ideas is actually quite old and originated with the Gestalt psychologists during the first two decades of this century. The attraction is that once you achieve insight you can readily transfer the solution to a related new problem—something not usually true with rote learning. It is the obvious goal of most educators (Rock and Palmer, 1990: 89); it is equally the obvious antithesis of standardized procedures and rigid rules.

Gestalt psychology has had an immense influence on the study of human perception in explaining how we make sense of what we see. It should, therefore, be of major importance to cartographers. We define ourselves as visual communication specialists. Our products are an integrated mix of graphic and textual techniques.

Gestalt is German for pattern or shape, although most people prefer "organization" as a more accurate interpretation because Gestalt principles are, in fact, principles of organization of the visual image. Its central tenant is well known, but seldom followed: "The whole is different from the sum of its parts". This idea that the nature of a complex whole can not be predicted by simply studying its component parts was first formulated by Wertheimer in 1912 in Frankfurt, in a paper concerned with a visual illusion called apparent motion. This effect is, of course, the one we rely on so extensively in the movies. He indicated that perception of the movement was radically different from the perception of each of the static images (Rock and Palmer, 1990: 84).

Gestalt theory maintains that the parts of an object INTERACT with one another and, in so doing, produce a "whole" which is very different from the sum of the various parts. Gestaltists talk about EMERGENT PROPERTIES produced by the interaction of the various components and give such examples as a melody in music, common table salt in chemistry, which is quite different from its poisonous and corrosive components, and the characteristics of a society which are often widely different from the individuals which comprise it. This is in direct contrast to ELEMENTARISM, a basic structuralist assumption that complex perceptions can be understood by identifying the elementary parts of the experience. The elementarist approach is often referred to as "mental chemistry" because it assumes perceptions can be analysed component-by-component, much as we study each atom, or even parts of an atom (Rock and Palmer, 1990: 84). Does all this sound familiar? It should; it is still common practise today, and many cartographic studies incorporate the notion without considering that it may not be valid.

Emergent properties are critical to Gestalt thinking, to their explanations of why we see the way we do and in a sometimes surprising manner. Gestaltists pointed out that what we "see" is not simply a matter of recording images with your eyes, but very much a result of how the visual system ORGANIZES or GROUPS the incoming information into units. Whole figures are primary—they are seen first and are visually dominant. The effect of the overall design is paramount and is more important than the detail comprising it, even on a highly complex product, and even though we can be shown to be capable of seeing huge amounts of small information. Tests have largely supported these notions; they would seem to be fundamentally true.

Given this introduction and quick review, it is very distressing to anyone who must struggle with cartographic design to note how little attention has been paid to overall layouts, proportions, aesthetics, cultural suitability, harmony, balance and coordination of the entire product, except for brief "motherhood" statements about their importance. Layout of a map is, for instance, only mentioned twice in Keates' *Cartographic Design and Production*, and as it is buried on pages 213 and 235 of a 250 page text, you might begin to wonder what you were supposed to be doing up until that point.

The stress on wholeness, integration, unity, consistency and, as the desktop publishers would say, remaining focused on the target, should be basic for cartographers. Everything else should be a means to that end. A map tends to be read in several distinct stages. Firstly, we obtain an OVERALL view of the image and this controls much of our initial reaction to it. We become aware of objects as FIGURES against BACKGROUNDS. We compare hues, values, brightness, size, contrasts, and so forth. As another stage, we then focus in on parts of the map image and buildup the detail by selective attention. The result of these studies is then combined mentally, and our brain tries to make sense of the result. At this point the aspects of memory and experience are introduced and become a major part of the visual system.

The creation of the FIGURE-GROUND relationship is the most essential one in perception and is subject to a huge amount of unconscious selectivity on the part of the brain—it cannot maintain objectivity. That is probably why so many people are surprised when they receive their snapshots back from the processor to discover just how messy the scene appears. The complex background was not seen by them, but was faithfully photographed. How many cartographers have unwittingly created virtually unreadable products because of a similar concentration on small details at the expense of the overall effect?

The classic cartographic illustration of the effects of figure-ground separation is, of course, the task of land and water differentiation, so well shown in Cuff and Mattson's (1982) book *Thematic Maps*. The control of contrast is the best tool we have in manipulating the graphic image and creating a number of visual levels. It is fundamental to text as much as it is to symbols.

The cartographer's task is to design a map that incorporates the user's visual perception to advantage, and not to create problems of detection and discrimination. To do this you must obviously understand the user's visual processes, background and abilities.

The Gestalt "Laws" explain some of the aspects of the complex images we create and help explain some of the techniques we have been using intuitively for years. They are not sufficient to explain everything, but they are a highly useful set of tools.

Psychologists have established that the mind will find the simplest possible meaning to fit the facts. This principle is called the LAW OF SIMPLICITY; it is the same as the LAW OF PARSIMONY in science. It is said that the best scientific explanation is the simplest one that fits the facts. Gestaltists applied this to the field of psychology. Details which are only approximately linear will be seen as a line; those roughly circular will probably be interpreted as forming a circle even though this shape has no bearing on the data being viewed. This means that you tend to see things, not as they are, but as your mind thinks they ought to be! Of course this can cause problems, such as when Schiaparelli "saw" the canals of Mars, touching off a time-wasting hunt that went on for years. It is also the reason why proof reading and marking student projects are such difficult tasks.

This has one result for designers of cartographic products that cannot be overstressed. It is the familiar symbol and text used in a normal way that is far easier to see and understand than the unfamiliar, exotic or, even worse, the familiar used in an unusual manner. A map is a highly complex graphic form. One must strive for simplicity in all aspects if it is to be read without mental gymnastics and errors on the part of the map user.

The most important of the Gestalt laws is known as the PRAGNANZ PRINCIPLE, first established by Koffka (1935). Known also as the Laws of Good Figure, the explanation states that, "the visual system converges on the most regular and symmetric perception consistent with the sensory information" (Rock and Palmer, 1990: 88). These laws, shown in Figures 18.1 and 18.2, include the individual laws of Good Continuation, Closure and Common Fate. Brief explanations and the most common text book illustrations are shown on the figures.

Pragnanz talks about a somewhat vague notion of a "good" figure, one which has a high degree of internal redundancy and for which any given part is predictable from previously seen parts. Recent testing has shown that the amount of information is critical to the perception of "goodness". Good figures contain relatively little information—they are simple and very basic—they can be matched more quickly (map legends), remembered better (map reading), and described more accurately (map interpretation) than poor ones. Poor figures are complex and hard to identify (Rock and Palmer, 1990: 88). Obviously, from a text viewpoint, the simpler the better—typefaces with complexities, decorations, or nonstandard proportions are defined here as "bad". We already know that they are difficult to read; here is the theoretical justification for that stand. Obviously we must avoid text with flourishes, complex serifs, high contrast letter forms, or erratic proportions. (Please, let us finally kill off Olde English!). The printer's maxim, "if you can't read it a glance, it's never read at all", is given some justification. The standard cartographic tradition of only using a single sans serif face on the map and a single, simple, low contrast, serif face for the surround and virtually everything else, seems appropriate.

The design of symbols, too, must follow the same principle. In fact, the whole idea of a symbol is to simplify the subject down to the basic, fundamental shape, removing all unnecessary detail. By its very nature, a symbol must be a good figure; the simple representation reminds us of what it represents, clearly and easily. A good symbol will illustrate its topic even to unsophisticated or inexperienced readers. In this it is similar to the work of an artist or cartoonist who can suggest a subject with only a few lines or a simple shape. This concept should make us avoid abstract or contrived symbols wherever possible. Because it is read by the same people at the same time with all the same habits, strengths and weaknesses, a symbol and a text label should be designed and used in a similar manner. This will also provide a consistency of approach which we now know to be highly important. One example of the visual link between text and symbols is shown on Figure 18.3, where the importance of the top half of both a letter and a symbol to its perception are illustrated. Once again it is our familiarity and competence with text that controls much of how we read a graphic symbol.

GESTALT "LAWS"

Proximity —Similarity—Continuity—Common Fate—Closure

PROXIMITY:

Visual groups are formed from elements which are spatially or temporally close to one another.

This example is seen to have three vertical rows of data because the vertical spacing is less than the horizontal.

SIMILARITY:

Groups are formed from the elements which are similar to one another.

Here we see horizontal rows because, although the dots are equally spaced, the horizontal ones are similar and the vertical ones are not.

GOOD CONTINUATION:

A trend in a set of elements will determine the direction in which the next element is seen.

The direction at the "cross-roads" obviously follows the trend already established.

CLOSURE:

Parts of a figure not present will be filled in visually to complete the figure.

The eye completes the figure—this also explains how we see lines of dots as a continuous shape.

COMMON FATE:

Objects which move or change together are seen as a unit or with a common fate. It reflects the great power of relative movement as an organizer for perception.

Here only one possibility of several is actually seen by the reader.

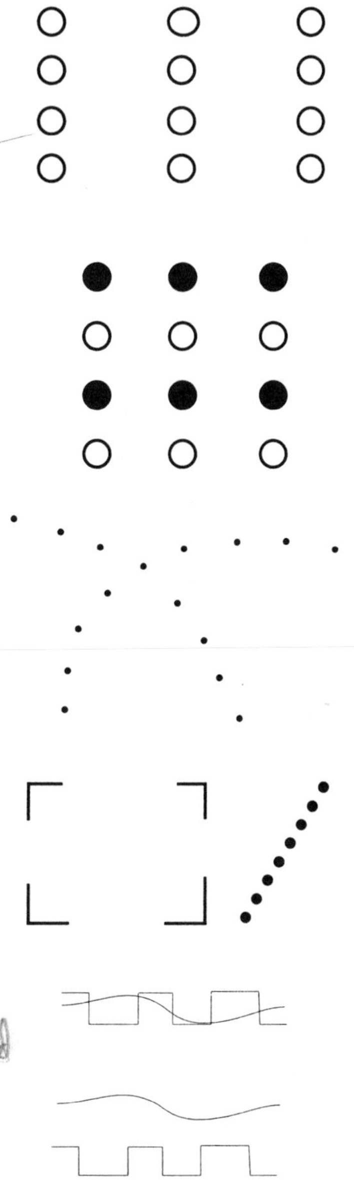

Figure 18.1 Gestalt "Laws" of proximity, similarity, continuity, common fate, and closure.

Gestalt breakdown

GROUPING LAWS

The visual system spontaneously organizes elements into larger groupings

Similarity

Good Continuation

Proximity

Common Fate

LAWS OF GOOD FIGURE

(Pragnanz Principle)

Visual systems converge on the most regular and symmetric perception consistent with the information

Good Continuation

Closure

Common Fate

Today these "laws" are viewed as descriptions of perception, rather than scientific laws of perception.

Figure 18.1 continued

The laws of continuation and closure have very obvious applications to name placements on maps. One major change that has occurred is the avoidance of extreme letter spacing, and sometimes of virtually any letter spacing at all. Much of this derives from studies that showed just how difficult letter spaced names are to read. Indeed, many older maps were essentially unreadable in the way they presented area names. Our explanation is obviously that they are no longer seen as a continuous unit. Some older maps allowed for the use of staggered names. These are likely to be a problem since we don't normally read this way and the visual link will probably not be established.

The law of common fate would seem mostly applicable to linear symbols, but it does explain nicely the visual link we create when we flow names together with complex lines, such as rivers, roads, railways, and boundaries. It shows the great power of relative movement as an organizing force in perception. It provides the vehicle for the major exception to our rule that lettering should be as normal (horizontal) as possible. Curved names on curved lines are seen as a single unit. If the trends are not very similar, a visual break is established and no obvious relationship

Gestalt samples of name placements

GOOD CONTINUATION

All alignments must be simple and spacings consistent; any parts of a single name must be seen obviously as a single, continuous structure without visual effort.

Modern text usege minimizes the number of times letter-spaced names can be used - the internal spacing must always be less than the spacing to other non-related text.

Split sections and simple alignments

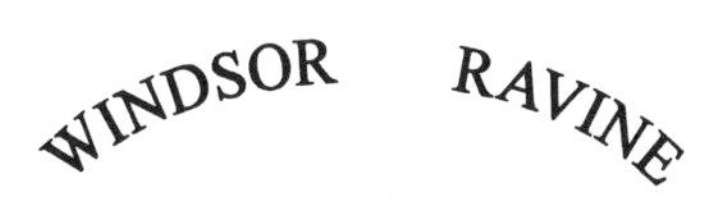

Simple and continuous curves

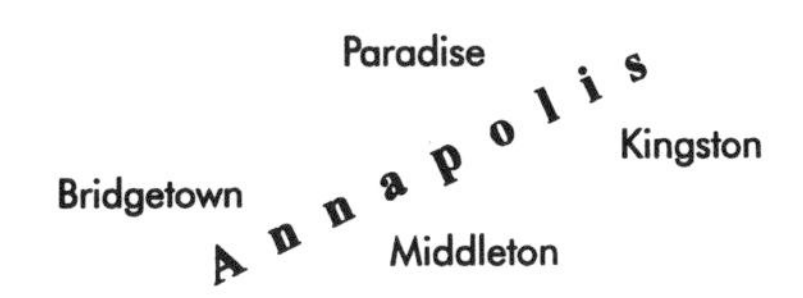

Letter-spaced, but visually unified

CLOSURE

Text which is placed over map detail and which covers part of it must be located so that the missing element can be inferred with accuracy and ease. Only place text over straight lines or very simple curves.

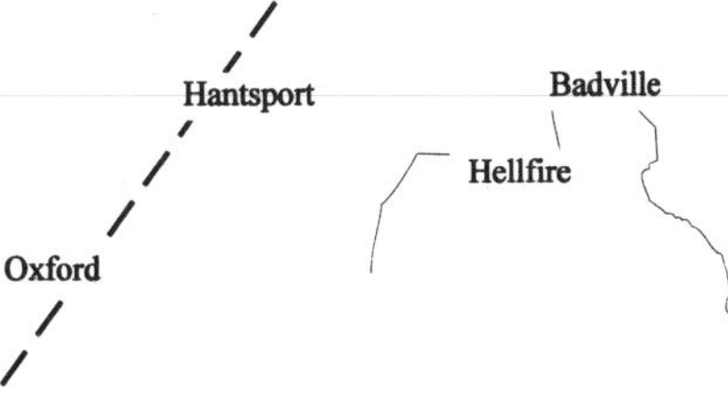

Boundary obvious Boundary missing

COMMON FATE

Names which are close to a symbol and flow in a visually similar manner will be seen as part of that symbol. Used a great deal for roads, contours, rivers and boundaries.

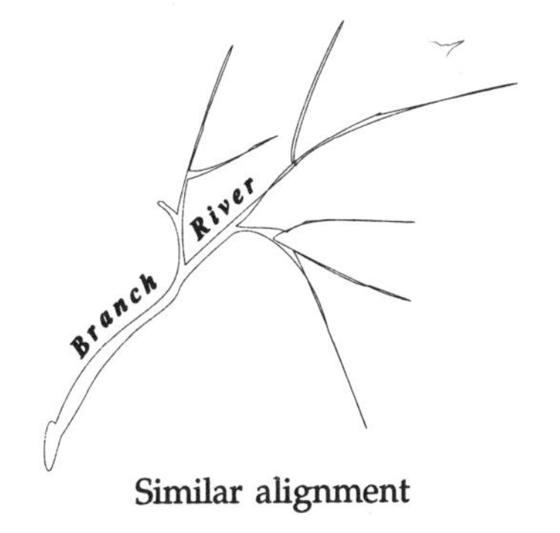

Similar alignment

Figure 18.2 Gestalt samples of name placements.

Gestalt samples of name placements

SIMILARITY

Most often used with symbols and to establish patterns and trends.

In type usage, similarity is a normal tool for the classification of names into categories. Similar type styles, colours, sizes or treatment are used to infer similar classification.

Conversely, a change in type appearance may be read as a change by the reader, even if you didn't intend it.

Note: this is an area where GRAPHIC REDUNDANCY is vital—change two aspects of the type (e.g. size and style) to clearly indicate an intended change.

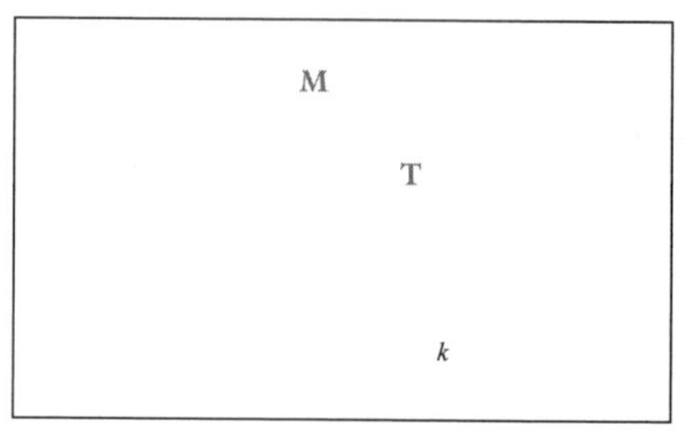

Classification clear and obvious—at least two graphic variables involved each time.

PROXIMITY

A major graphic tool for text placement.

It is essential that text be visually associated with its location. Point symbols must be very close to their own text, and separated visually by large amounts of space from non-related material.

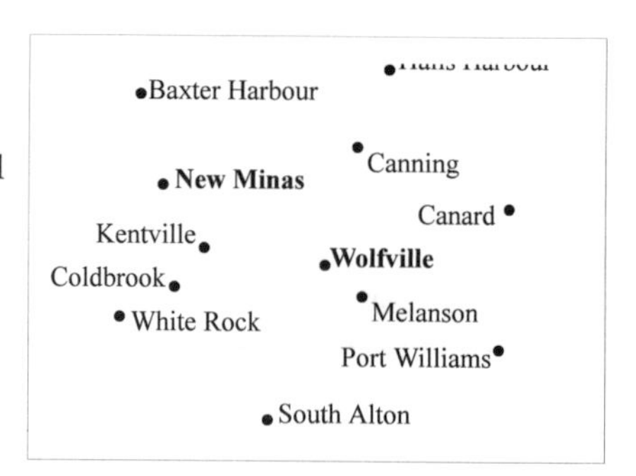

Figure 18.2 continued

is noted. Even here we must strive to be as simple and as normal as possible, and obviously any name that approaches the horizontal is to be preferred over one at an extreme angle. Upside down lettering is obviously to be avoided totally—it simply won't be seen or understood, and will vanish into the "background". Upside down lettering is never experienced in day-to-day reading; we have no familiarity or competence with it, and there has never been a good justification for using it, despite the habits of some mapping agencies.

Heider took the basic idea of Pragnanz further and developed ideas that should be of interest to cartographers and any other graphic workers. He developed the

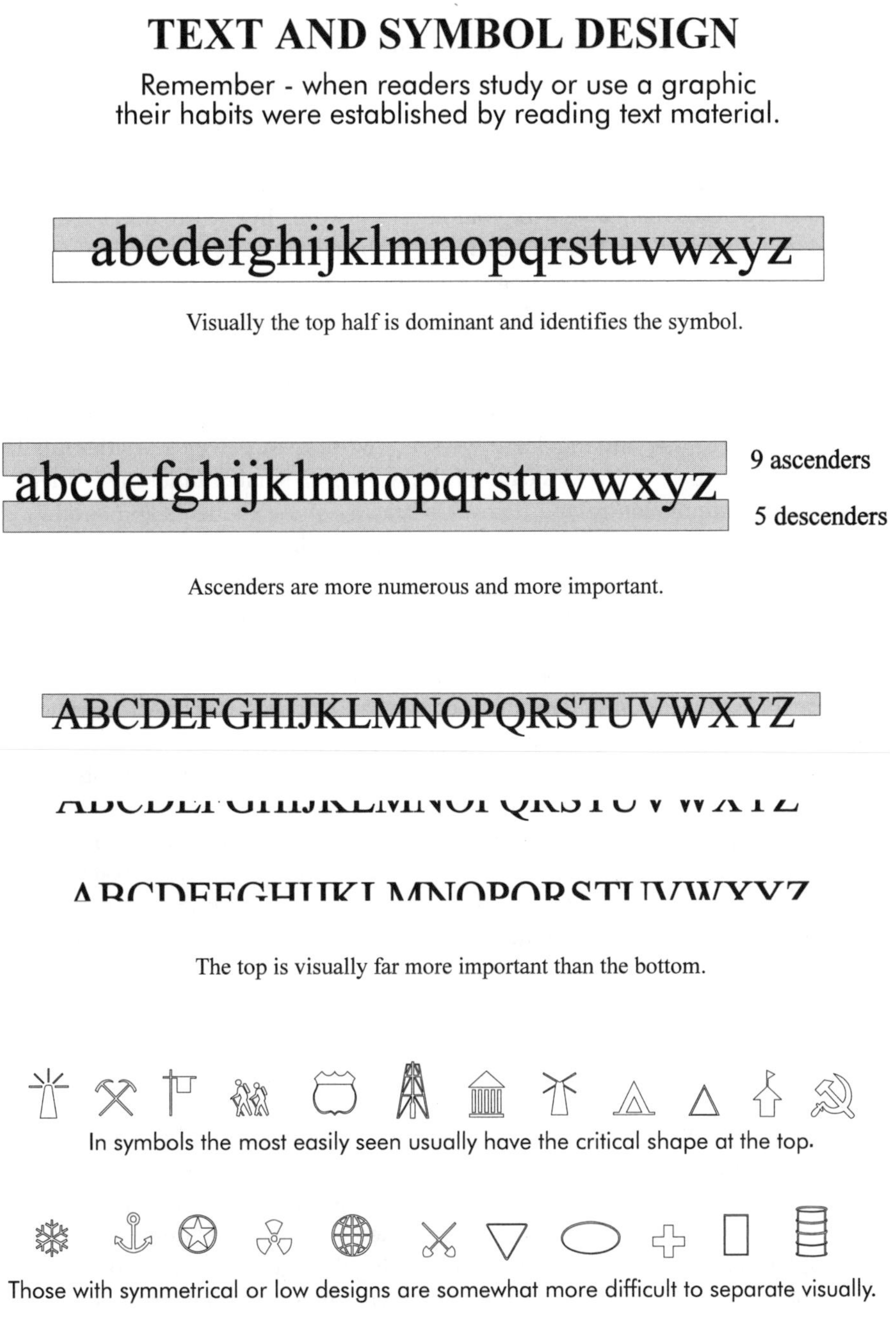

TEXT AND SYMBOL DESIGN

Remember - when readers study or use a graphic their habits were established by reading text material.

abcdefghijklmnopqrstuvwxyz

Visually the top half is dominant and identifies the symbol.

abcdefghijklmnopqrstuvwxyz

9 ascenders

5 descenders

Ascenders are more numerous and more important.

ABCDEFGHIJKLMNOPQRSTUVWXYZ

The top is visually far more important than the bottom.

In symbols the most easily seen usually have the critical shape at the top.

Those with symmetrical or low designs are somewhat more difficult to separate visually.

Figure 18.3 Some considerations for text and symbol design.

concept of BALANCE; the idea that individuals actually prefer HARMONIOUS relationships and actively seek them out (Rock and Palmer, 1990: 89). The contrasting function of map image and map textural material control its composition. Clarity is dependent on legible and well placed letters, and text placement requires that we create harmonious proportions and balances. Once again we see the importance of an integrated, overall design approach. Perhaps it will provide some pause for those people who suffer from "desktop fever" and are itching to try out those 600 typefaces that came with your latest software. In a word, don't! Neither should you approach the task of labelling a map on a piece-by-piece method—we are creating a single, integrated graphic communication, not a patchwork quilt!

The second group of Gestalt laws are known appropriately as the GROUPING LAWS, and they overlap the laws of good figure, as you might expect from a philosophy that is built around the notion of an integrated whole. Included once again is good continuation and common fate, while the concepts of PROXIMITY and SIMILARITY are added. Here we see how human perception attempts to organize data into simple and meaningful shapes from small, numerous elements. Note that a common control mechanism is spacing of the elements and another is similarity. Both are vital in the use of text on any publication.

Spacing of text and the use of consistent type styles has always been a major concern in the publishing industries as well as in cartography. Much has been published on how to achieve good communication by text manipulation, and information continues to arrive with the current popularity of desktop publishing. Some systems are highly complex; they are often contradictory in what they recommend. We have always needed a precise and easy to apply system.

Probably the most simple and useful system of achieving consistent and highly readable text material was developed by the late Newman Bumstead of the National Geographic Society in Washington for use in their publications. I am not aware if this was ever published, but it has proven so useful that it has been taught to a full generation of my students and greatly simplifies the design and analysis of the projects. Bumstead's system does not mention Gestalt, but the entire structure is based on Gestalt-like relationships.

The system is known as BUMSTEAD'S RULES, and after a brief introduction all students are assumed to use it consistently on everything, unless they can express an overpowering design constraint that forces them not to. Evaluation is equally simple—the expression "Bumstead violated" is attached to any type spacing error and no other explanation is necessary. Poor Bumstead by now must be the most violated human in history, but his memory certainly lives on, and he has been responsible for more competent projects being produced than anyone else.

Bumstead's laws, especially those concerned with the "surround space", are the simplest method we have found of controlling the figure/ground separation of text. They ensure the visual unity of a planned single topic and the visual separation of all non-related material. The basics of the system are shown on the accompanying Figures 18.4, 18.5, and 18.6. Hopefully it needs little explanation beyond what is shown there.

Bumstead's Rules

Rules of spacing relationships between text elements on any kind of graphic

1. The visual appearance of the space between letters of all words of the same style and size must be the same.

   Thus, all systems of mechanical or measured letter spacing are not acceptable. Courier and similar typewriter styles are to be avoided.

2. The space between letters must be seen to be less than the space between words.

   In order to create only the words intended, the separation of letters into visual groups by increased spacing is vital. Word space must be consistent and relatively small.

3. The space between words must be seen to be less than the space between lines of text in the information unit in which they appear.

   Word spacing must be seen to be less than that between lines in a paragraph, title, address etc.

4. The space between lines of a single text unit must be seen to be less than the space between lines separating units.

   Line spacing used within a paragraph, or any other closely related piece of information, must be seen to be less than the space used to separate paragraphs, different classifications, or other different information.

5. The space between lines separating different units of information must be less than the space that surrounds that textual information.

   The WHITE SPACE that surrounds the textual information must be seen to exceed any space used within that block. Don't cramp borders close to text blocks. Visually separate a text block from any other information that may surround it.

Figure 18.4 Bumstead's Rules of spacing relationships between text elements for graphics.

Bumstead's Rules

1. The visual appearance of the space between letters of all words of the same style and size must be the same.

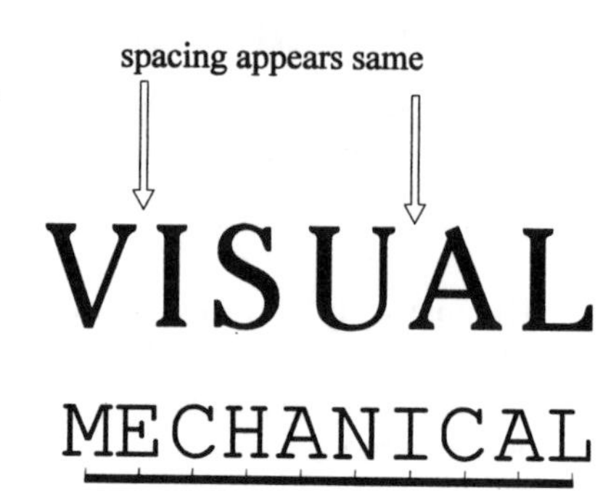

2. The space between letters must be seen to be less than the space between words.

good spacing

badspacing

3. The space between words must be seen to be less than the space between lines of text in the information unit in which they appear.

Normal reading is from left to right, and if you provide the correct line spacing, you ensure this.

You probably see this vertically

4. The space between lines of a single text unit must be seen to be less than the space between lines separating units.

CLASS 5 - Soils in this class have severe limitations that restrict the range of crops or require special conservation practices.

CLASS 6 - Soils in this class are capable of only producing perennial forage crops.

5. The space between lines separating different units of information must be less than the space which surrounds that textural information.

SURFICIAL GEOLOGY
Annapolis County

Figure 18.5 Applications of Bumstead's Rules on spacing relationships between text elements for graphics.

Spacing (Bumstead) Problems

Title block for a map of Seigniorial Settlement in 1709, Lake St. Peter, Quebec.

Note: Word spacing larger than line spacing; erratic word spacing; continuity of text destroyed by border; capital letters make message hard to read; border detracts from title message.

Railways
Principal Station
Station
Cutting
Double Track
Tunnel
Level Crossing
Embankment
Single Track
Bridge under
Bridge over
Footbridge
Roads
Two Way Bridge
Wooden
One Way Bridge
Concrete
Three Lanes or more wide
Sealed
Fenced
St Steel
Unfenced
Suspension
Metalled
Unmetalled
Vehicle Track
Foot Track
Electric Power Lines
Main Transmission Lines
Pylons actual positions
Wooden Poles
Distribution Lines
Bush Tramway
Telephone Line
Contours
Index
Supplementary
Intermediate
Approximate
Depression
Vegetation
Bush
Plantation
Orchard
Burnt or Fallen Bush
Hydrography
Drain
Drain beside Fence
Water Race
Stream
Indefinite Stream
Intermittent Stream
Dam; Ford; Waterfall
Cold Spring; Hot Spring
Trig Station
Windpump
Mine, underground
Rock Outcrop
Beacon; Buoy
Trig Station Beaconed
Footbridge; Gate
Mine, opencast
Cliff or Terrace
Spot Elevation in feet · 116
Fence
Gravel Pit; Cave
Stopbank
Sandhills
Wireless Aerial Masts
Pipeline
Quarry; Slip
Lighthouse

Part of legend from older O.S. 6 inch series (ink drawn)

Note: Spacing problems create tendency to read across columns; italics in volume are difficult.

Figure 18.6 Examples of violations of Bumstead's Rules.

This is a simple set of spacing relationships designed to ensure that information that should be visually linked is done so easily, and information that is distinct is visually separated. In so doing it incorporates the grouping laws as fundamental. It controls the white space or negative space surrounding each item. It conforms to normal reading techniques and the daily experience of readers. Of course, there is little that is new here; many designers do this quite intuitively in cartography and other applied arts. It is the simplicity and consistency of the system that makes it so useful. If followed carefully, it results in highly acceptable and, most importantly, highly readable graphic products. Our own normal teaching techniques include showing and obtaining samples from published advertising where Bumstead appears to be followed and others where his principles are violated. It is amazing how many very expensive ads by major corporations that don't seem to work well when first seen can be found to have a fundamental spacing flaw when analysed.

From an instructional and use standpoint, it has another decided advantage over many more complex systems. There are no formulas, no mathematics and no measurements required. The entire system is visual; if it seems to work it does, if it looks wrong it is. This is very much a reader-based application. It trains students very early on to look at their work and that of others critically, instead of following a formula approach. They learn to place themselves in the position of the map user instead of the map constructor; they become very critical, very quickly.

An example of poor spacing and several Bumstead violations are seen in Figure 18.6, the title block of a map on the Seigniorial Settlement of 1792 in the Lake St. Peter Region of Quebec. Notice how difficult this is to read, how inconsistent the spacing is, and how cramped the title itself is against the decorative borders. Do not think that this could not happen today. If you have "justification" set on many of our modern desktop software packages you can easily get word spacing that is even worse. It has another common error that is still seen today—a title that is far too long and lettered in all capitals. It is unreadable by all except the most dedicated user, guaranteed to put most of us to sleep before we've ever figured out what the map is all about.

One of the cartographic areas where spacing and unity problems often occur is in the design of legends, mostly because these items are overly complex, and the designer attempts to fit too much information into too small a space in the map surround. A common solution is the employment of multiple columns, where a symbol and its explanation are then followed on the same horizontal line by another symbol with another explanation. All too often the visual space between one explanation and the next symbol is less than the internal space used for a single item. The result is visual confusion—which symbol belongs to what explanation? People should not have to figure out this kind of puzzle. It is our job as designers to ensure that there is only one way to read the work—the correct way. The above example violates the grouping laws and violates Bumstead. It runs counter to one of the oldest guides in the printing business, which is to avoid 'Tombstoning", the unintentional reading across from one column to the next one. And yet we continually see this sort of thing in cartographic products.

Some national map series still don't pay much attention to the readability of their publications. An example is taken from one of the old style Ordnance Survey 6-inch plans. The conventional sign sheet shows lines visually running into one another and some potential confusion. It is not acceptable to say that people can figure this out in some short period, nor is it acceptable to say that experience limits this effect. With a very small change in design by the people who are supposed to be good graphic communicators, these potential errors or aggravations would not occur. The more people who can use our products easily, the more will use them and the more we will sell—it's simply good business not to antagonize the customers. This same legend illustrates another common habit of many cartographers. Why on earth would all the text in such a complex area be done in an italic face? Italics are designed to be poor communicators; that's how they draw attention to themselves when used for emphasis in a book. They should never be used for extensive amounts of reading; they simply don't work well, just as all capitalized legends are equally difficult to read.

A point-by-point analysis of all the various rules and guidelines used in the various mapping agencies and companies is obviously beyond the capabilities of what is supposed to be a short paper on the subject. It would be the size of a large text and would also be quite repetitive. Each mapping organization should investigate its own guidelines and standards to see if readability can be improved and consistency attained. This may not be a short task even at a single location. The guides for name placement given to students at the College of Geographic Sciences, for example, originated with the published articles on lettering by Imhof (1975), were supplemented by material from Keates (1982, 1989) and Robinson et al. (1978), and then rewritten. They have been expanded and changed virtually every year since. When a student or an instructor notes a problem that is not covered, a guide is developed to fit the situation. Much of the rewriting takes place in an attempt to maintain some sense of consistency and logic, hence the thrust of this chapter. Currently we have some 32 detailed guidelines for map face type alone, and each student receives some 10 pages of closely written sheets to which they are expected to adhere. We probably still have quite a way to go before this material can be considered comprehensive and integrated.

Figure 18.7 shows a few examples of considerations necessary for map face type. It includes a version of diagrams accompanying Imhof's original article on name placement from the *American Cartographer*, which we now believe to be in error. Some of our own ideas may prove to be erroneous, but that should only encourage us to look critically at what we do and take each opportunity to improve our techniques. Also, we must realize that cartography is one small branch of cultural communications. Styles and what is considered acceptable inevitably will change, and we must reflect that in much of our work. It is probably inevitable that ideas on how to communicate with text will remain fluid and the subject of endless discussion for the foreseeable future. It is to be hoped that this chapter can be considered a contribution to simplifying and bringing a sense of logic to a highly complex subject.

Study of some name placement conventions

3 • 1   Halifax • Halifax
4   2   Halifax   Halifax

Preferred Locations for spot names

Position #1 has capital and symbol very close, both designed to attract attention, therefore unity maintained. Bottom of text is more linear than top, therefore a preferred location. Proximity clearly better to right than to left. Position #4 clearly the worst option.

Halifax
•
Halifax

Two (less desirable) locations

Possible but not recommended. Proximity and unity problems because of visual separation between symbol and capital letter. Lower position is again usually the poorer choice.

Alternative, curved locations

Possible locations but effectiveness decreases rapidly with increased angle from horizontal and amount of curvature. Similarity and continuation both become a problem at distinct angles. Normal reading rarely calls for alignments distinct from the horizontal. The more variations used on a product the less unified it seems. Curves must be few, moderate and consistent.

ATLANTISCHER OZEAN   ATLANTISCHER OZEAN   ATLANTISCHER OZEAN

Good   **Poor**   **Poor**

Imhoff's curved text recommendations

Imhof avoids extreme curvature which is effective, but he allows text to go past a vertical position which must be the limit for readability of any kind. Note how in the first diagram only the last word can be read at all without distinct effort. All three of these options should be considered to be poor. Upside down lettering is to be avoided even on only partial words.In most cases re-positioning of names can avoid these problems.

Figure 18.7 Some name placement conventions.

REFERENCES

Birch, T.W. (1968). *Map and photo reading*. London: Edward Arnold.

Cuff, D., and Mattson, M. (1982). *Thematic maps, their design and production*. New York: Methuen.

Haber, R.N., and Hershenson, M. (1973). *The psychology of visual perception*. New York: Holt, Reinhart and Winston.

Imhof, E. (1975). Positioning names on maps. *The American Cartographer*, 2(2), 128-144.

Karssen, A. (1980). The artistic elements in map design. *The Cartographic Journal*, 17(2), 124-127.

Keates, J.S. (1981). *Understanding maps*. London: Longmans.

Keates, J.S. (1989). *Cartographic design and production*. London: Longmans.

Koffka, K. (1935). *Principles of Gestalt psychology*. New York: Harcourt, Brace and Co.

Robinson, A., Sale, R., and Morrison, J. (1978). *Elements of cartography*. New York: John Wiley and Sons.

Rock, I., and Palmer, S. (1990). The legacy of Gestalt psychology. *Scientific American*, 263(6), 84-90.

Trudel, M. (1973). *Atlas of New France*. Laval, Quebec: Les presses de L'Universite Laval.

Yasui, H. (1989). *Desktop publishing: Technology and design*. Eden Prairie, Minnesota: Paradigm Publishing International.

The Logic of Map Design

Phillip C. Muehrcke
Department of Geography, University of Wisconsin-Madison

Cartography happens at the design stage. Design is the creative heart and soul of our field. Long ago I was convinced that we learn more about ourselves and our environment in the process of mapping than by looking at the product of this process. This, more than anything else, is the gift of information age cartographic design. With rapidly growing frequency people are becoming their own map makers. This means they will have the benefit of experiencing the mapping process for themselves.

The burden of being one of the closing chapters of this book is that you might reasonably expect that I will bring together everything that has been written into a neat, coherent package. You might hope that I will add some final insight that transcends previous discussion. In fact, I imagine your expectations in reading this book were quite specific. You might have hoped, for example, to take away answers to the following key questions:

What do we mean by cartographic design?

What are the most basic principles of map design?

How do we distinguish well-designed maps from poorly designed ones?

Is it possible to teach a map design sense, or is it a talent some innately have and others don't?

Indeed, if design is the very core of cartography then, as mapping professionals, we are the ones others should legitimately look to for answers to these four questions (see also the discussion of the term "design" in Chapter 1). For a few moments lets look back at what the previous chapters have given us to work with. Several themes emerge.

(1) The long-standing split between practitioners and researchers persists today. Practising cartographers still say design research is of little value in their work. I hasten to point out that this does not mean design research has no value, however. It serves to educate students about the complexity of map design. The attempt to develop a theory of design also helps academics survive the tenure process and, thereby, be available to teach the next generation of practising cartographers. Although little significant design improvement may occur directly from design research, the net effect is to build a cartographic community and increase the pool of applied designers.

(2) There is a downward trend in the volume of map design literature, but an upward trend in mapping. The claim is that there is also an upward trend in poor map designs, implying that trained cartographers are losing control of mapping. I contend it is a myth that they ever had control to lose. Most maps are made by graphic artists and others not formally educated as cartographers.

I also wonder if it is true that a growing proportion of maps exhibit design failure. My sense is that the increased visibility of poorly designed maps may merely reflect the vastly greater number of maps being made, particularly with the aid of GIS technology.

(3) We now have the tools to move away from the practice of attempting to provide a single best map. Instead, we can now give the user a selection of maps and other representations. What makes this practical are fast desktop computers with a multiple window interface, the storage technology provided by CD-ROMs, and the cross-referencing links provided by hypertext software.

(4) We have the capability to incorporate more design dimensions or facets into our work. Effective integration of sound, animation, and interactions raises many new design challenges. At the same time, we can back our electronic maps up with vast databases, so that the map itself no longer needs to be designed to serve as the primary storage medium for geographical information. This unburdening of a map's data storage role should have a liberating effect on the map designer. We can now focus on the role of maps as graphical interfaces to the storehouse of geographical data contained in digital databases.

(5) The gender issue is most touchy. Beware the careless researcher in this arena, for emotion is intense. It might be good to remember that a map is a sophisticated hybrid representation. It involves simultaneous use of graphics, images, text, and number systems. Thus, even if significant gender-specific differences exist, they will be difficult to sort out through narrowly defined research projects.

It might be more fruitful to take the approach that, regardless of gender, individuals use different strategies to accomplish the same end. The strategy that is best for you is the one that you can make work. Rather than focus on potential gender differences, we might better expose map users to the range of useful strategies people are known to employ with respect to performing different tasks.

(6) The dominant key word reference to choropleth maps throughout this book is most shocking to me. GIS provides a convenient tool for integrating different variables (data layers) into a single map. With the aid of this technology we should expect to see dasymetric mapping displacing choropleth mapping. In fact, Ron Abler (former director of the Geography and Regional Science Program, National Science Foundation) has said that choropleth maps were a cartographic abomination that GIS will swiftly kill off (Abler, 1987). He made this statement a few years ago when soliciting proposals for the National Center for Geographic Information and Analysis (NCGIA). What happened?

(7) Much of the discussion in this book focuses on the importance of the map user. I would go even further to state that changes in the way maps are used in the

electronic age are probably far more significant than changes in how they are made. Except for the fact that I want a digital map for analytical and interactive purposes so the underlying database is accessible, it really makes little difference how a map is made.

I also question whether we, in practice, are true to our stated obsession with responsiveness to our customers. Or, do we ask them to see things our way? We can address this issue by considering the final theme I want to mention.

(8) There is a growing literature documenting a dissatisfaction with cartographers and their maps. Let's consider the public controversy we now face:

Peters (1983), the historian of Peter's projection fame, is a good place to start. His straightforward mission was to "right the wrongs" of hundreds of years of orthodox cartography. He would do this by offering a politically correct substitute for the many seriously flawed mapping options heretofore available. And his acclaim and wealth grew quickly. His message of outright misconduct on the part of professional cartographers found a large receptive audience. His claims got wide publicity (see Robinson, A.H., 1985, for one cartographer's reaction), and many embraced his message and bought his maps.

Next, let's consider the GIS challenge. This new technology is supposed to take us beyond maps. If we are to believe GIS proponents, then analog cartography is dead and digital cartography is dying. Most recently, the claim is that thematic cartography is dead. They say the paper map has been dead for a long time, cartographers have just not admitted its passing.

But site visits to GIS facilities reveal curious contradictions. It seems that nearly 90 percent of GIS development effort is in some way associated with digitizing or scanning existing maps. In other words, the knowledge base of GIS is overwhelmingly cartographic. It is even more obvious that roughly 90 percent of GIS products look suspiciously map-like to novices as well as professional cartographers. These "non-map maps" are displayed prominently and ubiquitously in GIS facilities, proudly announcing the accomplishments of the staff. Here we have the image of something that might be called GIS wallpaper.

The final insult to a dead cartography is that we get blamed for poor GIS product design. Apparently, to most people not schooled in the ways of GIS, these GIS products look enough like maps to be accepted as such.

Let's move on to Brian Harley (1988, 1989, 1990), who reminds us that our maps can have a dark and sinister side. Harley feels that maps could take on a meaning of their own, and become a territory in their own right. Thus, he concludes, maps have been used as an instrument of power throughout their history. But this should be no surprise. All forms of representation, indeed, all human tools, have been used to abusive ends. Maps merely belong to the club.

The question is whether we believe our maps are neutral, innocent products that merely reflect reality, or whether we understand the forces that control and limit map design and use. The cartographic literature, including our textbooks, does not seem to be as much at fault here as the cartographic literacy of those who use maps. I can accept our failure in education, but overcoming an abusive human nature seems beyond our grasp or responsibility.

Denis Wood (1992) carries on with the Harley "power of maps" message, but in a much more aggressive and accusatory way. He challenges the scientific objectivity of maps by stating that maps serve interests, that map agencies have agendas. He particularly attacks the sanitized character of institutional maps. He insightfully reveals the cultural/social context of mapping. But he also leaves the impression of loving mapping while not respecting the work of professional cartographers. There is an almost hostile anti-establishment theme, supported in part by ignoring widely available literature. Since we can have so little faith in official maps, we are encouraged to empower ourselves through personal mapping.

Monmonier (1991) adds a different flavour with his discussion of how maps lie and distort. His message is solid: distortion can't be avoided in mapping, so education is the way to prevent becoming a victim. The danger rests not in the message but in the packaging. His talk of lies appeals to a public fascination with trashing the establishment. Thus, his underlying message is easily lost.

Finally, let's consider Robert Kaplan's (1994) *Atlantic Monthly* article with its subtitle "The Lies of MapMakers." His message is that political maps reflect modernism—the era of nation-states. Now, in post-modern times, few nation-states exist and conventional political maps are a mockery of reality. These lies of out-of-touch mapmakers are deceptive and even dangerous. After making this direct attack on cartographers and their maps, he calls for a "new map"—a dynamic, multidimensional and multifaceted "final map." This final map is to capture the fluidity and chaos of current reality in most third world countries and even some cities in the United States. His conclusion is that if we can't have it all on a map it is better to have nothing. Apparently, some GIS blood flows through his veins!

In reviewing this public controversy, we can attribute some of the critique to marketing hype. Self-promotion is also a factor. And, to some degree, intellectual dishonesty is an issue. In each case there is an underlying truth or insight of significant value. But with the prevailing "grab the headline" mentality, the more balanced and truthful message is often buried so deep in the narrative that it can never quite overcome the catchy word (lies, distortion, power) or phrase that is so readily remembered.

But the critique also reflects a fundamental misunderstanding of what it means to map. The common link between these controversies is a conflict between the perceived nature of the map and the perceived nature of the reality mapped.

The issue is not design aesthetics. By this I mean it is not map readability in the strictly graphic sense of clarity and legibility. It is not a matter of line weights, black/white or colour, symbol dimensionality (1, 2, or 3), map orientation, type selection or placement, and so forth.

The issue is the appropriateness or comprehensibility of the very idea or concept of a map's design. The problem is related directly to what it means to map. Stated most directly, the issue is the logic of cartographic design. To explore this issue -- to answer our critics -- I would like you to consider several points.

WHAT IT MEANS TO REPRESENT

My first major point is that a map is a form of graphic representation and, as such, shares much with other forms of representation such as words and numbers. To represent is to abstract. Abstraction frees us from the tyranny of our physical existence. It lets us transcend the me-here-now constraint and the total reliance on direct sensory information.

Cartographic representation lets us visualize the environmental "What if . . . ?" Individual maps fall along a gradient of abstraction, from large scale and detailed to small scale and general. We can just as legitimately cover the floor of a large room with a map of a small region as we can map the entire surface of the earth on a postage stamp. Accuracy is a relative concept that is determined by available mapping resources. The idea of "truth to scale" is more meaningful than an absolute standard of truth.

Since all representation is not reality, it is unreal or fake. Representation is only possible by distorting the real. We are left with a highly selective and biased representation. If you want perfection, go with reality, since there can be no perfect representation. By being imperfect, however, representation lets us see things that otherwise would be impossible. For example, mapping lets us see the structure of the whole earth's surface simultaneously.

Thus distortion is not necessarily bad. It can be our friend if we know how to use it effectively. Our privilege to take advantage of abstraction in the form of cartographic representation does carry with it certain responsibilities, however. Namely, we must understand the forms distortion potentially takes in our map designs, and we must learn to compensate for or neutralize the possible negative side of these distortions.

Some of our critics seem to have missed the point here—the issue of user responsibility. No amount of safety features can prevent the harm caused by misusing tools. Maps are no exception. Ultimately, users must learn to handle mapping tools responsibly.

CAUSE VERSUS EFFECT IN MAPPING

My second major point is that there seems to be confusion over cause and effect in mapping. There are two aspects of this confusion. On one hand, some people seem to have trouble understanding what came first. Was it the map or was it the concept the map illustrates? We are being accused of inventing such concepts as the nation-state, forest, climate zone, soil class, and so on. Thus, we are being blamed when difficulties arise in applying these abstractions.

However flattering it may be to have these abstractions attributed to us, those who use maps would benefit from realizing that maps reflect in visual form the way people think. In fact, I am willing to go further and say that there is something very natural about our maps. That is why they are so believable. We have made our maps reflect our conception of things and the environment.

Related to this first issue is the idea that our maps are optimized for drafting tools. Therefore, since our tools are changing, we can abandon traditional mapping methods. The debate centres on the use of lines in mapping, especially for making hard edges or boundaries between areal classes (regions). At some point in our distant history someone picked up a stick and drew a line in the sand, and said "let this line stand for river or forest edge." I would argue that drawing the line constituted a great conceptual leap but was a minor technological breakthrough. The idea of "edge" and its representation is the significant part of this scenario.

GIS experts would like us to embrace the myth of cartographic realism. But I have already pointed out that the biggest danger with representation is that it can become too true to be useful. We call this design clutter. It violates the cartographic principle of clarity/legibility.

GIS technology may help us to quickly view a larger number of design variations, or make it possible to flexibly alter the map design on command using hot buttons/features to get a more thorough perspective—but it will not be possible to put everything on a single map. Those that advocate overcoming map clutter by zooming in on areas of special interest do not understand the basic relationship between scale and abstraction.

DUAL INFORMATION PROCESSING STYLES

My third major point is that we must remember the role played by our dual information processing styles. People really are of two minds. We use intuition, a holistic form of unconscious thought. And we use analysis, a picking-apart and weighing form of conscious thought. The most creative/productive thinkers—epitomized by inventors/scientists/designers—use a natural mix of intuition and analysis. This is also true of map designers. The reason we are so inarticulate about our work is that intuition occurs in the unconscious. But we know far more than we can tell (see M. Wood, Chapter 6). This tacit knowledge is what informs intuition and makes "gut feelings" so effective.

The use of tacit knowledge causes problems for our analytical mind, however, since we are not able to verbalize how or why we came to a particular decision. It just felt like the right thing to do given the circumstances.

We can learn a lot from sports trainers about this information processing duality. Their "inner sports revolution" literature makes it clear that the novice characteristically is limited by a conscious focus on rules and procedures. In contrast, the expert functions intuitively in a holistic way. Ask the tennis or golf expert the key to success in their sport and they will say "just hit the ball" or "watch me." They will emphasize practice until it feels right. Then just do it. They might be pressed into giving some rules or procedures but, rest assured, such analytical thinking becomes unconscious when they are playing the game.

For map designers, dual information processing styles imply that it is difficult, if not impossible, to derive a rule-based system. We are trying to provide an analytical solution to a holistic problem.

For map users, dual information processing styles imply that the burden is on the map user to go beyond individual maps and complementary representations (see M. Monmonier, Chapter 7). It clearly is not reasonable or fair to ask one map, or one analytical slice, to do it all. Hypertext tools, which open the user to many context-sensitive options, may prove useful here. But ultimately, things still must "click" for the map user. We must go beyond analysis to a more holistic state if map use is to be most effective.

So, how do we get these three points across to our critics. And, more importantly, in the future how do we minimize this type of confusion about mapping.

POWER OF EDUCATION

As several authors have suggested in this book, the solution seems to lie in better education of map users. We must do a better job of teaching users to read, analyse, and interpret our maps. What did you expect me to say!

The ugly truth of our profession is that we already know how to use maps far more effectively than the vast majority of people are using them. We also know how to design maps far more effectively than most people are designing them. Psychophysical, cognitive, and visualization research may raise our intellectual status and get us tenure, but the advances in map design improvement made through this approach have been disappointingly small. If we are really concerned about the map user, the basis for making much bigger and quicker gains is already within our grasp. We merely need to catalogue and teach the strategies practised by expert map makers and users. Only after that foundation education is established does it make sense to attempt a fine-tuning of map design parameters to achieve maximum effectiveness.

Unfortunately, in view of the current state of our educational institutions and the many competing demands on student time and energy, it is unlikely we can do much more than we are doing now to train people to use maps. And, as most studies have shown, the level of user sophistication is dismal. Apparently, less than half of the adult population in the United States can perform even the most basic tasks related to using maps.

This leaves you—the map designer—to do a better job of conveying a sense of map logic through the map itself. Monmonier is on the right track here with his *map complementarity* message. Better titles and legends will help. Multiple map perspectives will also help, as many of you have mentioned. Mixed-mode (tables/graphs/explanatory text blocks/voice overs/etc.) representations will also help.

But we must also be more forthcoming in conveying a sense of map accuracy. The current National Map Accuracy Standards (NMAS) statement is sterile. It is not at all appropriate for the bulk of our maps, especially those representing composites of thematic variables. We might add a *stability index* to these maps. This index would provide a statement of the confidence a person can have that the map reflects something real and not merely the artifacts of map design decisions.

CONCLUSION

In these above remarks, I have addressed the recent public critique of cartography from the perspective of the map designer. I have shown that it is easy to make excuses and pass blame. But it is not very satisfying to be so misunderstood by the consumers of our product.

It is clear that we must do a better job of explaining and demonstrating the value of mapping and the importance of good map design. The previous chapters have presented ideas on this matter. I have little quarrel with what is written. It was insightful and stimulating.

But design research requires tight experimental control, which calls for a narrowing of focus. I am merely reminding the readers not to lose sight of the broader perspective associated with mapping logic. This seems to be where we are facing the greatest challenge to our discipline.

To achieve our educational goal I am asking you to focus directly on the idea of mapping. This means to stress mapping logic and the relation between map and reality over all else. Our critics in large part misunderstand what it means to represent, especially in the cartographic mode.

This seems like something we can address more effectively as map designers.

REFERENCES

Abler, R.F. (1987). The National Science Foundation National Center for Geographic Information and Analysis. *International Journal of Geographical Information Systems*, 1(4), 303-326.

Harley, J.B. (1988). Maps, knowledge, and power. In D. Cosgrove and S. Daniels (Eds.), *The iconography of landscape: Essays on the symbolic representation, design and use of past environments* (pp. 289-290). Cambridge: Cambridge University Press.

Harley, J.B. (1989). Deconstructing the map. *Cartographica*, 26(2), 1-20.

Harley, J.B. (1990). Cartography, ethics, and social theory. *Cartographica*, 27(2), 1-23.

Kaplan, R.D. (1994). The coming anarchy. *The Atlantic Monthly*, 273(2), 44-76.

Monmonier, M. (1991). *How to lie with maps*. Chicago: University of Chicago Press.

Peters, A. (1983). *The new cartography*. New York: Friendship Press.

Robinson, A.H. (1985). Arno Peters and his new cartography. *The American Cartographer*, 12(2), 103-111.

Wood, D. (1992). *The power of maps*. New York: The Guilford Press.

Advances in Cartographic Design

C. Peter Keller

Department of Geography, University of Victoria

Clifford H. Wood

Department of Geography, Memorial University

So what is new in map design? What makes today's professional cartographer judge one map a successful design and another a failure? Does the cartographic dilettante agree with the professional cartographer's judgement? How are design evaluation criteria changing and what is initiating the change? The preceding chapters have explored answers to some of the above questions. What have we learned?

THE NEED FOR MAP DESIGN INVESTIGATION

Cartography is about communication; the map is but another form of language. Any language is bound by rules, yet contains considerable diversity and is dynamic. The preceding chapters have shown that this certainly holds true for the cartographic language. The chapters have re-confirmed that there exist many different forms of cartographic expression and that the cartographic language has changed and will continue to change through time. Some of the chapters have focused on rules of cartographic communication. We have learned that rules of the cartographic language do not form absolute law and that some of the rules are still poorly understood. Indeed, it would appear that questioning, re-examination and unintentional or deliberate violation of established rules is part of what keeps the cartographic language dynamic.

Map design is both the practice and the law of cartographic language. The practice of cartographic design is the process of combining and applying cartographic rules to create maps. But map design also defines cartographic language; repetition of a particular map design creates the rules and systems of representation that make up cartographic language. Map design, therefore, is fundamental to cartographic language. Every language attracts individuals who make it their profession to study and to influence it, while others (the majority) use it to communicate. This holds true, as well, for the cartographic language. There have always been academic cartographers that have made it their business to study the process of, to critique and to search for innovations in map design, and there have always been those who practise map design to communicate spatial data and associated information.

The introductory chapter in this book pointed out that academic cartographers had lost interest in investigating map design in recent years, leading to a neglect of this subject. Did this trend imply that the advancement of map design and the dynamics of the cartographic language had reached a point of stagnation? Throughout this book we have learned that the answer is a resounding 'no'. It should have become obvious that it is a misconception to believe that it is only academic cartographers who conceptualize and shape innovation in map design. On the contrary, practising cartographers play a vital, if not leading role in the advancement of cartographic language. Every practising cartographer is involved in shaping the evolution of map design. Every deliberate or unintentional move, however slight, on behalf of a practising cartographer to break from the accepted norms of map design, and every move to embrace functionality of new technology in map design constitutes advancement. And this is exactly what has happened in the last few years.

We have learned again and again throughout this book that the last decade has seen a trend in cartography in which the advent and popularization of digital graphics software, computer drafting, GIS and the information superhighway have changed the playing field of map design. It would appear that practising cartographers have been given little choice but to accept this change. They have been forced by technological circumstance to explore and to adjust to a new environment in which to conduct map design.

Not only has the playing field on which to conduct map design changed around established practising cartographers, but graphic software and GIS also have attracted a new breed of players wishing to communicate via cartography. This new set of players includes those who are very familiar and comfortable with the digital playing field, but who lack the cartographer's training in map concepts and graphic design. Established cartographers have had to face considerable challenge and competition from this new group.

Given the above noted trends, it becomes clear that it is the practising cartographers and the computer literate cartographic novices that have manoeuvred the transition of map design from the analog to the digital world. It is these two groups, therefore, that have to a large extent been responsible for keeping the dynamics and diversity of the cartographic language alive in recent years.

The role and importance of the practising cartographer in the map design agenda has been recognized in M. Wood's chapter on the practitioner's view. Wood raises our awareness that the academic community would do well to consult with and to listen to the practising cartographer when studying map design. Perhaps the practising cartographers' membership in the map design community is given too little attention at present. Much can be learned from their contribution to the literature, often in the form of technical notes in newsletters such as the Canadian Cartographic Association's *Cartouche*.

Reading the preceding chapters should have made it obvious that the impacts of the digital revolution on map design have been considerable and exciting. However, the introductory chapter has reminded us that not all has been positive. The standard of map design appears to have declined in recent years, evidenced by a notable deterioration in the quality of some of the recently published map products. Both academic and practising cartographers have begun to take note of this deterioration, and interest in the study of map design is rekindling.

MAP DESIGN AND CHANGING SOCIETY

Has the most recent shift in the nature and quality of cartographic communication been due solely to the technological revolution, or do transformations in the cartographic language also reflect a change in societal values? Language reflects values of society at the time, and cartography is no exception. Map design does not advance in isolation, and the relationship between cartographic language and society has been explored by a number of authors in this book and elsewhere. The study of the relationship between maps and society should make for fascinating reading. Today's society is obsessed by technology and information, two variables already noted to be closely linked to map design. Today's society is also increasingly environmentally conscious and is busy breaking down barriers in the search for the global village. The map has a role to play in all this. Today's high technology map analysis and map design capabilities increasingly are used by special interest groups to explore and to state opposing viewpoints. Perhaps more than ever before, map design is used to communicate society's contemporary thinking and to bring out divergent viewpoints.

The perception has always been that there is a close relationship between the discipline of geography and the map. Assuming that the geographical discipline has been keeping up with changing society, then what about the association between advances in geographic thought and cartographic communication? Krygier's chapter investigates whether geographers understand and conceptualize maps and other visual representations as geographic methods. He suggests that there should be a close association between geographic thinking and cartographic design, and that paradigm shifts in geography should include a critical analysis of maps and their function. He concludes that cartographers have failed to realize and study the direct relationship between geographic thought and map design. There is more to cartographic design than maps, pictures and multimedia; beyond communication, cartographic design is about exploration, synthesis, analysis and confirmation. Geographers and cartographers alike should take note and make use of the powers of the cartographic language.

There exists a fashion in contemporary society to distrust experts and to question packaged information. This trend is of relevance to cartographic communication, a theme picked up in Huffman's chapter where we are challenged to follow and participate in a debate on postmodernism and contemporary design theory. Huffman is not satisfied with map design solely to communicate factual information. He advocates a style of map design that encourages critical thinking. Maps should be designed and map readers should be educated to view maps critically and, in the process, map readers should learn to raise their awareness about the world that the maps construct and represent. A well known form of cartographic expression that stimulates critical thinking is the cartogram; but new technologies offer opportunity to explore other forms of cartographic expression that encourage critical thinking. Map design research just has to find and popularize them.

Today's society is trying hard to be increasingly inclusive while catering to everybody's differences, that is, we try to do our best to address the needs of all members of society and to eliminate barriers and inequalities. Those who set the map design research agenda are aware of this trend in society and are trying to address the issues. Ongoing research is exploring how to bring the cartographic language to all members of society and is seeking to identify forms of cartographic expression most suited to individual groups. The preceding chapters show that the search continues for map designs that make the cartographic language accessible to the blind and other disadvantaged groups, and that we are beginning to investigate the relationship between map design and children. We also are beginning to explore the role of gender in map design, although Kumler and Buttenfield's chapter informs us that remarkably little work has been completed here to date.

The relationship between map design and changing societal values forms an area of investigation that we hopefully will hear more about as cartographers make it their business and priority to investigate it. But map design has to learn to walk in our new high technology society before attempting to run and to address the needs of a changing society. Those in charge of the map design research agenda must not forget to learn more about and to explore further the relationship between map design and the new technologies.

MAP DESIGN AND THE NEW TECHNOLOGIES

Much has been written to remind us that advances in technology and information management are radically changing the general rules of communication. This includes cartographic communication, a fact brought out again and again in a number of chapters in this book. We are reminded that cartographers have an important role to play in the new technologies and that where cartographic design is going in the new information era should depend on direction given by the cartographic profession. Of course, the discipline of cartography has to agree to accept and to live up to that challenge. Cartographers must realize that the cartographic language will change under the influence of new technologies even if cartographers fail to take charge. Without the cartographers' guidance, new directions will simply be shaped by the computer literate, but cartographically naive.

A number of chapters in this book have picked up on the latter line of thinking, stressing the need for map design specialists to get more actively involved in the high technology agenda. The call is made for cartographers to become more directly involved in the design of software that supports cartographic composition. The time has come for cartographers to get directly involved in the process of firmly embedding traditional knowledge of map design in the functional capabilities of software that supports the creation of maps, regardless of whether the software is marketed as a graphics, GIS, or visualization tool. We need more research along the lines of enquiry reported by Mersey. Her approach, to proceed by investigating map designs done in the past, and to follow on by exploring whether today's software packages can support these designs, has identified serious software limitations. We need these types of map design studies to push technology by exposing its limitations.

However, it is not sufficient simply to automate traditional cartographic knowledge. Cartographic design must take advantage of the new technologies to break beyond automation of traditional cartographic techniques. This point is raised in Mackaness's chapter on automated cartography and the human paradigm when he prompts us to address the specific cartographic design needs of today's GIS, exploratory data analysis (EDA), and general digital graphics software. The good news is that work appears to be progressing in these directions, as should have become evident when reading some of the chapters in this book. But more work is needed. We need more research along the lines of McGranaghan's investigation into map readers' responses to technological limitations. This type of investigation is important for two reasons. First, we need to comprehend the limitations and constraints imposed on map design by today's technology. Second, we must identify technological shortcomings that seriously hinder effective cartographic communication and we must push the technology industry to remedy them.

Among the great potentials of the digital revolution are the opportunities to take cartographic communication from a static to a dynamic environment, to offer multiple views of data using map complementarity (see Monmonier's chapter), and to generally move cartographic communication closer to the field of visualization and exploratory analysis. Indeed, the relationship between map design and scientific visualization is one that should be at the forefront of the map design research agenda, as has been advocated, for example, by MacEachren and Ganter (1990), MacEachren et al. (1992), MacEachren and Taylor (1994), and some of the chapters in this book.

The most important issue to investigate immediately appears to be the search for innovative designs to support the growing capability for users to interact with the actual design process using graphics software. The questions of how dynamic and complementarity capabilities of the digital environment can be utilized to design dynamic maps and/or multiple representations, and how effective these new forms of representation would be in communicating cartographic messages have not been answered today, but research clearly is progressing. In the process, an academic question appears to have surfaced. What should be the emphasis of a dynamic map, an interactive map and/or simultaneous multiple representations? Should it be communication or data exploration? Patton and Cammack note that some proponents of the data exploration and visualization paradigm wish to distance these new approaches from a more narrowly defined concept of the communication paradigm—but it should be obvious that each can benefit from the other. Returning to a point of Krygier's made earlier, there is more to cartographic design than communication using maps, pictures and multimedia; cartographic design is also about exploration, synthesis, analysis, and confirmation. Cartographers would be wise to take note of, and to treat all the above as equally important aspects of the cartographic language.

Cognitive studies have always formed an important component of the map design research agenda. Chapters in this book confirm that cognitive research is alive and advancing. But, as Patton and Cammack note, the goals of cognitive studies in cartography and the search for optimal cartographic communication may shift with a movement from a strict communication paradigm to a visualization paradigm. Cartographers will need to continue cognitive research into cartographic communication, but it is becoming obvious that the value of this line of enquiry increasingly will be to guide and evaluate the design of effective visualization tools.

BRIDGING THE GAP BETWEEN THEORETICAL AND APPLIED RESEARCH

An old dichotomy that has resurfaced again in this collection of chapters is the rift between theoretical and applied map design research. Belbin, in his chapter on Gestalt theory, is most outspoken about this rift. He observes that neither side appears to communicate much with the other, and that practising cartographers pay little attention to the theories developed by academic enquiry. He suggests that this is perhaps because theoretical research concentrates too much on small detail at the expense of the overall effect, while practising cartographers must look at the whole which is different from the sum of its parts.

The preceding collection of chapters has tried to pull us in both directions. Some have argued that advances in map design can be gained by listening to practising cartographers and by studying existing map designs, and that we should advance new technology to allow us to do old things with them. Others have made the point that we should not try to automate traditional cartographic know-how. Instead, we should focus on exploring new capabilities offered by the computer, and we should start to address the cartographic design needs of today's digital drafting and GIS software users. Some have tried to steer us strictly towards map design for communication, while others are encouraging us to chart new territories of map design for exploration, synthesis and analysis. Some are trying to convince us that the map design research agenda should be technology driven, while others see the need for it to be directed by philosophy and societal values.

Surely the map design research agenda should not be a debate between theoretical and applied research; nor should it allow for a separation between academic and practising cartographers. It should also avoid drawing a hard distinction between traditional and computer cartography. All viewpoints expressed in the preceding chapters merit investigation. There are no rights and wrongs. The chapters in this book present a summary of some of the recent reactionary and proactive thinking. What is important is that the different research efforts do not proceed in isolation of each other; what we need is ongoing constructive dialogue.

REFERENCES

MacEachren, A.M., and Ganter, J.H. (1990). A pattern identification approach to cartographic visualization. *Cartographica*, 27(2), 64-81.

MacEachren, A.M., Buttenfield, B.P., Campbell, J., DiBiase, D., and Monmonier, M. (1992). Visualization. In R. Abler, M. Marcus, and J.M. Olson (Eds.), *Geography's inner worlds: Pervasive themes in contemporary American geography* (pp. 99-137). New Brunswick, NJ: Rutgers University Press.

MacEachren, A.M., and Taylor, D.R.F. (1994). *Visualization in modern cartography*. Oxford: Pergamon.

Subject Index

A

B

C

D

E

F

G

H

I

K

L

M

N

O

P

O

R

S

T

U

V

W

Z

Author Index